Computational Statistical Methodologies and Modelling for Artificial Intelligence

This book covers computational statistics-based approaches for Artificial Intelligence. The aim of this book is to provide comprehensive coverage of the fundamentals through the applications of the different kinds of mathematical modelling and statistical techniques and describing their applications in different Artificial Intelligence systems. The primary users of this book will include researchers, academicians, postgraduate students, and specialists in the areas of data science, mathematical modelling, and Artificial Intelligence. It will also serve as a valuable resource for many others in the fields of electrical, computer, and optical engineering.

The key features of this book are:

- Presents development of several real-world problem applications and experimental research in the field of computational statistics and mathematical modelling for Artificial Intelligence
- Examines the evolution of fundamental research into industrialized research and the transformation of applied investigation into real-time applications
- Examines the applications involving analytical and statistical solutions, and provides foundational and advanced concepts for beginners and industry professionals
- Provides a dynamic perspective to the concept of computational statistics for analysis of data and applications in intelligent systems with an objective of ensuring sustainability issues for ease of different stakeholders in various fields
- Integrates recent methodologies and challenges by employing mathematical modeling and statistical techniques for Artificial Intelligence

Edge AI in Future Computing

Series Editors:

Arun Kumar Sangaiah, SCOPE, VIT University, Tamil Nadu

Mamta Mittal, G. B. Pant Government Engineering College, Okhla, New Delhi

Soft Computing Techniques in Engineering, Health, Mathematical and Social Sciences
Pradip Debnath and S. A. Mohiuddine

Machine Learning for Edge Computing: Frameworks, Patterns and Best Practices
Amitoj Singh, Vinay Kukreja, Taghi Javdani Gandomani

Internet of Things: Frameworks for Enabling and Emerging Technologies
Bharat Bhushan, Sudhir Kumar Sharma, Bhuvan Unhelkar, Muhammad Fazal Ijaz, Lamia Karim

Soft Computing: Recent Advances and Applications in Engineering and Mathematical Sciences
Pradip Debnath, Oscar Castillo, Poom Kumam

Computational Statistical Methodologies and Modelling for Artificial Intelligence
Edited by Priyanka Harjule, Azizur Rahman, Basant Agarwal and Vinita Tiwari

For more information about this series, please visit: www.routledge.com/Edge-AI-in-Future-Computing/book-series/EAIFC

Computational Statistical Methodologies and Modelling for Artificial Intelligence

Edited by
Priyanka Harjule, Azizur Rahman, Basant Agarwal
and Vinita Tiwari

CRC Press is an imprint of the
Taylor & Francis Group, an informa business

Designed cover image: Shutterstock

First edition published 2023
by CRC Press
6000 Broken Sound Parkway NW, Suite 300, Boca Raton, FL 33487-2742

and by CRC Press
4 Park Square, Milton Park, Abingdon, Oxon, OX14 4RN

CRC Press is an imprint of Taylor & Francis Group, LLC

ISBN: 978-1-032-17080-0 (hbk)
ISBN: 978-1-032-18142-4 (pbk)
ISBN: 978-1-003-25305-1 (ebk)

DOI: 10.1201/9781003253051

Typeset in Times
by Newgen Publishing UK

Dedication

To everyone who has contributed and supported this book

Contents

Preface

Computational statistical science is a progressive field that employs advanced computing strategies to solve and understand real-life problems. In widespread usage, computational models are integrated with Artificial Intelligence (AI) algorithms, which is the soul of the digital computer or computer-controlled robot to perform intelligent tasks. AI is an interdisciplinary field and is frequently applied to developing systems endowed with humans' intellectual processes, including the ability to reason, identify meaning, generalize, or learn from memories. AI has multiple approaches, and advancements in machine learning and deep learning are creating a metamorphosis in every industry sector.

Applications of computational methods for AI are all around us, such as personal assistants, search engines, image reorganization tools, etc. In the last decade, AI has seen innumerable successes with significant societal benefits and have contributed to the dynamic economy of the world. Deep learning is another machine learning dimension that runs inputs through biologically inspired neural network architectures.

With the availability of lots of data all around us, computational statistical techniques have evolved and become an integral part of AI. The research trends in AI are highly influenced by mathematical modelling and computational statistics modifications. Hence, the editors and authors aim to bring together the recent research on the computational statistical methods and models applied to new datasets through this book.

In recent years, powerful computational models based on deep learning and machine learning approaches have shown significant success in dealing with a massive amount of data in unsupervised settings. This book is a culmination of efforts from researchers and academicians across the globe to present various computational and statistical methods and models for the field of AI.

AI has seen tremendous growth in its impact on human life over the past decades. With each passing day, AI's sphere of influence is expanding, and today it is ubiquitous. The editors and contributors have considered the dynamically changing and ever-evolving nature of AI while compiling this book. This book provides a dynamic perspective on computational statistics for data analysis and applications in intelligent systems to ensure sustainability issues for the ease of different stakeholders in various fields. An in-depth explanation and description of several methodologies are also covered in the book. It is curated to cater to the needs of both beginners and advanced learners.

The book mainly focuses on the five major themes: Theme 1: *Statistics and AI methods with applications*, Theme 2: *Machine learning-adopted models*, Theme 3: *Development of the forecasting component of decision support tools*, Theme 4: *Socio-economic and environmental modelling*, and Theme 5: *Healthcare and mental disorder detection with AIs*. Each of these themes is covered by a range of relevant chapters. A brief description of the themes follows.

Theme 1 consists of three chapters and covers the broad spectrum of statistics and AI methods with applications. In particular, the first chapter focuses on introducing the contemporary techniques for computational statistics in AI. In addition, the chapter describes recent research in computational statistics and its application to AI. Chapter 2 presents an improved random forest for classification and regression using a dynamic weighted scheme. This chapter explores the relationship between test samples and decision trees, based on which weightage of the decision tree is computed, and the aggregation/weighted voting is performed. Thus, as the test sample change, the calculated weights also change. The proposed method is tested over various heterogeneous datasets compared to state-of-the-art competitors. The authors claimed that it showed improvement for both regression and classification tasks tested over the UCI datasets. Chapter 3 presents the study of computational statistical methodologies for modelling the evolution of COVID-19 in India during the second wave. In their research, the authors studied two approaches, namely deterministic and stochastic, to calculate the value of the basic reproduction number. Further stability analysis of the solution is also discussed. Results in the study demonstrate that the case numbers can be controlled through social distancing, use of face masks, regularly sanitizing, and washing hands.

Theme 2 covers applications of computational statistical models employed in machine learning and deep learning algorithms in various fields such as transfer learning, ride-sharing, nowcasting, and stress-level detection. It consists of three chapters dealing with data issues, and statistical methods comprise the foundation of machine learning and deep learning algorithms. More specifically, Chapter 4 presents an efficient and accurate model to detect drivers' behaviour from their captured images while driving. The deep learning model is used as a feature extractor trained using images, from which the foreground section is segmented using the Grab Cut. Then, those extracted features are used to classify the behaviours of the driver and predict the label of distinct classes as output. Chapter 5 aims to improve the performance of machine learning models that classify fake reviews. In this research, Decision tree, Random Forest, Gradient Boosting, AdaBoost, and Bi-LSTM were implemented to get the best performance on data collected from the current ride-sharing applications review section. Chapter 6 focuses on the nowcasting of selected imports and exports of Bangladesh. Nowcasting, a combination of "now" and "forecast", estimates a target variable's current state, or a close approximation of it, either forwards or backwards in time, utilising information available on time. It has a wide range of applications, attempting to supplement and assist users' decisions. This research applied autoregressive moving average, neural network models, and support vector regression for modelling and nowcasting selected imports and exports, considering annual data from 1976 to 2020 in an Asian country. The findings revealed that support vector models had superior performance compared the other models considered in this study.

Theme 3 covers the latest developments in the forecasting component of decision support tools. Forecasting is critical for successfully executing an organization's strategic operations and functions. There are four chapters in this theme with time-series data in computational statistical models. For example, Chapter 7 focuses on analysing oversampling and ensemble learning methods for credit card fraud detection. The increasing number of online transactions have resulted in more security measures taken to protect the data of the transactions and the customers. Companies have consulted various data scientists to provide better frameworks for fraud detection – this chapter analyses the ensemble model of prominent machine learning algorithms. The ensembles were created using multiple algorithms such as Logistic Regression, Random Forest, KNN (K Nearest Neighbours), Naive Bayes, and SVM (Support Vector Machine). It was observed that random forest and KNN performed the best among the methods.

Chapter 8 aims to predict the stock market trends; the authors propose a hybrid model that uses daily news, online news data, and time series analysis. The numerical data and textual data are used to train models separately. They came up with a hybrid ensembled model that is used to predict stock trends more accurately. Chapter 9 proposes an AI bot, which is helpful for industry hiring strategy. The authors claim that the time and services spent on hiring people can be reduced by using the proposed recruiter chatbot. The proposed bot can conduct interviews, and can judge candidates' self-confidence, anxiety, fear, and behaviour using dynamic analysis. The facial expression is recognized with convolution neural networks, and the speech recognition is implemented with Google APIs. The authors reported facial emotion recognition accuracy of 71% and speech recognition accuracy of 91%. On average, they claim, their model works with 81% overall accuracy. Chapter 10 contributes to the literature on the use of advanced statistical techniques applied to intelligent data analysis scenarios within the retail IT project management industry against an underlying phenomenological Cynefin framework. It undertakes factor and cluster analysis via pattern recognition to measure the assignment of projects and project managers to respective Cynefin domains. Correlation to project success outcomes are analysed using regression analysis techniques to predict the optimal combination of variables leading to successful project outcomes. The proposed model uses advanced statistical methods to identify and align patterns and dimensions across disparate datasets, seeking varieties that are not easily identifiable to human respondents.

Theme 4: One of the integral aspects of computational statistical methods is data analysis using statistical tools and techniques. Data analysis is key to applying statistical techniques and mathematical modelling in guiding policy decisions and decision-making. This theme incorporates the development of regional economic models that imitate the macroeconomic model. The authors collected real-world case studies and data in Chapters 11, 12, and 13. Under this theme, intelligent data analysis is inferred. Chapter 11 aims to describe socio-economic development in Palestine. Mathematical modelling is used as the primary tool to explain the main factors that affect the development of the Palestine economy. The research establishes the idea of using and applying mathematical modelling for regional socio-economic development in Palestine. The authors proposes a number of modelling equations and equilibrium points, and bifurcation is calculated and measured for stability options by analysing the Jacobin Matrix. Chapter 12 presents the computational statistical methods for uncertainty assessment in geoscience. Since complex data sets require computational statistics to analyse the data, two such computational statistical methods are compared in this chapter. Geostatistical algorithms and computational tools are used to predict block and point grades (mineral resource estimates). Artificial intelligence and machine learning techniques are applied to those models and three-dimensional representations of mineralization for uncertainty assessment.

Chapter 13 compares computational models for validating the geospatial distribution of the water quality index. This chapter aims to identify the best geospatial predictive model for the spatial distribution of WQIs for coastal water quality. Eight widely used interpolation techniques such as local polynomial interpolation (LPI), global polynomial interpolation (GPI), inverse distance weighted interpolation (IDW), radial basis function (RBF), simple kriging (SK), universal kriging (UK), disjunctive kriging (DK), and empirical Bayesian kriging (EBK) are utilized for WQIs. This study was carried out in Cork Harbour, Ireland, as a case study for assessing coastal water quality. According to the cross-validation results in the chapter, the UK (RMSE = 6.0, MSE = 0.0, MAE = 4.3, and R2 = 0.8) and the EBK (RMSE = 6.2, MSE = 0.0, MAE = 4.6, and R2 = 0.78) methods performed excellently in predicting WQIs, respectively. The results show that the EBK computational interpolation method could effectively reduce uncertainty in geospatial predicting WQIs.

The final Theme 5 consists of three chapters focusing on the theme of healthcare and mental disorder detection with AI. In particular, Chapter 14 presents the computational study of pattern discovery of autism spectrum disorder using tensor decomposition models based on eye-blinking data. Using tensor dimensionality reduction and feature extraction algorithms, it proposes a new algorithm to discriminate subjects with autistic spectrum disorder (ASD) from typically developed (TD) children based on their eye-blink patterns. It also achieves classification results with an error rate lower than 3%. Moreover, Chapter 15 presents the case study of Mohalla clinics in Delhi, India. The authors aimed to identify the determinants for the use of primary care. A questionnaire survey was carried out in patients visiting Mohalla clinics (government-run accessible primary care facilities) in selected districts of Delhi, India. The exploratory factor analysis was conducted to assess the satisfaction of patients of Mohalla clinics. Further, the Chi-square test of association was used to explore relationships among various categorical variables of social importance.

Lastly, Chapter 16 discusses stress-level detection using smartphone sensors. It is complicated to measure an individual's stress because of its personalized nature. Therefore, the authors use smartphone sensors to identify stress levels. Data was collected from 48 users using the "Sensors Recorder" app over six months. It collects data from the accelerometer, gyroscope, ambient light sensors, call data, and app usage in the background. In the foreground, it provides information about the user's stress level by giving daily notifications. The Long Short-Term Model was used for prediction, and an accuracy of 82.8% was achieved.

In conclusion, various contemporary issues in methodological innovations, programming, and applications are covered in this book with diverse arenas of current popularity and state-of-the-art research in computational statistics, advanced level modelling with data science, and AIs.

Jaipur, Rajasthan, India Priyanka Harjule
Wagga Wagga, NSW, Australia Azizur Rahman
Jaipur, Rajasthan, India Basant Agarwal
Jaipur, Rajasthan, India Vinita Tiwari
15 May 2022

Acknowledgments

Our efforts to produce a book in the evolving field of computational statistics-based approaches for AI is a culmination of efforts from several scientists, academicians, and practitioners across the world. The invited call for contributions in this fast-growing and rapidly evolving field has received an overwhelming response from the fraternity around the globe. A total of 54 proposals were received for this book. Each of these proposal manuscripts was thoroughly checked and peer reviewed by at least three reviewers who are renowned experts in the respective topic. Based on the peer review outcomes and the suitability of the revised manuscript on the book's theme, 16 high-quality works were chosen for final publication. We express our sincere thanks to all our reviewers for sparing their valuable time and sharing their expertise. We also express our sincere gratitude and best wishes to all the authors for contributing quality research work related to the book. We are also thankful to the Charles Sturt University, Australia, Indian Institute of Information Technology Kota, and Malviya National Institute of Technology Jaipur, India for extending their support in every way possible for this book. Finally, we are also very thankful to the publishers' team of CRC press for their support at every stage, which helped us in the timely production of this book. We hope that this book will be handy for both students and expert researchers working in the relevant domains.

Priyanka Harjule
Azizur Rahman
Basant Agarwal
Vinita Tiwari

About the Editors

Priyanka Harjule is currently working as Assistant Professor in the department of mathematics, Malviya National Institute of Technology Jaipur. She holds a Ph.D degree from MNIT Jaipur in applied mathematics. Er research areas include mathematical modelling and computational statistical methodologies in machine learning. She has published more than 30 research papers in reputed journals and conferences. Dr. Harjule has established national and international connections with different sponsored projects and is an active participant of several research and developmental activities.

Azizur Rahman, PhD, is an applied statistician and data scientist with expertise in developing and applying novel methodologies, models and technologies. He is the Leader of the "Statistics and Data Mining Research Group" at Charles Sturt University, Australia. Prof. Rahman is able to assist in understanding multi-disciplinary research issues within various fields, including how to understand the individual activities that occur within very complex behavioural, socio-economic and ecological systems. Prof. Rahman develops "alternative methods in microsimulation modelling technologies", which are handy tools for socio-economic policy analysis and evaluation. His 2016 and 2020 books have contributed significantly to the area of small area estimation, microsimulation modelling, data science and policy analysis. His research interests encompass issues in simple to multi-facet investigations in various fields ranging from the mathematical sciences to the law and legal studies. He has more than 135 scholarly publications, including a few books. Prof. Rahman and his collaborative teams have received more than $2.00 million in external grants for research, including funding from the Australian Federal and State Governments. He serves on a range of editorial boards, including the *International Journal of Microsimulation (IJM) and Australasian Journal of Regional Studies (AJRS).* He obtained several awards, including a Research Excellence Award and the CSU-RED Achievement Award.

Basant Agarwal is currently working as an Assistant Professor in the field of Computer Science and Engineering at the Indian Institute of Information Technology Kota (IIIT-Kota), India, which is an Institute of National Importance. Dr Basant Agarwal holds a M.Tech. and Ph.D. in Computer Science and Engineering from Malaviya National Institute of Technology (MNIT) Jaipur, India. He has more than 9 years of experience in research and teaching. He has worked as a Postdoc Research Fellow at the Norwegian University of Science and Technology (NTNU), Norway, under the prestigious ERCIM (European Research Consortium for Informatics and Mathematics) fellowship in 2016. He has also worked as a Research Scientist at Temasek Laboratories, National University of Singapore (NUS), Singapore. He has published more than 60 research papers in reputed conferences and Journals. His research interests include Machine Learning, Natural Language Processing, and Intelligent Systems.

Vinita Tiwari is currently working as an Assistant Professor in the department of electronics and Communication at IIIT-kota. She received her B.Tech degree in Electronics and Communication Engineering from M.B.M Engineering collage, Jodhpur followed by Master's and Ph.D degree in Communication Engineering and High Speed Optical Communication respectively under the faculty development program of BITS-Pilani, Pilani Campus. Her research interest includes high speed optical link designs, advanced modulation techniques, Soliton communication, mathematical modelling, blockchain etc.

Contributors

Faruq Abdulla is currently working as Junior Lecturer at RTM Al-Kabir Technical University, Bangladesh. He has completed his B.Sc (Honors) and M.Sc degrees in Statistics from Islamic University, Kushtia, Bangladesh in 2013 (held in 2014) and 2014 (held in 2016), respectively. He was ranked 1st position at both B.Sc and M.Sc levels with Gold Medal for securing the highest marks in the Faculty of Applied Science & Technology in the M.Sc final examination. He has more than twenty research articles (peer-reviewed) in the field of immunoinformatics, time series, econometrics, environmental statistics, sampling, and numerical methods in different renowned international journals including SCOPUS & SCI indexing journals with high impact factor. Now, he is continuing his research in the field of data mining, machine learning, biostatistics, public health, and bioinformatics.

Basant Agarwal is currently working as Assistant Professor in the field of Computer Science and Engineering at the Indian Institute of Information Technology Kota (IIIT-Kota), India, which is an Institute of National Importance. Dr Basant Agarwal holds a M.Tech. and Ph.D. in Computer Science and Engineering from Malaviya National Institute of Technology (MNIT) Jaipur, India. He has more than 9 years of experience in research and teaching. He has worked as a Postdoc Research Fellow at the Norwegian University of Science and Technology (NTNU), Norway, under the prestigious ERCIM (European Research Consortium for Informatics and Mathematics) fellowship in 2016. He has also worked as a Research Scientist at Temasek Laboratories, National University of Singapore (NUS), Singapore. He has published more than 60 research papers in reputed conferences and Journals. His research interests include Machine Learning, Natural Language Processing, and Intelligent Systems.

Soumyajit Bal is currently working as an Assistant Software Engineer in Accenture Bengaluru, India. He received B.Tech. Graduate degree in Information Technology from College of Engineering and Technology, Bhubaneswar, an autonomous College under BijuPatnaik University of Technology (BPUT), Odisha, India. His research interests lie in Data Management and Visualization, Computer Vision.

Lokesh Bansal is a final-year BTech student in the Department of Computer Science and Engineering at the Indian Institute of Information Technology in Kota, India.

Jacqui Coombes is the Managing Director & CEO of Amira Global. Jacqui also serves as an Independent Director on the Board of ICRAR, is on the Advisory Board of Start Up WA, and the Advisory Board of State of Play. A Statistician by training, Jacqui has experience across the mine value chain, commodities, and across the globe. Jacqui has a Masters in Geostatistics and is author of "The Art and science of Resource Estimation" and "I'd Like to be OK with MIK, UC?" Jacqui has a PhD in Statistical Reasoning for JORC Code Competency Development and a Masters in Commercial and Resources Law.

Sanjit Kumar Dash is currently working as Assistant Professor in the Department of Information Technology in College of Engineering and Technology, Bhubaneswar, India, an autonomous and constituent college under the BijuPatnaik University of Technology, Odisha, India. He obtained his Ph.D. degree in Computer Science Engineering from the Utkal University, Odisha, India. He is the life member of CSI. His research interests include Wireless Mobile Network, Wireless Sensor Network, and Machine Learning.

Sidhartha Bibekananda Dash is currently working as Systems Engineer in TCS Bangalore, India. He received B.Tech. Graduate degree in Information Technology from College of Engineering and Technology, Bhubaneswar, an autonomous College under Biju Patnaik University of Technology (BPUT), Odisha, India. His research interests lie in Image processing, Deep learning, Computer Vision, Data Visualization.

Mir Talas Mahammad Diganta is a PhD candidate in Civil Engineering at the National University of Ireland Galway. He works in arena of water quality modeling, aquatic ecosystem, and environmental analytical chemistry. He has published several scholarly publications. He acts as a Research Associate for the EcoHydroInformatics Research Group at NUI Galway.

Rohit Bhaskar Rao Gurijala completed his B. Tech in Electronics and Communication Engineering from Indian Institute of Information Technology, Kota in the year 2021. In the same year, he joined MDI Gurgaon to pursue his higher studies in PGDM-HRM.

Priyanka Harjule is currently working as Assistant Professor in the department of mathematics, Malviya National Institute of Technology Jaipur. She holds a Ph.D degree from MNIT Jaipur in applied mathematics. Er research areas include mathematical modelling and computational statistical methodologies in machine learning. She has published more than 30 research papers in reputed journals and conferences. Dr. Harjule has established national and international connections with different sponsored projects and is an active participant of several research and developmental activities.

Md Zobaer Hasan is currently a lecturer with the School of Science, Monash University Malaysia. He obtained his BSc (Hons) and MSc in Statistics from Shahjalal University of Science & Technology, Bangladesh. He then undertook his PhD in Financial Statistics at Universiti Sains Malaysia, Malaysia. During his Ph.D., he focused his research on the Bangladesh stock market. Currently, he is engaging himself in some multidisciplinary research works which will be analyzed using statistical tools.

Jennifer Hayes is Doctor of Information Technology candidate at Charles Sturt University with industry expertise in the delivery of large-scale transformational IT projects and programs across the Education, Energy and Retail sectors. Jennifer is a passionate proponent of improving the delivery of complex projects and programs with research interests in evidencing the existence of chaos in IT project management in Retail. Jennifer holds Master's degrees in both IT Management and Project

Management, receiving the IT Masters Award for the graduating student with the highest GPA in 2011. She has also received the Executive Dean's Award in 2019 and 2020 during her coursework as a doctoral candidate.

Most Hasna Hena completed her B.Sc and M.Sc degree with First Class Second position in the Department of Information and Communication Engineering from the University of Rajshahi, Bangladesh. She received gold medal Award and Merit Scholarship from Rajshahi University for her good result in the B.Sc (Honors) Examination. She also Received NSICT-2010 fellow award from Ministry of Science, Information and Communication Technology (MOSICT), Bangladesh, for her M. Sc. Research work. Her research area is Network Security, Machine Learning, Deep Learning and Communication.

Ana Horta is a geospatial scientist with a research background in geostatistics. She is currently Senior Lecturer and discipline lead in Geospatial Sciences at Charles Sturt University, Australia. Ana's research work contributed to the field of spatial modelling and geostatistics through the development of new methodologies for environmental, agricultural and health applications, for which she received funding from the Australian Research Council and NSW government. Besides her research, Ana has provided environmental consultancy services to private and public companies concerning contamination assessments and environmental strategic assessments.

Md Moyazzem Hossain is working as Associate Professor in the Department of Statistics, Jahangirnagar University, Bangladesh. Hossain is an applied statistician and a data engineer with expertise in both developing and applying recent methodologies, models, and techniques. In addition to teaching, Hossain is enjoying research activities. As a result, he has published more than 100 research articles in peer-reviewed journals published by world-renowned publishers. He is also engaging in different research activities and passionate about collaborative and multidisciplinary research. Moreover, he served as a volunteer reviewer and member of the Editorial Board of different international journals.

Harshit is a postgraduate student at the department of Mathematics in Malviya National Institute of Technology Jaipur. He did his graduation from University of Delhi with Mathematics, Statistics and computer science. His research interests include Mathematical Modelling and stochastic simulations. He is interested in higher studies in the field of Mathematics. This work is a part of his masters project which is a study of computational statistical methodologies for modelling the evolution of COVID-19 in India during the second wave.

Taminul Islam was born on June 4, 2000 in Jamalpur city, Mymensingh Division, Bangladesh. He received his Bachelor of Science (B.Sc.) with first class in the field of Computer Science and Engineering (CSE) from Daffodil International University, Ashulia, Dhaka, Bangladesh, in 2022. During his student life, he worked as a Student Prefect and Junior Research Assistant at Daffodil International University. He achieved a Full Free Scholarship in his B.Sc. program from Daffodil International University. His research interest lies in the area of Artificial Intelligence and Human-Machine Interaction. He has completed his B.Sc. thesis in the field of Machine Learning and

Deep Learning. He has collaborated actively with researchers in several other disciplines of Computer Technology.

Vikas Jain (Student Member, IEEE) completed his bachelor's (B.E.) degree in computer science engineering from Rajiv Gandhi Proudyogiki Vishwavidyalaya, Bhopal, Madhya Pradesh, India, in 2009, and the M.Tech. degree in information and communication technology from Dhirubhai Ambani Institute of Information and Communication Technology (DAIICT) Gandhinagar, India, in 2011. He is currently a Ph.D. Research Scholar with the Indian Institute of Information Technology Vadodara, Gandhinagar, India, since 2016. Prior to this, he served as an Assistant Professor for over five years in Bhopal, Madhya Pradesh, India. His research interests include machine learning, hyperspectral image classification, and deep learning. He also serves as a reviewer for the IEEE journals and Transactions in his area of work.

Pooja Jain is woking with Indian Institute of Information Technology, Nagpur in the department of Computer Science and Engineering. She has close to 17 years of experience in industry and academia. She is IEEE senior member and has been an expert/keynote in almost 75 workshops/conferences.

Nour Jamal was awarded high honor degree in mathematics applied to economics from Birzeit University and master degree in mathematical modeling from Palestine Technical University – Kadoorie, West Bank, Palestine. She is currently working on mathematical modeling research at Palestine Technical University – Kadoorie. Her research interest includes mathematical modeling, numerical analysis, bifurcation, and stability analysis.

Anupam Kumar (Member, IEEE) received the M.Tech. degree in system modeling and control and the Ph.D. degree from the Electronics and Communication Engineering Department, IIT Roorkee, India, in 2012 and 2018, respectively. He currently serves as an Assistant Professor with the Department of Electronics and Communication Engineering, IIIT Kota, Rajasthan, India, since 2019. He has published many papers in journals, conferences, book chapters, and one patent granted. His current research interests include fuzzy logic systems (type-1 and type-2), robotics, control and automation, fractional-order controllers, biomedical engineering, healthcare applications, deep learning, and machine learning. He is a recipient of the Best Paper Award from NIT Kurukshetra. He is serving as a reviewer for journals like IEEE TRANSACTIONS and Elsevier, Springer, Taylor Francis, and Wiley journals. He also served as a reviewer for many conferences. His current Google scholar citation is around 330.

Rahul Kumar is currently pursuing a Ph.D. from the Faculty of Management Studies, University of Delhi. He has M.Sc. and M. Phil degrees in Operational Research from the Department of Operational Research, Faculty of Mathematical Sciences, University of Delhi. He has published research papers in several journals of repute including International Transactions in Operational Research, Journal of Energy Sector Management, International Journal of Reliability and Safety, etc. He is currently pursuing research on Distribution Network Design in Public Health.

Tapan Kumar is currently working with Indian Institute of Information Technology, Nagpur in the Department of Electronics and Communication Engineering. He has close to 20 years of experience in industry as well as academia.

Kaushal Kumar is working as Assistant Professor in the Department of Operational Research, Faculty of Mathematical Sciences, University of Delhi. He has a Ph.D. degree from the Faculty of Management Studies, University of Delhi. He has professional and research experience of more than 10 years. Previously, he worked in Cognizant Technology Solutions India Pvt. Ltd. and Department of Food Civil Supplies and Consumer Affairs, Government of Punjab. One of his articles has won the best paper award at a conference organized by the Indian Institute of Science, Bangalore in 2018.

Amit Kumar received his engineering degree in Computer Science and Engineering from The Institution of Electronics and Telecommunication Engineers, New Delhi, in 2008. He received his MTech and PhD in Computer Science and Engineering from PDPM Indian Institute of Information Technology Design and MAnufacturing Jabalpur, in 2010 and 2016, respectively. Presently, he is working as an assistant professor in Computer Science and Engineering Department in Indian Institute of Information Technology, Kota, India. His research interest includes robotics, computer vision, machine learning and multi-robot systems.

Arindom Kundu was born on November 5, 2000 in Jhenaidah city, Khulna Division, Bangladesh. He has completed his Bachelor of Science (B.Sc.) in the Department of Computer Science and Engineering from Daffodil International University, Ashulia, Dhaka, Bangladesh, in 2022. After his graduation, he is collaborating with researchers in the field of Computer Technology. His research interest is in Artificial Intelligence, Human-Machine Interaction, Computer Technology and Data Mining.

Rishalatun Jannat Lima was born on June 20, 1999 in Jamalpur city, Mymensingh Division, Bangladesh. She has completed her Bachelor of Science (B.Sc.) in the Department of Computer Science and Engineering from Daffodil International University, Ashulia, Dhaka, Bangladesh, in 2022. After her graduation, she is collaborating with young researchers in the field of Data Mining and Machine Learning. Her research interest is in Data mining, Artificial Intelligence, and Computer Networks.

Scott McManus is Lecturer in Spatial Statistics & Science, Analytics, Big Data Analytics and Data mining. He is currently researching methods of assessing uncertainty in spatial domains using Bayesian methods to improve public reporting of mining results. He has over 25 years of industry experience as a consultant. Scott is a member of "Statistics and Data Mining Research Group". Other research interests include big data analytics in MNC Local Area Health District, and Resilience of Mangroves in Watson Taylor Lake after, Drought, Fire and Flood.

Prateek Meena is a final-year BTech student in the Department of Computer Science and Engineering at the Indian Institute of Information Technology in Kota, India.

Champake Mendis is an adjunct lecturer and industry supervisor. He is a member of "Statistics and Data Mining Research Group" at Charles Sturt University. He has more than 20 scholarly publications including a book. Champake is an experienced Data Scientist with a demonstrated history spanning over 20 years in the ICT, five years in financial services superannuation/pension disciplines. Industries worked include Finance, Defence, Education, Transport, Telecom, Construction and Insurance. He holds a PhD in Computing and Information Systems and M.Eng.Sc. in Mechatronics from University of Melbourne and was a member of one of the best AI research groups in Australia. He is currently the Chief Data Scientist of Triple A Super. He is also the Assistant Secretary of IEEE VIC/TAS Section. His current research interests include the areas of complex networks, project management, wireless sensor networks, and AI and ML algorithms for data fusion, with emphasis on numerical modelling, simulation, and performance analysis.

Anubhuti Mittal is a postgraduate student at the department of Mathematics in Malviya National Institute of Technology Jaipur. She did her graduation University of Delhi. Her research interests include computational statistics.

Stephen Nash is a senior lecturer in hydraulic/water engineering at NUI Galway. His research expertise is the numerical and physical modelling of surface water bodies and associated infrastructure with particular focus on the modelling of tidal currents, waves, water quality, coastal flooding/ersosion, tidal/wave energy and aquaculture. He co-leads the Marine Modelling Group at NUI Galway and is an SFI MaREI funded investigator in Marine Renewable Energy Technologies and in Coastal and Marine Systems. He has 80+ peer-reviewed publications and has won a number of national and international awards for research excellence including the Engineers Ireland excellence award for technical innovation".

Agnieszka I. Olbert is a lecturer in Civil Engineering and Programme Director for Project and Construction Management at NUI Galway. She has a vast experience in hydrodynamic and water quality modelling of surface waters and specializes in marine/coastal climate change forecasting, flood protection, water quality analysis, data assimilation and marine renewable energy assessment. She leads the EcoHydroInformatics Research Group at NUI Galway and collaborates on numerous projects funded by the EU, Irish and industry sources. She acts as an expert advisor to the Irish Government and state agencies. Dr. Olbert has over 50 publications and serves as guest editor and editorial board member. She has received a number of awards for research excellence including the "Irish Marine Industry Award for Excellence in Marine Research".

Rekha Pal is a B.Tech undergraduate in the Department of Computer Science and Engineering at Indian Institute of Information Technology, Nagpur. Her current research focuses on utilizing smartphone sensors to detect stress among working professionals.

Kourosh Parand received his Ph.D. in Applied Mathematics from the Amirkabir University of Technology, Iran in 2007. Professor Parand is the academic staff at the Shahid Beheshti University, Tehran, Iran, and at the University of Waterloo. He has an outstanding and unique curriculum vitae with experience in numerical analysis, cognitive science, and neural networks. Kourosh Parand is a world leader in applying methods of spectral methods as well as machine learning approaches to nonlinear dynamical models.

Ashish Phophalia (M'13) holds Ph.D. and M.Tech. degree from Dhirubhai Ambani Institute of Information and Communication Technology (DAIICT) Gandhinagar, India in 2016 and 2010, respectively, and B.E. in Computer Science & Engineering from Rajasthan University, India in 2008. Dr. Phophalia is currently working as an assistant professor at the Indian Institute of Information Technology Vadodara (IIITV), Gandhinagar, India, since 2016. He is a member of IEEE since 2013. His area of research is image processing, machine learning, and medical image analysis. Dr. Phophalia also serves as a reviewer in IEEE journals of his area of interest.

Hamidreza Pouretemad is an Iranian neuropsychologist who is the founder and the Dean of the Institute for Cognitive and Brain Sciences at SBU. He received his Ph.D. from the Department of Psychological Medicine, Institute of Psychiatry, London University in 1998. Pouretemad's research interests center on neurodevelopmental disorders and he is a frequent contributor to advanced research lines including the application of new technologies for the diagnosis and treatment of autism and dyslexia. He has coauthored several books and is the author of many articles on autism and dyslexia, mainly devoted to the Persian-speaking population.

Jamal Amani Rad received a Ph.D. in applied mathematics from the Department of Computer Science at Shahid Beheshti University (SBU) in October 2015. Following his Ph.D., he started a one-year computational cognitive modeling postdoctoral fellowship at Institute for Cognitive and Brain Sciences in May 2016 where he is currently midway through his fourth year as an assistant professor. In addition to having at his relatively young age 81 publications in quality journals/conferences, he has worked independently during his five years after graduation.

Azizur Rahman, Ph.D. is an applied statistician and data scientist with expertise in developing and applying novel methodologies, models and technologies. He is the Leader of the "Statistics and Data Mining Research Group" at Charles Sturt University, Australia. He can assist in understanding multi-disciplinary research issues within various fields, including understanding the individual activities that occur within very complex scientific, behavioural, socio-economic, and ecological systems. Prof. Rahman develops "alternative methods in microsimulation modelling technologies", which are handy tools for socio-economic policy analysis and evaluation. He has more than 120 publications, including a few books. His 2020 and 2016 books have contributed significantly to the fields of "data science and policy analysis" and "small area estimation and microsimulation modelling", respectively. The Australian Federal and State Governments fund his research, and he serves on a range of editorial boards, including the International Journal of Microsimulation (IJM) and Sustaining Regions. He received several awards, including the SOCM Research Excellence Award 2018 and the CSU-RED Achievement Award 2019.

Utkarsh Kumar Rai received the B.Tech degree in Electronics and Communication Engineering from the Indian Institute of Information Technology, Kota in 2017. In 2021, he joined Deloitte Consulting India Private Limited, where he is currently working as a Business Analyst .

Negar Sammaknejad received her Ph.D. in Cognitive Psychology from the University of California, Irvine in 2012. Following her Ph.D., she started a one-year postdoctoral fellowship at Institute for Cognitive and Brain Sciences where she is currently midway through his fifth year as an assistant professor. Her research interest is visual perception, face perception, and neuromarketing.

Omar Sharif is currently a lecturer (senior scale) with mathematics and statistics department, Universal College Bangladesh (Monash University pathway program). In the past, he was a lecturer (senior scale) with the General educational development, Daffodil International University (DIU). He obtained his BSc (Hons) in Mathematics and MSc in Applied Mathematics from University of Dhaka, Bangladesh. He then undertook his MSc(Research) in Financial Statistics at Universiti Tunku Abdul Rahman, Malaysia. During his MSc., he focused his research on the Data science, AI, mathematical modeling in health science, and Malaysian stock market. After that, he has started to do research on Malaysian stock market data analysis as well as the stock markets efficiency measuring in developing countries. Additionally, He has collaborated actively with researchers in several other disciplines of data science. Currently, he is engaging himself in some multidisciplinary research works which is related to adaptive mathematical model.

Ashish Sharma is presently working as Assistant Professor in Indian Institute of Information Technology Kota. He received the Ph.D degree from MNIT Jaipur. His research interests include Computer Architecture, VLSI and Embedded System, System-Level Design and Cyber-Physical System.

Shashank Shekhar is a final-year BTech student in the Department of Computer Science and Engineering at the Indian Institute of Information Technology in Kota, India.

Shashwat Singh is currently pursuing Masters in Data Science from Columbia University. He completed his undergrad in Computer Science and Engineering from Indian Institute of Information Technology. He is deeply passionate about building products from abstract ideas and his current research focuses on financial and HEALTHCARE DOMAIN.

Ankit Kumar Singh received the B.Tech degree in Electronics and Communication Engineering from the Indian Institute of Information Technology, Kota in 2017. In 2021, he joined Verizon India Private Limited, where he is currently working as a Full Stack Developer

Vinita Tiwari is currently working as Assistant Professor in the department of electronics and Communication at IIIT-Kota. She received her B.Tech degree in Electronics and Communication Engineering from M.B.M Engineering collage, Jodhpur followed by Master's and Ph.D degree in Communication Engineering and High Speed Optical Communication respectively under the faculty development program of BITS-Pilani, Pilani Campus. Her research interest includes high speed optical link designs, advanced modulation techniques, Soliton communication, mathematical modelling, blockchain etc.

Aditya Prasad Tripathy is currently working as a Systems Engineer in TCS Bangalore, India. He received B.Tech. Graduate degree in Information Technology from College of Engineering and Technology, Bhubaneswar, an autonomous College under Biju Patnaik University of Technology (BPUT), Odisha, India. His research interests lie in Image processing, Machine Learning, Statistical Analysis and Data Visualization.

Debabrata Tripathy is currently working as an Assistant System Engineer-Trainee in TCS Hyderabad, India. He received B.Tech. Graduate degree in Information Technology from College of Engineering and Technology, Bhubaneswar, an autonomous College under BijuPatnaik University of Technology (BPUT), Odisha, India. His research interests lie in Robotic Process Automation, Machine learning and Statistical Analysis

Md Galal Uddin is the PhD candidate (Final year) in Civil Engineering. He has six years of experience in hydrodynamic and water quality modelling, water quality analysis with expertise in machine learning, deep learning, and AI, and data assimilation. He also has experience in environmental chemistry. He acts as a Research Director for the EcoHydroInformatics Research Group at NUI Galway and collaborates on numerous MS/PhD projects. He has two years' experience in environmental analytical chemistry in a national chemical metrology reference institute. Uddin's significantly contributes to the field of coastal water quality modelling and he developed a weighted quadratic mean (WQM) water quality index model for assessing coastal water quality through his PhD program. He has published several scholarly publications".

Rohith Reddy Vangal is a B.Tech undergraduate in the Department of Computer Science Engineering at Indian Institute of Information Technology, Nagpur. His current research interests include stress detection in working professionals using smartphone sensors

Yashita Watchpillai is a B.Tech undergraduate in the Department of Computer Science Engineering at Indian Institute of Information Technology, Nagpur. Her current research focuses on utilizing smartphone sensors to detect stress among working professionals.

Sonal Yadav is Assistant Professor at the Department of Computer Science and Engineering, National Institute of Technology Raipur. She has completed her post-doctoral from the Indian Institute of Technology (IIT) Guwahati, India in 2020. She received her M.Tech and PhD degree from the Malaviya National Institute of Technology (MNIT) Jaipur, India. Her research interests lie in Networks-on-Chip, Computer Architecture, Deep Learning and Artificial Intelligence. Her PhD thesis work awarded travel grant in DATE 2019. She is the recipient of the best paper award by TCVLSI in IEEE-SES 2018.

Ayu Yaduvanshi is a final-year BTech student in the Department of Computer Science and Engineering at the Indian Institute of Information Technology in Kota, India.

Theme 1

Statistics and AI Methods with Applications

1 A Review of Computational Statistics and Artificial Intelligence Methodologies

Priyanka Harjule[*1], *Azizur Rahman*[2], *Basant Agarwal*[3] *and Vinita Tiwari*[4]
[1]Department of Mathematics, Malaviya National Institute of Technology Jaipur Campus, India
[2]School of Computing, Mathematics and Engineering, Charles Sturt University, Wagga Wagga, Australia
[3]Department of Computer Science and Engineering, Indian Institute of Information Technology Kota, MNIT Jaipur Campus, India
[4]Department of Electronics and communications, Indian Institute of Information Technology Kota, MNIT Jaipur Campus, India
*Corresponding author email: priyanka.maths@mnit.ac.in

CONTENTS

1.1 INTRODUCTION

Statistics derive its livelihood from numbers, and computational statistics depends mainly on how these numbers are utilized to gain valuable insights from the data [1, 2]. Computational statistics is an evolutionary field that employs advanced computing strategies to understand and solve real-world problems [3]. In widespread usage,

DOI: 10.1201/9781003253051-2

computational models integrated into Artificial Intelligence (AI) algorithms provide the ability for a computer to perform intelligent tasks similar to what a human being can do. The machine learns by mimicking the characteristics of humans, such as the ability to reason, discover meaning, generalize, or learn from previous experience. AI is an interdisciplinary field of research with multiple approaches. However, advancements in machine learning (ML) and deep learning are being significantly used in real-world problems such as misinformation detection [4], hate detection [5], sentiment analysis [6], medical image analysis [7], biomedical health informatics [8], malware detection [9], etc.

Applications of computational methods for AI are all around–d us [10–11]. With its focus on performing specific tasks, AI has experienced numerous breakthroughs in the last decade that have had significant societal benefits and have contributed to the economic vitality of the world. A few examples of applications of computational statistics in AI include image recognition software and Siri, Alexa, and other personal assistants. With the availability of lots of data all around us, computational statistical techniques have evolved and become an integral part of AI. The research trends in AI are highly influenced by the modifications in mathematical modelling and computational statistics [12]. In recent years, powerful computational models based on deep learning and ML approaches have shown significant success in dealing with a massive amount of data in unsupervised settings.

The process of moving from data to scientific understanding is often challenging [13–14]. Statistical science provides a principled framework for this process, contrasting with pure prediction central to numerous ML procedures. However, as the size and complexity of modern data sets continue to increase and the ambition of the understanding sought from the data also expands, fresh challenges emerge from implementing formal statistical approaches. The 20th-century computational statistical methodologies (often based on Markov chain and sequential Monte Carlo methods and their variants) have brought complex problems within the grasp of classical and Bayesian model-based paradigms. Today's complex issues pose immense challenges for principled statistical methods. Computational statistics combines analytical capabilities and statistical workflow to gain insights from data [15]. Computational and statistical methods can identify the latest problem-solving strategies that might be useful in economics, mathematical modelling, intelligent systems, and software engineering. Computational and statistical methods flawlessly analyse and interpret big data [16]. Statistics have vital uses in our daily lives, including biomedicals and genomics research, weather forecasting, stock marketing, and quality testing [17–19]. Data science fields without statistics are very challenging. For example, we use correlation analysis, a handy component of statistics, to find the relation between two or more variables. Statistical methods deal with unstructured data, which is massive, in a proficient manner. Advancements in computational statistical methodologies have become an integral part of a more comprehensive, more affluent, and extensive field known as "Artificial Intelligence."

AI is a multidisciplinary subject with various approaches. AI is the simulation of the functions that receive codes from surroundings [20]. It is the intelligence of computers to do work better and faster way than human intelligence [21, 22]. Though AI also requires human intelligence. Alexa, self-driving cars, and robotic works are examples of AI. Most of the AI methods, the main algorithms applied in

Computers are Artificial neural networks, fuzzy logic systems, genetic algorithms, particle swarm optimization, simulated annealing, and evolutionary computing. AI has enhanced engineering in a variety of areas. Many new frameworks to demonstrate the role of technology powered by AI have been proposed, and research continues.

Figure 1.1 presents a flowchart of typical relationships between statistics, computing, and data in the context of modelling and applications. It reveals that statistics and computing have interacted with each other through a series of traditional and contemporary innovations, including computational methods, computing algorithms, AI, and ML. These advanced techniques are employed in data integration, pre-processing, and data analysis to build simple to complex models that generate new insights from the data. We learn from the data through fresh insights and then decide and implement our local, state, national, and international strategies in all domains of our life.

This introductory chapter presents an overview of the applications of computational statistical methodologies in AI through a critical literature review. It demonstrates the findings and strategies adopted by various researchers in studying and investigating different computational statistics methodologies in AI, ML, and deep learning. The chapter's primary focus is to present a collection of state-of-the-art approaches, concentrating on the high-performing, innovative solutions to the most prevailing and demanding challenges faced in AI research.

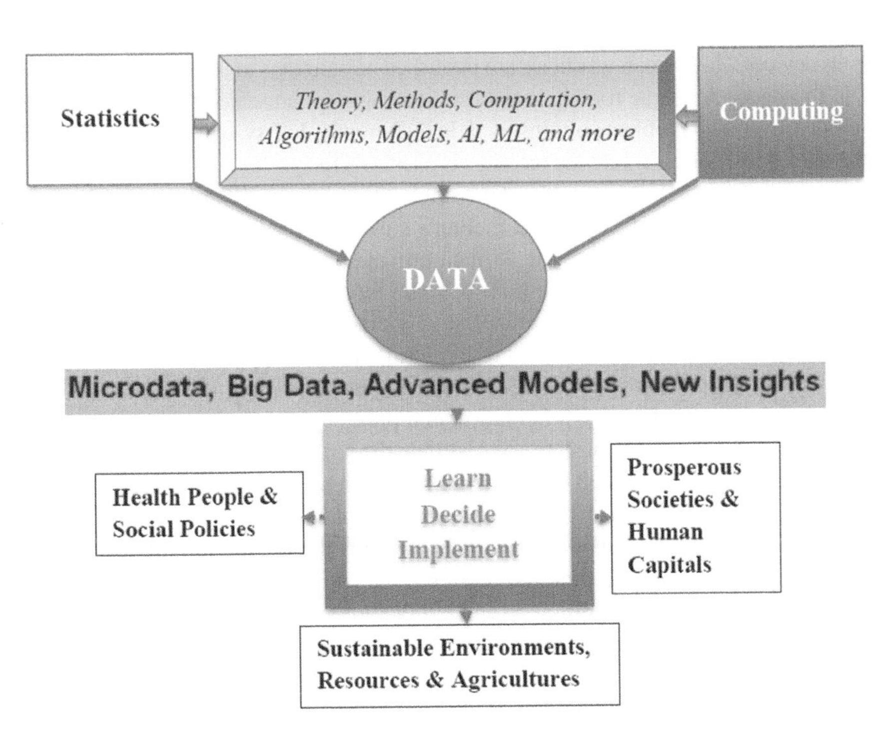

FIGURE 1.1 Interlinks between statistics, computing, and data in the lens of CS, models, and AIs.

Source: authors work.

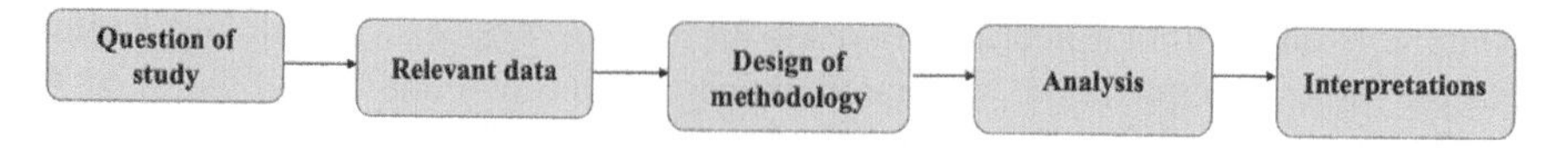

FIGURE 1.2 Flow of computational statistical method.

1.2 CURRENT METHODOLOGIES

This section provides an overview of current methodologies in computational modelling and AI with the relevant sub-headings. The flow of a typical computational statistical method is presented in Figure 1.2. The main focus here is to introduce some key methods with their relevance to applications.

1.2.1 Related Work on Computational Statistics

In the past several years, much practical work has been done with the help of computational statistical (CS) methods in various fields of AI. The contribution of CS methodologies is majorly in planning and design of the studying, analysing data collected and examining the data quality, comparison and investigation of causality and associations, and finally, assessing the uncertainty of results. This section presents a literature review of applications of CS methodologies in AI. Computational and statistical methods have a wide range of utilizations. For instance, effective gateway systems for interdomain routing constitute a significant problem in connecting Internet of Things (IoT) heterogeneous devices. However, few articles are published on solutions to the scalability problem of IoT systems. A unique routing scalable system work is presented by Shen et al. [23], considering energy problems of sensors in IoT. Currently, IoT systems dealing with non-interoperability issues require the message broker system.

For large-scale MIMO systems, the complexity minimization of the matrix inversion is investigated [24]. Statistical methods solve potential hazardous text identification, which is an essential point in data processing and analysis research. Zhang et al. [25] presented statistical methods of spam filtering techniques. In the pedagogy field, research has been done on quantitative methods. Meghanathan, N. offers uses of correlational analysis of decay centrality in mathematics applications in intelligent systems [26]. Migration system forecasting using mathematical and simulation modelling is described in Ref. [27]. Schmertmann presented research work on a population model when the stable birth rate is less than the fertility rate [28]. Communication systems can be upgraded with the help of mathematical-statistical methods in natural languages. Syntactic structures of complex sentences were made as simple sentence structures [29]. Social networking analysis methods help in community monitoring and analysis, influence propagation, and expert identification presented community mining problems and methodologies and applications to handle them. Further, some key literature review information is presented in tabulation form in Table 1.1.

To solve scalability complications in IoT systems by perfect planning of resources hazardous text related to terroristic, fanatical, and suicidal organizations could be recognized using statistical methods. In managing smart cities, urban data streaming emerges from IoT systems assistance. Using a robust message broker data analysis of these intelligent systems would be facile. Uses in pedagogical aspects of

TABLE 1.1
Summary of Information from Key Literature Review of Applications of Computational Statistics

Author	Research Aims	Methodology	Significant findings	Strategies
A. Bhavana and A.N. Nandha Kumar [33]	To solve scalability issues in IoT. Enhance interdomain routing among heterogeneous devices with maximum communication in IoT. To decrease latency or any other communication degradation during heavy traffic.	A new IoT **Ecosystem Map arranges** which has all the transactional information. This system gave all information about the terminal nodes. There are two algorithms implemented for scalability improvement – in primary algorithm inputs are Device span and selected IoT device that gives output of scalable communication vector. In the secondary algorithm Input is cost of route and output is optimised path.	It was found that transmission errors are associated with IoT. Ecosystem Map. The rise of cost factor is minimal without an IoT Ecosystem Map. The impact of IEM is that in any network condition Cost per packet with increasing number of nodes and packet is lesser with the IEM system.	Minimise the cost per packet within the optimum path.
Inas Abuqaddom and Amjad Hudaib [34]	To detect software defects by **hybrid** SMOTE-**Ensemble** approach with use of **Cost-Sensitive learner (CSL)** to handle uneven distribution.	Ensemble techniques improve classifiers for better prediction of defects. A Best classifier has Higher probability of Detection and lower probability of false alarm means maximum g-mean that gives more accurate classification of defect and non-defect modules. Different classifiers on four data bases tested and analysed results, the best classifier found with maximum g-mean.	Cost-sensitive learner – SMOTE Adaboost classifier gives best results. It performs very well in high imbalance datasets. Prediction of defects in software can be done more precisely using multiple techniques together.	Using multiple ML techniques on different databases.

(continued)

TABLE 1.1 (Continued)
Summary of Information from Key Literature Review of Applications of Computational Statistics

Author	Research Aims	Methodology	Significant findings	Strategies
B. Archana and T.P. Surekha [35]	To simplify channel estimation and detection for MIMO-OFDMA receiver with three element-based antenna. Minimizing the computational cost without altering radio frequencies.	MIMO-OFDMA receiver used with three-element steering antenna. Firstly, modulation is applied on the selected image to change its resolution and contrast Then it is sent to OFDMA channel. Then at receiver terminal demodulation is applied and channel estimation done with parameters least square mean square error and compressive sensing methods. Received data by channel estimation analysed with signal to noise ratio and bit error rate.	Inverse relation between SNR and BER established. Output from this MIMO OFDMA system with steerable antenna gives high accuracy data.	Break the signals into channels. Channel estimating with compressive sensing.
Vladislav Babutskiy and Igor Sidorov [36]	To identify potentially hazardous texts not relative to their hashtags that are posted on social sites. To know the text domain by data processing approach. Identify potential hazardous text impacts on people's mindsets.	Every text on social sites contains some hashtags, hashtags are analysed and bring out keywords from text. Keywords and hashtags are compared and characteristic of relations found between these. A statistical analysis is applied on the neutral text and finds out which feature presents in what strength.	Statistical research method gives results of text identification better than other methods under theme uncertainty. It identifies hazardous text precisely.	Count the frequencies of keywords.

Elarbi Badidi [37]	To construct a flawless scalable system of urban IoT data and its processing without revealing its privacy. Make use of a message broker-based system to make better interconnection between IoT devices.	A MQTT broker used as a gateway of data received from IoT devices. Then these data sets are sent into kafka cluster These data sets are analysed in a Hadoop cluster by batch processing.	This MQTT and kafka-based structure helps in streaming data from IoT devices. This detects real-time events. Kafka cluster systems can easily deal with large data amounts.	MQTT used for device communication and kafka for data consuming.
Thomas Barot and Radek krpec [38]	To apply an alternative approach of Fisher's Exact test when general statistical methods are used by software, because when a statistical software solution is used this test limits only to a manual solution.	A transformation of the second categorical variable to integer numbers is applied. Frequencies rewritten in the form of the first categorical variable. Then modified the hypothesis and tested it.	This study makes a comparison between computational simplification of statistical methods with Fisher's exact test. For unfulfilling chi-square test conditions proposed method applied for testing of hypothesis for significant dependencies between categorical variables.	Transformation of categorical variables, modification of hypothesis.
V. V. Bystrov, M. G. Shishaev, S. N. Malygina and D. N. Khaliullina [39]	To manage the interregional migration process developing the prototype of a polymodel complex for estimating the migration process.	Two subsystems are considered in each zone Economy and Population to know the migration model. Agents are created to represent each sector in these subsystems. Output collected from the population subsystem is used to calculate the supply of labour. This data is transferred into the second subsystem economy,	Migration model estimated by these subsystems: Population and economy. The forecasting component evaluates situations in interregional migration. Functioning of the forecasting component depends on the analytic and simulation modeling.	Computational experiments, organization and simulation of previous models.

(continued)

TABLE 1.1 (Continued)
Summary of Information from Key Literature Review of Applications of Computational Statistics

Author	Research Aims	Methodology	Significant findings	Strategies
Co Ton Minh Dang [40]	To patternising syntactic structures of Vietnamese complex sentences by experimental approach depends on DCG and subject-predicate grammar pattern.	A table of syntactic patterns created based on C-V combination. Construct syntactic rules for different types of phrases. Divide the patterned syntactic structures into different classes. Two ML support vector machines and Naive Bayes.	Six syntactic structures have been patterned. Experiment of 300 complex Vietnamese sentences by patternizing syntactic structure has been done successfully. If DCG syntactic rules and vocabulary are modified this proposed system approaches higher accuracy.	Classification of C-V structure.
Igor O. Datyev and Andrey M. Fedorov [41]	To analyse real territorial features of communities based on their virtual activities. Construct methods to interpret communities based on social networking data.	An analysis of identification of the home region of VKontakte users has been performed. Correlation of territoriality of the VK user with his friends and community groups identified. Content and context analysis of the user on VK used as a better way with high accuracy for finding their territoriality.	Direct observation of social network accounts would not give accurate results about territoriality. Indirect method of analyse profile attributes, analysis of content and context, user activity are specified the territoriality with reliability.	Using all available data on social networking sites of users.
V. V. Dikovitsky and M. G. Shishaev [42]	To extract knowledge from legislative documents by experimental evaluation making use of semantic analysis.	Some universal dependencies that come by automatic parsing were skillfully interpreted and used. After analysing statements, the resulting table is represented as a sequence of lexemes, properties, and relations in the universal	Stable syntactic constructions and universal dependencies of text were obtained. Relation between universal dependencies of high accuracy gives automatic	Semantic network, pattern of deontological statement.

	Developed multilevel analysis of text by merge of linguistic and semantic methods.	dependencies. Percentage of syntactic role of the elements of deontological statements shows the possibility of automatic search of deontological statements in a sentence.	search of deontological statements with high accuracy in a text.	
Zdena Dobesova and Jan Pinos [43]	To recognise main causes that enforce high school students to pursue in the department of geoinformatics Palacky University in Olomouc by decision trees.	A questionnaire was performed for two years with the various secondary students that contained 12 possible responses. This collected data transformed into a binary matrix. Use this data as an input in decision trees. Decision trees calculate the influence of GIS week, attend an open day, lecture, friend, teacher on study.	Highest influence factors are attendance on open day, recommendation of any friend, teacher's advice on students for choosing geoinformatics studies. GIS week's attendance also has not any great impact on students. Internet advertising influence is very low.	Comparing decision trees.
Denis V. Gruzenkin, Anton S. Mikhalev, Galina V. Grishina, Roman Yu. Tsarev, and Vladislav N. Rutskiy [44]	To identify blunders in log N-version software by block chain technology. Upgrade functions of N-version software and enhance its dependability.	A classification of log error types done and divides error messages into six types. Probability of total error detection obtained by probability of error detection of these all types of error. Higher probability error would be eliminated firstly.	Block chain technology successfully diagnosed residual error by volume of messages in logs using probability of error detection. During software operation higher probability error can be detected and eliminated. That upgrades the reliability of multi version softwares.	Probability of errors

(continued)

TABLE 1.1 (Continued)
Summary of Information from Key Literature Review of Applications of Computational Statistics

Author	Research Aims	Methodology	Significant findings	Strategies
Yang Chen Hui L, Mei Chen, Zhenyu Dai, Huanjun L., and Ming Zhu [45]	To find an approach that can identify each cluster's typical features with high accuracy. Collect similar features using clustering by the DBSCAN algorithm.	Construct a similarity matrix having entries of euclidean distance of features in the data set. Classified features based on available number of features within the neighbourhood. FESIM applied and small clusters are made. Use the K-means algorithm to identify clustering done successfully.	Fesim obtained results with better accuracy compared to other methods with respect to the number of selected features in the Mice Protein Expression dataset. Accuracy is less when number of selected features is small and improved as increased in scadi data set and same in Epileptic seizure recognition dataset. Experiments shows that accuracy goes upward as dimension of features increases.	Farthest euclidean distance, tree classifier.
Li Tang, Hui L, Mei Chen, Zhenyu Dai, and Ming Zhu [30]	To present an ETL task schedule algorithm based on an altruism method to manage ETL tasks. This algorithm should upgrade the utilization of resources and task regulation.	In Altruism-based ETL Tasks Schedule Algorithm (AETSA) all jobs are submitted in a scheduler and paused then all jobs settled based on their IO costs. finds out which task is supreme and resumes that task. Timer is monitoring specified time to start the task and record the cpu-utilization and follow an altruistic schedule.	AETSA is performing better scheduling than FCFS and meets SJF methods efficiency. Efficiency of FCFS decreases as CPU upper bound reaches higher. At cpu bound 85% SJF is better than AETSA. Factor of improvement of AETSA is higher, more than 12%, than FCFS.	Factor of improvement, CPU utilization

Zuzana Majdisova, Vaclav Skala, and Michal Smolik [47]	To find out stationary points of a sampled surface by RBF interpolation. And form an algorithm that detects the shape of the curve on which stationary points lie on.	To dictate stationary points of the data set piecewise RBF interpolation is used. For the efficiency of interpolation shape parameter α should be large. Make four subdomains for stationary points and reduce them according to the maximum possible distance between any two points to find isolated stationary points.	Proposed approach experimented on four sample data sets. For the first function curve is found as a line segment and the result of the proposed approach coincides with the analytical approach. For all four data sets total seven curves are found and one isolated point is also found. Result is the same as the analytical approach.	Gauss RBF, distance between stationary points.

the chi-square test and Fisher's test in quantitative research are commonly acceptable. Decision support systems can monitor population changes, migration models, and birth and mortality rates in a specific region. Patterning syntactic structures of complex sentences to expand communication and grasp information monitoring of social network sites to determine the social territoriality of virtual communities. AETSA schedules extract, transform and load (ETL) tasks to reduce response time and enhance the efficiency of job execution [30]. This approach could manage multiple tasks at a time. RBF interpolation methods can determine stationary points of a sampled data set and the curve shape made by those points. Computational statistical techniques have also been used to model the evolution of the COVID-19 pandemic [31,32,46].

1.2.2 Related Work in AI

There are three important methodologies in AI: supervised learning, unsupervised learning, and reinforcement learning [17]. Supervised learning makes use of labelled data for training AI systems by learning a function $f: U \rightarrow V$ that describes the underlying relation between input and output. Input information consists of matrix $n X m$of given features $x \subset X$ and label vector $Y = (y_1, y_2, \ldots, y_n) \subset Y$. Here, n is the total number of observations, m is the number of features, and U and V are the input and output space, respectively. Examples of supervised learning methodologies are support vector machine, logistic regression, k-means clustering and decision trees, etc. [48]. In unsupervised learning patterns are extracted from unlabelled data. Examples of unsupervised learning involve principal component analysis, dbscan, hierarchical clustering, etc. [17–19]. Finally, in reinforcement learning, the model is trained using its mistakes (i.e., trial and error approach). It uses the concepts of the Markov chain decision process from probability. The data to an AI algorithm is in the form of values from stock market prices [49], audio signals, texts, climate data, and data from complex systems such as chess games. The following paragraph presents a few examples of applications in AI.

Remarkable progress and evolution have occured in AI in the past decade. The applications are in the fields of automatic face recognition, speech recognition, language processing [50], self-driving cars, and games (such as chess) where a computer beats a human player [51,52]. Due to the ability to represent grammar, the models of hidden Markov chains from statistics are used successfully in speech recognition, text translation, and analysis [53,54]. Today, language translation apps are top-rated as they enable translating languages such as Chinese or any European languages in real-time [55]. Another popular area of research and applications of AI is in medicine. AI is used for the early detection of many diseases, accurate diagnosis, or predicting acute events [56–59, 60]. The future scope in this area is the development of personalized and customized medicines, leading to tailor-made treatments for individuals or sub-groups [61–63]. Further economics uses many AI methods, especially in macroeconomics, to develop models for individual consumer behaviour [64–65].

Regardless of the progress and developments achieved in AI, precaution needs to be taken while applying AI algorithms. There have been examples and reports related to the limitations of AI [66]. Furthermore, careful consideration of AI systems is required given the severe consequences of false-positive and false-negative decisions [66]. For example, there have been cases in which the identification of offenders using automated video surveillance has been reported to be 0.67% [67]. Similarly, the applications of AI in medicine cannot afford to have wrong decisions or decisions with low efficiency [68]. Further, Table 1.2 presents some more related works in AI.

1.2.3 A Comparison of CS and AI

Statistics is a discipline that works with data to hold data and make predictions about more extensive data. It works by visualizing, wrangling, and analysing. CS is using statistics with computers. This involves applications in areas of computational biology, computational linguistics, computational material sciences, and many more. AI is also an extensive and advanced branch of computer science. CS provides a solid foundation for all the applications in AI, and contains types that are reactive machines, theory of minds, and self-awareness. Siri, Alexa, self-driving cars, and robots are outstanding examples of AI. AI makes computer systems able to do tasks that need human intelligence.

CS are much needed with the growing volume of sensed data. Collecting vast amounts of data and making choices from them has been on the rise, especially in the unprecedented times of the COVID-19 pandemic. As a result, advanced statistical analysis is the need of the hour. With its applications in various fields, the scope of CS in AI is different for different problems, as is seen from the literature review in Section 1.2. In intelligent systems, statistical methods are used to deal with vast amounts of urban data of IoT systems in monitoring cities. CS methods are needed in solving and analysing scalable problems of IoT. An amalgamation of AI and IoT has the power to develop industries, economics, the standard of living, and businesses and can upgrade technologies. The task of IoT is that all devices connect efficiently, and AI deals with device learning. Statistics is the nucleus element of AI. The applicability of statistical methods in AI development has been shown in Ref. [20].

Decision support systems help companies make confident decisions. Intelligent decision support systems that are AI-powered include robo-advisors in financial technology, which use AI to make decisions based on previous learning and outcomes. In health care, AI-powered image processing is used to detect disease by images. In communication systems, computational and statistical methods are used to enhance the scope of natural languages and do the computational analysis of texts. Therefore, it is clear that CS methodologies and AI must go hand in hand to provide solutions to many challenging big data problems arising in today's global world. The research community is continuously developing advanced statistical analysis methods and technologies that go with AI applications.

1.3 DISCUSSION AND CONCLUSION

Research in AI has been playing a significant role in every field, including economics, political science, and engineering. Throughout this chapter, it was observed and

TABLE 1.2
Summary of information from Key Literature Review of Applications of Artificial Intelligence

Author	Research aim	Methodology	Significant findings
Agata Blasiak, Jeffrey Khong, and Theodore Kee [66]	To increase the implementation of personalised medicine using CURATE AI. Discussion about the CURATE AI and its applications.	Recognise the best combination of a particular drug and its doses. CURATE. AI examines the adjustable patient specific profiles and checks optimal doses throughout the whole treatment of individuals.	This study covered the path of clinical implementation of dose optimization. CURATE AI present dose optimization for proposed treatment of multiple myeloma and immunosuppression
Jafferey Dustin [69]	To examine algorithm of hiring in amazon and discrimination of amazon's recruitment tool based on Artificial intelligence.	Examine past resumes with computer model, check how recruiting tool react on the gender discrimination words present on the resumes.	Amazon's recruitment tool puts down resumes that include the word "women's,." It puts down graduates of women's colleges.
Sarah friedrich et al. [20]	To provide applications and strategies of AI\ML methods in cardiovascular research. A systematic review	Firstly a literature search is done by using PUBMED and EMBASE restricted by clinical studies. Extract and analyse data on study population and qualities. Categorized AI/ML methods and further sub categorized. Comparison of AI/ML methods with classical methods.	AI/ML methods applied in majority (87%). In supervised learning and in unsupervised learning AI/ML methods mostly used in preprocessing. AI/ML algorithms are grantee for clinical practice. Variation in quality of method is noticed because there is no harmony for which disease which method should be applied.
Eirini Ntoutsi et al. [70]	To Examine the bias and fairness of decision-making systems based on AI by doing surveys of recent technical methods. Discussion of challenges and direction for AI solutions for social welfare.	For mitigating bias, the first method focuses on the data. For fairer result modification of original data done and make balanced in terms of discrimination. Second method of altering the model of ML.	There is no preferable method of mitigating bias for each category. Fairness of the result cannot be acquired by only a mathematical approach.

		Accounting for bias is described by proactively and retroactively.	Biases are involved densely in our society so the bias problem in AI can not be removed by only a technical approach.
Sumit das et al. [71]	To present a general view of real-world applications of ML. Discuss applications of AI in ML.	Applications of types of ML such as unsupervised learning, supervised learning, reinforcement learning, and recommender systems are described. Artificial agents in ML learn by cooperate environment and training data and can deal with large data set.	AI agents are general problem solvers in ML. Virtual doctors can be created using machine learning and AI that can diagnose the symptoms. ML can be proven to be a helpful aspect in process of information time machine.
Jahanzaib Shabbir, and Tarique Anwer [72]	To analyse expansion of AI in various fields and what it is capable of doing today. In comparison of AI with human intelligence what are the challenges, similarities, and pros and cons of AI.	Turing test and Eugene Goostman Test are used to test how AI is differ from human intelligence. When computers have enough algorithms to perform tasks then AI can represent humans in an intellectual way. Cognitive letter recognition model is used to understand text.	Advancement of AI will affect the society in economic and social ways. It is considered favourable in war, health, and poverty problems. But there are challenges that humans are totally dependent on technology and loss of employment, and it challenges human thinking. By new computing structures of the cloud AI becomes inexpensive to any organization.

concluded that CS serve as a backbone for theoretical and practical understanding of AI. As noted in Ref. [73], According to weak AI, the principal value of the computer in the study of the mind is that it gives us a potent tool. [...] But according to strong AI, the computer is not merely a tool for studying the reason; instead, the appropriately programmed computer is a mind [...]. Thus, computational statistics plays an integral role in weak AI. That is, applications of computational statistics are mainly on data-driven problems of AI. It is also observed that statistical analysis enhances AI systems' safe and successful development.

The aim of this book is to investigate the recent research trends in CS using deep learning/ML, which will contribute to the conceptual knowledge network of researchers across the globe. Fundamental concepts of gathering, processing, and analysing the dataset from a batch process, related exhaustive literature review, rigorous experimentation results, and application-oriented approaches are demonstrated in chapters spanning the book. This book presents applications based on real-world problems and experimental research in computational statistics and mathematical modelling for AI. The book describes these transpiring research fields, which will be a valuable resource for researchers and practising engineers.

REFERENCES

1. James E. Gentle, Wolfgang Karl Härdle, Yuichi Mori (2012) "Handbook of Computational Statistics Concepts and Methods," Springer Berlin, Heidelberg, https://doi.org/10.1007/978-3-642-21551-3, 2012.
2. Rahman, A. (2020). Statistics for Data Science and Policy Analysis. Springer.
3. Wang, C., Chen, M. H., Schifano, E., Wu, J., & Yan, J. (2016). Statistical methods and computing for big data. Statistics and its interface, 9(4), 399–414. https://doi.org/10.4310/SII.2016.v9.n4.a1.
4. B. Agarwal, A. Agarwal, P. Harjule & A. Rahman (2022) "Understanding the intent behind sharing misinformation on social media," Journal of Experimental & Theoretical Artificial Intelligence, DOI: 10.1080/0952813X.2021.1960637
5. PR Vishnu, B Agarwal, P Vinod, KA Dhanya, Alice Baroni (2021) "A Deep Learning-Based Model for an efficient Hate-speech Detection in Twitter," in Book chapter in the book titled "Securing Social Networks in Cyberspace," CRC Press., pp:245.
6. Agarwal, B. Financial sentiment analysis model utilizing knowledge-base and domain-specific representation. *Multimed Tools Appl* (2022). https://doi.org/10.1007/s11042-022-12181-y.
7. V. N. Manjunath Aradhya, Mufti Mahmud, D. S. Guru, Basant Agarwal & M. Shamim Kaiser One-shot Cluster-Based Approach for the Detection of COVID–19 from Chest X–ray Images. *Cognitive Computation* 13, 873–881 (2021). https://doi.org/10.1007/s12559-020-09774-w.
8. B. Agarwal, V.E. Balas, L. Jain, R. Poonia, M. Sharma (Eds.), Deep Learning Techniques for Biomedical and Health Informatics (1st ed.), Elsevier (2020), 10.1016/C2018-0-04781-7
9. Chowdhury, M.M.H., Rahman, A., & Islam, M. R. (2018). Protecting data from malware threats using machine learning technique. In Proceedings of the 2017 12th IEEE Conference on Industrial Electronics and Applications (ICIEA) (pp. 1691–1694). IEEE, Institute of Electrical and Electronics Engineers. https://doi.org/10.1109/ICIEA.2017.8283111.

10. Das, S., Rahman, A., Ahamed, A., & Rahman, S. T. (2019). Multi-level models can benefit from minimizing higher-order variations: An illustration using child malnutrition data. Journal of Statistical Computation and Simulation, 89(6), 1090–1110. https://doi.org/10.1080/00949655.2018.1553242
11. Rahman, A., Harding, A., Tanton, R., & Liu, S. (2013). Simulating the characteristics of populations at the small area level: new validation techniques for a spatial microsimulation model in Australia. Computational Statistics and Data Analysis, 57(1), 149–165. https://doi.org/10.1016/j.csda.2012.06.018
12. Sharif, O., Hasan, M. Z., & Rahman, A. (2022). Determining an effective short term COVID-19 prediction model in ASEAN countries. Scientific Reports, 12(2022), 1–11. [12:5083].
13. Rahman, A., & Kuddus, M. A. (2021). Modelling the transmission dynamics of COVID-19 in six high-burden countries. BioMed Research International, 2021, 1–17. [5089184]. https://doi.org/10.1155/2021/5089184.
14. Abdulla, F., Nain, Z., Karimuzzaman, M., Hossain, M. M., & Rahman, A. (2021). A non-linear biostatistical graphical modeling of preventive actions and healthcare factors in controlling COVID-19 pandemic. International Journal of Environmental Research and Public Health, 18(9), [4491]. https://doi.org/10.3390/ijerph18094491
15. Rahman, A., & Harding, A. (2016). Small area estimation and microsimulation modeling. Chapman and Hall/CRC.
16. Rahman, A. (2019). Statistics-based data preprocessing methods and machine learning algorithms for big data analysis. International Journal of Artificial Intelligence, 17(2), 44–65.
17. Sutton RS, Barto AG (2018) Reinforcement learning: An introduction. MIT press, Cambridge, MA.
18. Harjule, P., Rahman, A., & Agarwal, B. (2021). A cross-sectional study of anxiety, stress, perception and mental health towards online learning of school children in India during COVID-19. Journal of Interdisciplinary Mathematics, 24(2), 411–424. https://doi.org/10.1080/09720502.2021.1889780.
19. Agarwal, B., Agarwal, A., Harjule, P., & Rahman, A. (2022). Understanding the intent behind sharing misinformation on social media. Journal of Experimental and Theoretical Artificial Intelligence, 1–15.
20. Friedrich, S., Antes, G., Behr, S. *et al.* (2021) Is there a role for statistics in artificial intelligence?. *Adv Data Anal Classif* 16, 823–846.
21. Moor J (2006) The Dartmouth College artificial intelligence conference: The next fifty years. AI Magazine 27(4), 87–87.
22. Solomonoff RJ (1985) The time scale of artificial intelligence: Reflections on social effects. Human Syst Manag 5(2), 149–153
23. Shen, J., Wang, A., Wang, C., Hung, P.C.K., Lai, C.F.(2017) An efficient centroid-based routing protocol for energy management in WSN-assisted IoT. IEEE Access 5, 18469–18479
24. Wu, M., Yin, B., Vosoughi, A., Studer, C., Cavallaro, J.R., Dick, C. (2013).: Approximate matrix inversion for high-throughput data detection in the large-scale MIMO uplink. In: 2013 IEEE International Symposium on Circuits and Systems (ISCAS), pp. 2155–2158
25. Zhang, L., Zhu, J., Yao, T. (2004): An evaluation of statistical spam filtering techniques. ACM Trans. Asian Lang. Inf. Process. (TALIP) 3(4), 243–269
26. Meghanathan, N. (2017).: Correlation Analysis of decay centrality. In: 6th Computer Science On-line Conference: Cybernetics and Mathematics Applications in Intelligent Systems, pp. 407–418. Springer. ISBN 978-3-319-57263-5.

27. Bystrov, Vitaliy V. et al. (2018). "Development of the Forecasting Component of the Decision Support System for the Regulation of Inter-regional Migration Processes." *Advances in Intelligent Systems and Computing* 859, 60–71. https://doi.org/10.1007/978-3-030-00211-4_7
28. Schmertmann, C.P. (2012): Stationary populations with below-replacement fertility. Demogr. Res. 26, 319–330, Article 14.
29. Ha, L.Q.T. (2009).: Xây dựng công cụ tìm kiếm tài liệu học tập bằng các truy vấn ngôn ngữ tự nhiên trên kho học liệu mở tiếng Việt "Developing a search engine for learning materials using natural language queries in the Vietnamese Open Learning Materials." Master thesis, VNUHCM.
30. Tang, Li, et al. (2018) "Enhancing Concurrent ETL Task Schedule with Altruistic Strategy." *Proceedings of the Computational Methods in Systems and Software*. Springer, Cham, pp. 201–212.
31. Harjule, P.; Poonia, R.C.; Agrawal, B.; Saudagar, A.K.J.; Altameem, A.; Alkhathami, M.; Khan, M.B.; Hasanat, M.H.A.; Malik, K.M. An Effective Strategy and Mathematical Model to Predict the Sustainable Evolution of the Impact of the Pandemic Lockdown. Healthcare 2022, 10, 759.
32. Priyanka Harjule, Vinita Tiwari and Anupam Kumar; (2021) "Mathematical models to predict COVID-19 outbreak: An Interim Review," Journal of Interdisciplinary Mathematics DOI: 10.1080/09720502.2020.1848316
33. Bhavana, A., and A. N. Nandha Kumar. (2018) "An Analytical Modeling for Leveraging Scalable Communication in IoT for Inter-Domain Routing." *Proceedings of the Computational Methods in Systems and Software*. Springer, Cham, pp. 1–11.
34. Abuqaddom, Inas, and Amjad Hudaib. (2018) "Cost-sensitive learner on hybrid SMOTE-ensemble approach to predict software defects." *Proceedings of the Computational Methods in Systems and Software*. Springer, Cham, pp. 12–21.
35. Archana, B., and T. P. Surekha. (2018) "A compressive sensing based channel estimator and detection system for MIMO-OFDMA system." *Proceedings of the Computational Methods in Systems and Software*. Springer, Cham, pp. 22–31.
36. Babutskiy, Vladislav, and Igor Sidorov. "A novel approach to the potentially hazardous text identification under theme uncertainty based on intelligent data analysis." *Proceedings of the Computational Methods in Systems and Software*. Springer, Cham, 2018, pp. 32–38.
37. Badidi, Elarbi. "Towards a message broker based platform for real-time streaming of urban IoT data." *Proceedings of the Computational Methods in Systems and Software*. Springer, Cham, 2018, pp. 39–49.
38. Barot, Tomas, and Radek Krpec. "Alternative approach to fisher's exact test with application in pedagogical research." *Proceedings of the Computational Methods in Systems and Software*. Springer, Cham, 2018, pp. 50–59.
39. Bystrov, V. V., et al. "Development of the forecasting component of the decision support system for the regulation of inter-regional migration processes." *Proceedings of the Computational Methods in Systems and Software*. Springer, Cham, 2018.
40. Dang, Co Ton Minh. "Modeling syntactic structures of vietnamese complex sentences." *Proceedings of the Computational Methods in Systems and Software*. Springer, Cham, 2018, pp. 81–91.
41. Datyev, Igor O., and Andrey M. Fedorov. "Information monitoring of community's territoriality based on online social network." *Proceedings of the Computational Methods in Systems and Software*. Springer, Cham, 2018.

42. Dikovitsky, V. V., and M. G. Shishaev. "Automated extraction of deontological statements through a multilevel analysis of legal acts." *Proceedings of the Computational Methods in Systems and Software*. Springer, Cham, 2018, pp 102–110.
43. Dobesova, Zdena, and Jan Pinos. "Using decision trees to predict the likelihood of high school students enrolling for university studies." *Proceedings of the Computational Methods in Systems and Software*. Springer, Cham, 2018, pp. 111–119.
44. Gruzenkin, Denis V., et al. "Using blockchain technology to improve N-version software dependability." *Proceedings of the Computational Methods in Systems and Software*. Springer, Cham, 2018, pp. 132–137.
45. Chen, Yang, et al. "Enhancing feature selection with density cluster for better clustering." *Proceedings of the Computational Methods in Systems and Software*. Springer, Cham, 2018, pp. 138–150.
46. Priyanka Harjule, Basant Agarwal, Ashish Burdak, Satvik Gupta, Saurav Singh and Shivdeep Singh,"Forecasting and Seasonal Analysis of air quality index using machine learning models during COVID-19 pandemic,"4th International Conference on Computer Networks, Big Data and IoT (ICCBI 2021).
47. Majdisova, Zuzana, Vaclav Skala, and Michal Smolik. "Determination of stationary points and their bindings in dataset using RBF methods." *Proceedings of the Computational Methods in Systems and Software*. Springer, Cham, 2018, pp. 213–224.
48. Barrachina S, Bender O, Casacuberta F, Civera J, Cubel E, Khadivi S, Lagarda A, Ney H, Tomás J, Vidal E, Vilar JM (2009) Statistical approaches to computer-assisted translation. Comput Linguistics 35(1), 3–28.
49. Basant Agarwal, Priyanka Harjule, Lakshit Chauhan, Upkar Saraswat, Parth Agarwal (2021) "Prediction of dogecoin price using deep learning and social media trends." EaI Endorsed Transactions on Industrial Networks and Intelligent Systems. https://dx.doi.org/10.4108/eai.29-9-2021.171188
50. Richa Sharma, Sudha Morwal, Basant Agarwal, "Named Entity Recognition using Neural Language Model and CRF for Hindi Language," In Computer Speech & Language. Volume 74, July 2022, 101356, doi: https://doi.org/10.1016/j.csl.2022.101356, 2022
51. Koch C (2016) How the computer beat the go player. Sci Am Mind 27(4):20–23.
52. Silver D, Hubert T, Schrittwieser J, Antonoglou I, Lai M, Guez A, Lanctot M, Sifre L, Kumaran D, Graepel T, Lillicrap T, Simonyan K, Hassabis D (2018) A general reinforcement learning algorithm that masters chess, shogi, and Go through self-play. Science
53. Kozielski M, Doetsch P, Ney H (2013) Improvements in RWTH's System for Off-Line Handwriting Recognition. In: 2013 12th international conference on document analysis and recognition, IEEE,
54. Juang BH, Rabiner LR (1991) Hidden markov models for speech recognition. Technometrics 33(3), 251–272.
55. European Commission (2020a) https://ec.europa.eu/info/resources-partners/machine-translation-public-administrations-etranslation_en#translateonline, accessed 8.05.2022
56. Chen H, Engkvist O, Wang Y, Olivecrona M, Blaschke T (2018) The rise of deep learning in drug discovery. Drug Discovery Today 23(6),1241–1250
57. Chen CLP, Liu Z (2018) Broad learning system: an effective and efficient incremental learning system without the need for deep architecture. IEEE Trans Neural Netw Learn Syst 29(1), 10–24.

58. Burt JR, Torosdagli N, Khosravan N, RaviPrakash H, Mortazi A, Tissavirasingham F, Hussein S, Bagci U (2018) Deep learning beyond cats and dogs: recent advances in diagnosing breast cancer with deep neural networks. British J Radiol 91(1089):20170545
59. Burton A, Altman DG, Royston P, Holder RL (2006) The design of simulation studies in medical statistics. Stat Med 25(24), 4279–4292
60. Priyanka Harjule, Manva Mohd. Tokir, Tanuj Mehta, Shivam Gurjar , Anupam Kumar, and Basant Agarwal. Journal of Computational Biology.ahead of print http://doi.org/10.1089/cmb.2021.0267
61. Blasiak A, Khong J, Kee T (2020) CURATE.AI: optimizing personalized medicine with artificial intelligence. SLAS TECHNOLOGY: Trans Life Sci Innov 25(2), 95–105
62. Hamburg MA, Collins FS (2010) The path to personalized medicine. N Engl J Med 363(4):301–304
63. Schork NJ (2019) Artificial intelligence and personalized medicine. In: Von Hoff D, Han H (eds) Precision medicine in cancer therapy, cancer treatment and research. Springer, Cham
64. Ng S (2018) Opportunities and challenges: lessons from analyzing terabytes of scanner data. In: Honore B, Pakes A, Piazzesi M, Samuelson L (eds) Advances in economics and econometrics, Cambridge University Press, pp 1–34,
65. McCracken MW, Ng S (2016) FRED-MD: a monthly database for macroeconomic research. J Business Econ Stat 34(4), 574–589.
66. AInow (2020) https://ainowinstitute.org/, accessed 08.05.2022
67. Bundespolizeipräsidium Potsdam (2018) Abschlussbericht Teilprojekt 1 "Biometrische Gesichtserkennung." www.bundespolizei.de/Web/DE/04Aktuelles/01Meldungen/2018/10/181011_abschlussbericht_gesichtserkennung_down.pdf?__blob=publicationFile=1, accessed 07.05.2022
68. FDA (2019) www.fda.gov/media/122535/download, accessed 8.05.2022
69. Dastin, Jeffrey. "Amazon scraps secret AI recruiting tool that showed bias against women." *Ethics of Data and Analytics*. Auerbach Publications, 2018. 296–299.
70. Ntoutsi, Eirini, et al. "Bias in data-driven artificial intelligence systems – An introductory survey." *Wiley Interdisciplinary Reviews: Data Mining and Knowledge Discovery* 10.3 (2020), e1356.
71. Das, Sumit, et al. "Applications of artificial intelligence in machine learning: review and prospect." *International Journal of Computer Applications* 115.9 (2015), 31–41.
72. Shabbir, Jahanzaib, and Tarique Anwer. "Artificial intelligence and its role in near future." *arXiv preprint arXiv:1804.01396* (2018).
73. Searle J (1980) Minds, Brains and Programs. Behavioral Brain Sci 3(3), 417–457.

2 An Improved Random Forest for Classification and Regression Using Dynamic Weighted Scheme

Vikas Jain and Ashish Phophalia*
Indian Institute of Information Technology, Vadodara, India
E-mails: 201671001@iiitvadodara.ac.in
Ashish_p@iiitvadodara.ac.in

CONTENTS

2.1 INTRODUCTION

Classification and regression are the two most crucial tasks in the field of data mining, computer vision, pattern recognition, and machine learning [1]. There are several approaches, such as perceptron learning, naive Bayes classifier, support vector machine, *k*-nearest neighbor, logistic regression, linear regression, decision

DOI: 10.1201/9781003253051-3

tree, neural network, clustering, principal component analysis, and many more for classification and regression [1–4]. These approaches can broadly be classified into supervised and unsupervised categories. Researchers have heavily exploited all of these algorithms over the last two decades with applications to various fields [5–8]. In 2001, Breiman [9] proposed an ensemble of decision tree approach known as random forest (RF).[1] It is one of the most popular choices in the fields of classification and regression. It is a supervised machine learning approach that uses decision trees as the base learners. Decision trees have been around for several years [10–12]. They have become more popular because the ensemble of randomized trees tends to produce better predictions than individual tree predictions. The decision trees are worked as a rule-based system, which answers the question in terms of "yes" or "no" and accordingly moves either left or right side of the tree. In order to be independent, decision trees are trained over the sample generated by bootstrap sampling. Therefore, they can be easily implemented as parallel threads, which results in increasing their operating speed. Hence, decision trees are more prevalent in various domains such as machine learning [9, 11, 12], medical applications [6], and computer vision [2, 13–16].

Decision trees involve several essential parameters like splitting criterion, the number of candidate features selected at a node for splitting, tree depth, and the minimum number of samples required for splitting at an internal node. However, thus far no one has determined the best choice of these parameters [2,17]. Although several attempts have been made in the direction to decide and design the best splitting criterion [7, 18–20], no one has come up with the generalized choice. However, several heuristic approaches have been designed for building the decision tree and forest. A lot of work has been done in the past to improve the performance and consistency of the RF [21–25]. For example, Biau et al. [22] theoretically proved the consistency of the RF constructed by randomly selecting the splitting feature and its value. Further in [25] Biau et al. randomly select the subset of candidate features, and the split point is selected as the midpoint of the selected feature value. The feature with the most significant decrease in impurity is selected as a splitting one. Denil et al. [23] proposed a new variant of RF that differs in terms of feature and splitting point selection as compared to conventional RF. It used Poisson distribution for feature selection and the only subset of randomly selected data samples used for each feature. In [24] Wang et al. proposed two independent Bernoulli distributions to control the splitting feature and splitting point selection process of tree construction. Hence, the study gains the advantages of conventional RF and increases randomization in the construction of decision trees.

The conventional RF decision is based on the decisions made by base classifiers (i.e., decision trees). It considers equal weights to the votes cast by each decision tree [9], and final prediction is based on the majority voting. It should be noted that in real-life applications, datasets consist of a large number of features. However, all the features may not be equally informative; hence the decision trees constructed using less informative features will not contribute significantly. Since randomness is injected during the construction of the decision trees during the selection of data samples and features, decision trees may be constructed on data samples populated by less informative features. In such cases, the equal weightage assigned to the decision tree will lead to poor performance; hence weighted voting is performed. The weight assigned to the decision tree can be categorized as static and dynamic. The static weight is computed during the training phase and is kept fixed for all decision trees throughout the testing phase. On the other hand, the weight factor is computed during the testing phase in the dynamic weighting scheme and keeps varying for all the decision trees. Winham et al. [26] proposed an approach to computing the weight based on the performance of the decision tree calculated using out-of-bag (OOB) samples. Paul et al. [27] initialized the weight factor with one and kept reducing the value based on the decision tree performance based on the OOB samples. Ref. [28] used the entropy or Gini score to compute the confidence level as a weight factor for the decision trees. Note that all these methods do not talk about the relationship between decision trees and test samples. It should be noted that a decision tree may be suitable for classifying some test samples but bad for some other test samples. So, instead of static tree weight, decision tree should have a dynamic weight score for each test sample to achevie a better outcome.

In [29], we introduced the Exponentially Weighted Random Forest (EWRF) classifier that improves the classification and regression accuracy of the forest. In particular, instead of assigning a static weight to each decision tree, a weighing score is calculated based on the similarity between the test sample and the decision tree. The exponential distribution is used to compute the final weight score; hence it is called an EWRF. This weight is not static for decision trees since the weight score keeps changing for the same decision tree; that is why it is a dynamic weighing scheme. Note that the existing weighting schemes do not take into account dynamic voting [26–28, 30]. The EWRF approach has been tested for the classification of hyperspectral images (HSI) from the Indian Pines and Pavia University [31] and Kennedy's Space Center (KSC)). Further, the proposed EWRF approach is applied for the classification of objects over Caltech-101 and Caltech-256. In addition to this, the handwritten digit classification is performed using the EWRF approach.

2.1.1 Proposed Work

The overall elements of this chapter can be summarized as:

- We propose a dynamic weighting mechanism for the RF. It calculates the exponential weight score based on the similarity between the test sample and decision trees.
- The efficacy of the proposed approach is tested over three HSI classification tasks. Earlier, the approach was tested over two Indian Pines and Pavia University images, and results were found to be motivational and published [31]. To this end, further, the work was extended for the classification of KSC hyperspectral images and compared to state-of-the-art methods.
- The EWRF approach was applied over the soil moisture prediction task in remote sensing as a regression application.
- Finally, the EWRF approach was used for the classification of the objects and handwritten digits over the Caltech-101, Caltech-256, and MNIST images.

The outline of this chapter is as follows: we introduce the RF as a classifier and regressor in Section 2.2. Section 2.3 presents the EWRF approach and shows the dynamic weight computation. Section 2.4 describes the experiments conducted for the classification of hyperspectral images, objects, and digit classification. Finally, we present the conclusion of the work in Section 2.5.

2.2 RANDOM FOREST

We briefly discuss here the conventional RF [9] for completeness. It is composed of a pre-defined number of binary decision trees. Each tree in the forest is grown using a bootstrap sample from the training dataset [32]. The decision tree keeps growing until the stoppage criterion is reached. The prediction is based on the class distribution probability computed at the leaf node of the decision tree. In the regressor tree, the output is the average of the outcome associated with data samples present at the leaf node. Thus, the prediction of the RF depends on the outcome of the base classifier (i.e., decision and regressor tree).

Let $D \in \mathbb{R}^{M \times N}$ be a dataset, having a M number of data samples with N attributes. Let the dataset can be categorized into $y_i \in \{y_1, y_2, \ldots, y_c\}$ class labels. The dataset is divided into training set $D_1 \in \mathbb{R}^{M_1 \times N}$ such that $M_1 < M$ and testing set $D_2 \in \mathbb{R}^{(M-M_1) \times N}$.

In the next section, we describe the computation of class distribution probability for the prediction of class labels. Furthermore, it describes the prediction process of RF. It also describes the weight factor that affects the overall predictive power of the RF.

2.2.1 Random Forest as Classifier

Decision trees constructed over the sub-sampled dataset are generated using bootstrap sampling performed over the training set D_1 [32]. During the testing phase, the test sample guided either the left or right side of a decision tree based on the classifier value present at the internal nodes until it reached the leaf node. The class probability is computed at the leaf node to predict the class label by the decision tree.

The computation of the class distribution probability, $p^t_{j,h}(\mathbf{x}_k)$, for the y_j class at the terminal node h in the decision tree t, corresponding to test sample $\mathbf{x}_k$, can be given as Eq. 2.1:

$$p^t_{j,h}(\mathbf{x}_k) = \frac{1}{n_h} \sum_{i \in h} \mathbb{I}(y_i = y_j) \tag{2.1}$$

here: n_h is total number of samples present at the node node h, and $\mathbb{1}()$ is an indicator function.

The decision tree assigns the class label based on maximum class distribution probability. Therefore, Eq. 2.2 computes the maximum class distribution probability out of c number of classes corresponding to the test sample $\mathbf{x}_k$:

$$\hat{y}^t_j = \max_{1 \le j \le c} \{p^t_{j,h}(\mathbf{x}_k)\} \tag{2.2}$$

Every decision tree assigns the class label based on the maximum class probability. The RF predicts the final class label based on the majority voting by assigning equal weight (i.e., $\mathbb{1}$ to each decision tree). Eq. 2.3 to count each decision tree's votes or count the number of votes corresponding to y_j class. Let T be the total number of trees present in the forest; hence the total number of votes against each class label can be computed as

$$V(y_i = y_j) = \sum_{t=1}^{T} \mathbb{1} \cdot \mathbb{J}(\hat{y}_j^t) \tag{2.3}$$

Here: $\mathbb{J}(\cdot)$ is an indicator function, and T is the total number of trees. Finally, the RF performs the majority voting to assign the class label; i.e., it assigns the class label of the class that got the maximum votes:

$$\hat{y}_k = \max_{1 \le j \le c}\{V(y_i = y_j)\} \tag{2.4}$$

2.2.2 Random Forest as Regressor

In the regressor task, instead of classifying the data, decision trees have to predict the output value; that's why they are known as regressor trees. In regression, each data point is associated with a single real value as a output (i.e., $y_i \in \mathbb{R}$). In order to construct the regressor tree, mostly Mean Squared Error (MSE) is used as the splitting criterion, and the rest of the tree construction process remains the same as the decision tree. During the testing, the test sample is passed to each regressor tree and guided either to the left or right side of the tree based on the internal node value. In the end, the prediction is performed at the leaf node of the tree. The predicted value is the mean of the outcome of data samples available at the leaf node.

The output for the test sample $\mathbf{x}_k$, at the leaf node h, by the regressor tree t, can be computed as the mean of output values of data samples available at the leaf node. It can be given as Eq. 2.5:

$$\hat{y}_h^t = \frac{1}{n_h} \sum_{i \in h} y_i \tag{2.5}$$

The final output of the RF is the average of output values predicted by each decision tree. These trees are equally contributing. Hence, the computation of the final output using forest can be given as

$$\hat{y}_k = \frac{1}{T} \sum_{t=1}^{T} \mathbb{1} \cdot (\hat{y}_h^t) \tag{2.6}$$

Note that the conventional RF assigns equal weight to each decision tree. However, a decision tree may not always correctly classify all the data samples. Therefore, instead of assigning equal weight, dynamic weight should be assigned. It results in

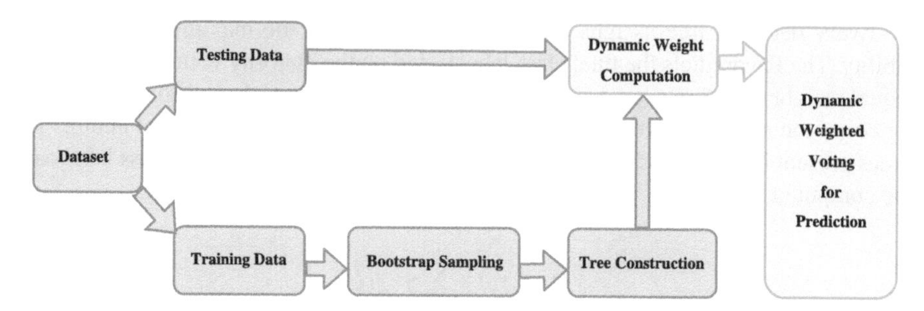

FIGURE 2.1 The pipeline architecture of the proposed method to show the overall trees construction and dynamic weight computation phase.

an increase in the overall performance of RF. To this end, in the next section, the proposed method is described in detail, and how the dynamic weight is computed for each decision tree with respect to each test sample is described in detail.

2.3 PROPOSED METHOD

The proposed approach computes the weight based on the similarity between the test sample and the decision tree. The weight computation is done during the testing phase; for example, Figure 2.1 shows the pipeline architecture of the proposed approach. Initially, the dataset is divided into training and testing set. The training data is used for the decision tree construction using bootstrap sampling. Later, the dynamic weight score is computed during the testing between test sample and decision tree. Finally, the dynamic weighted voting is performed for the prediction.

2.3.1 Dynamic Weight Score Computation

In the RF, it should be noted that a decision tree may not always correctly classify all test samples; hence instead of assigning static weight, dynamic weight should be assigned. Hence, once all decision trees are constructed, the dynamic weight score is computed for each decision tree with respect to the test sample. It is based on the euclidean distance between the test sample and the decision tree. Therefore, this weight gets changed as the test sample gets changed. This dynamic weight computation considers the similarity between the test sample and the decision tree.

At an internal node of a decision tree, t, the data is split at the F_i feature value. A test sample $\mathbf{x} = \{a_1, a_2, ..., a_j, ..., a_N\}$ is guided either to the left, if the corresponding attribute value of the test sample, $\mathbf{x}$, is less than or equal to split point value τ, i.e., $(a_j^{\mathbf{x}} \leq \tau)$ or right subtree otherwise, as $(a_j^{\mathbf{x}} > \tau)$, and move down until it reaches one of the leaf nodes of the decision tree. The similarity measure is computed as the sum of the squared distance between corresponding attribute values in the test sample, $\mathbf{x}$, and participating internal nodes value, F_i, in the path of the decision tree t. Let there be k number of internal nodes in the path traversed by the test sample $\mathbf{x}$ in the decision tree t; hence the similarity can be computed as

$$d = \sum_{l=1}^{k} \| F_i - a_j^{\mathbf{x}} \|_2; \quad \forall F_i \in t; a_j \in x \tag{2.7}$$

Thus, $\{d_1, d_2, ..., d_T\}$ distances are computed for each test sample, with respect to $\{t_1, t_2, ..., t_T\}$ decision trees, respectively. Note that distance and the exponential weight are inversely proportional (i.e., the smaller the value of d, the more the weight score for the decision tree). In other words, the more similarity between the test sample and the decision tree. Hence, the weight computed for each decision tree is directly proportional to the similarity measure between the test sample and decision tree. Therefore, as the test sample changes, the similarity measurement keeps changing; hence the corresponding weight associated with a decision tree is changing. This results in a change in the overall weight score. Note that we are using the traversal path of the test sample in each decision tree, which would give the idea that leaf nodes would have similar samples. The overall similarity measurement between the test sample and the decision tree is shown in Figure 2.2. The final weight score is computed using an exponential distribution applied over the similarity measure. Therefore, the exponential weight score is computed using Eq. 2.8.

$$W_{\mathbf{x}}^{t} = \frac{1}{Z} \exp\left\{ -\frac{\sum \| F_i - a_j^{\mathbf{x}} \|_2}{\alpha} \right\} \tag{2.8}$$

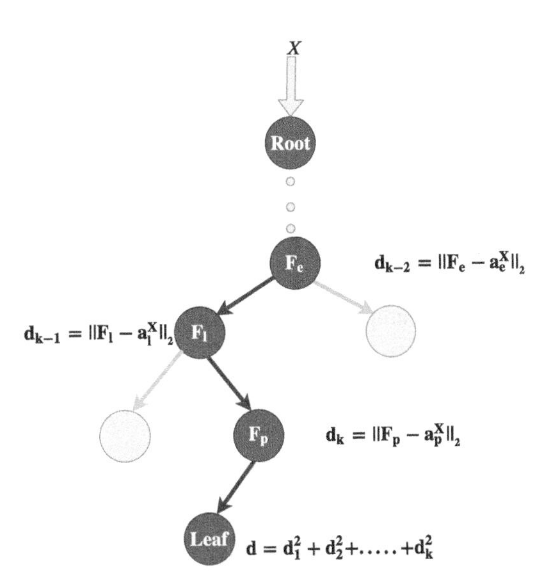

FIGURE 2.2 An example to show the calculation of similarity measure during testing. In this example an test sample **x** follows the path marked as bold blue lines (F_e, F_l and F_p) to reach up to leaf node. The distance is calculated at nodes appearing in the path followed by the test sample. All distances are sum up at the leaf node and use it to compute the exponential weight score for the decision tree.

Here Z is the normalizing term, which is the sum of weights of all decision trees. The $\alpha \in [0,1]$ is one of the hyper-parameter used to adjust the weight score. Similarly, an exponential weight score can be computed for all the decision trees in the forest. Once the weight factor is computed, Eq. 2.3 can be modified as Eq. 2.9 in the proposed EWRF approach.

$$V(y_i = y_j) = \sum_{t=1}^{T} (W_{\mathbf{x}}^t) \cdot \mathbb{J}(\hat{y}_j^t) \tag{2.9}$$

Similarly, for the regressor tree, Eq. 2.6 in the proposed EWRF approach turns out to be

$$\hat{y}_k = \frac{1}{T} \sum_{t=1}^{T} (W_{\mathbf{x}}^t) \cdot \hat{y}_h^t \tag{2.10}$$

Hence, in the proposed EWRF approach, weighted voting is performed using Eq. 2.9 and Eq. 2.10 for the prediction of class value in the classification and output value in the regression tasks, respectively. The overall architecture of the proposed EWRF approach is shown in Figure 2.3. The overall working steps for the proposed EWRF are described in algorithm 1. Note that the distances and hence the corresponding weight do not affect the classification and regression task directly. It is only a way to distinguish between trees and test samples, which helps to improve the decision capability of the forest. The conventional RF is the particular case of the proposed approach. If we assign equal weights to each decision tree, the exponentially weighted forest acts as conventional RF.

Algorithm 1: Exponential Weighted Random Forest

Input: Dataset $D \in \mathbb{R}^{M \times N}$, T: # trees

Output: Predicted class or output value

Step 1: Divide the dataset into training set $D_1 \in \mathbb{R}^{M^1 \times N}$ and testing set $D_2 \in \mathbb{R}^{(M-M^1) \times N}$

Step 2: Perform bootstrap sampling T times and construct decision or regression tree

Step 3: Testing over D_2:

Step 3.1: for *each* $\mathbf{x} \in D_2$ **do**

Step 3.1.1: for *tree* $t = 1$, **to** T **do**

Step 3.1.2: Compute the distance between test sample and decision tree as:

$$d = \sum_{l=1}^{k} \| F_i - a_j^{\mathbf{x}} \|_2; \quad \forall F_i \in t; a_j \in x$$

Step 3.1.3: Compute exponential weight score as:

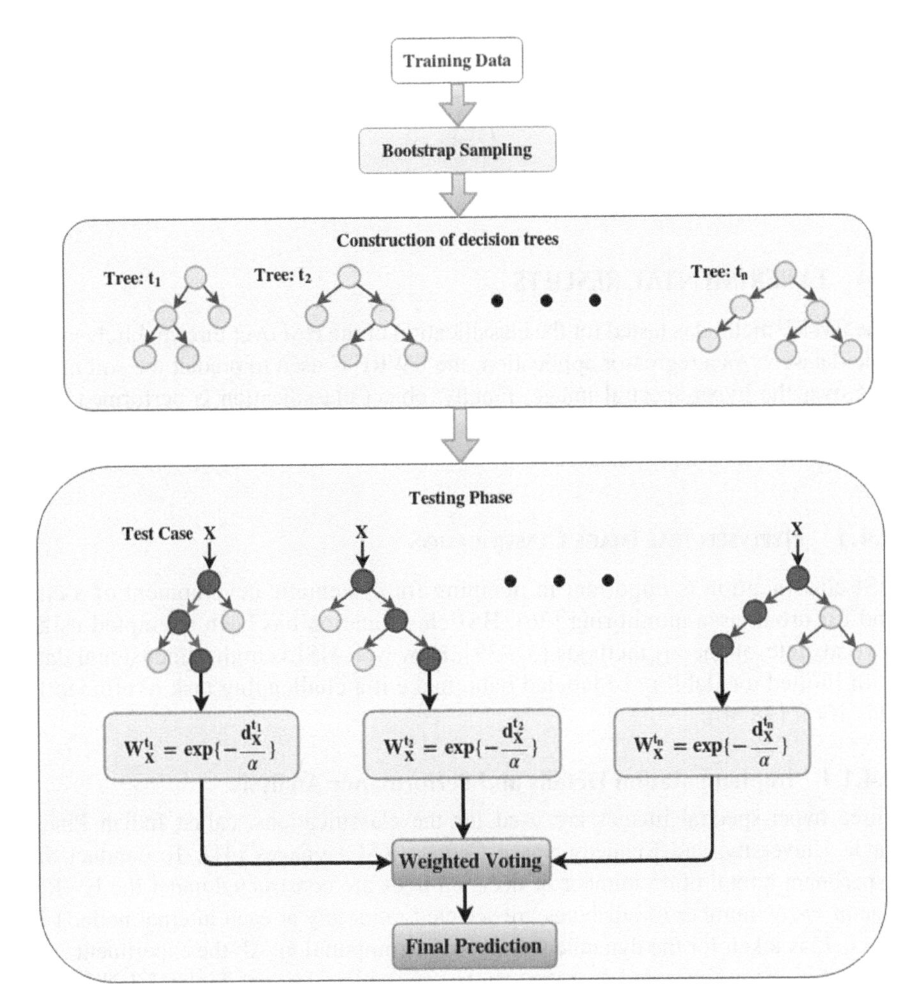

FIGURE 2.3 The overall architecture of the proposed Exponential Weighted Random Forest (EWRF) approach.

$$W_x^t = \frac{1}{Z} \exp\left\{ -\frac{\sum \| F_i - a_j^x \|_2}{\alpha} \right\}$$

end

Step 3.2: Normalize the weight score computed for each tree

Step 3.3: In classification task, predict class value as:

$$V(y_i = y_j) = \sum_{t=1}^{T} (W_x^t) \cdot \mathbb{J}(\hat{y}_j^t)$$

(OR) In regression task, predict the output value as:

$$\hat{y}_k = \frac{1}{T}\sum_{t=1}^{T}(W_x^t)\cdot\hat{y}_h^t$$

end

2.4 EXPERIMENTAL RESULTS

The EWRF method is tested for the classification of the HSI over three publicly available datasets. As a regressor application, the EWRF is used to predict the soil moisture over the hyper-spectral image. Finally, object classification is performed over the Caltech-101 [33], Caltech-256 [34], and handwritten digit classification over the well-known MNIST [35] dataset.

2.4.1 Hyperspectral Image Classification

HSI classification is important in planning for systematic development of a city and for urban area monitoring [36]. HSI classification has been attempted using several state-of-the-art methods [37–39]. However, HSI is high-dimensional data with limited availability of labeled data, make it a challenging task to efficiently classify it [38,40].

2.4.1.1 Implementation Details and Performance Analysis

Three hyper-spectral images are used for the classifications, called Indian Pines, Pavia University, and Kennedy Space Center (KSC) images [41]. To conduct the experiment a total of 45 number of decision trees are constructed under the EWRF. The $m = \sqrt{N}$ number of attributes are selected randomly at each internal node. The $\alpha = 0.75$ is taken for the dynamic weight score computation. All the experiments are iterated over ten times and average results are presented in the Tables 2.1–2.3. The train-test ratio and the other parameters for implementing the Enriched RF (ERF) [42], SVM-3DG [43], Boosted RF (BRF) [44], and Cascaded RF (CRF) [45] are taken from the corresponding references.

To analyze the performance of the state-of-the-art methods, three measures – overall accuracy (OA), average accuracy (AA), and kappa coefficient (κ) [46] – are used. The EWRF method is compared with the RF [9], ERF [42], BRF [44], SVM-3DG [43], and CRF [45]. In the experiment, we did not extract any features from the given datasets. Instead, we considered each pixel value as a feature in our approach.

The results for Indian Pines are shown in Table 2.1. One can see that the EWRF method shows improvement for eight classes out of sixteen classes in terms of class-specific accuracies. The OA, AA, and κ coefficient values of EWRF are significantly better than the state-of-the-art methods. Table 2.2 describes the obtained results for Pavia University. The AA and κ coefficient values obtained by the EWRF method are better than using all approaches except the SVM-3DG [43] method. These results

TABLE 2.1
Classification Accuracies Computed over Indian Pines Image Using State-of-the-art Methods and EWRF Method

Class	RF [5]	ERF [38]	BRF [40]	SVM 3DG [39	CRF [41]	EWRF
	58.9	65.1	66.9	**96.8**	70.8	85.1
	64.2	70.7	71	58.5	79.7	**88.1**
	53.6	52.4	59.6	**93.4**	69.9	81.2
	97.5	98.3	97.9	96.4	**98.8**	93.7
	80.3	83.1	86.4	86.1	91.4	**97.9**
	95.3	93.9	95.2	95.8	94.7	**99.1**
	41.7	59.1	53.9	**100**	83.6	85
	37.1	41.4	45.1	**100**	53.2	99.8
	68	84	79	**100**	90	69.1
	66	71.2	70.9	68.9	81.9	**93.7**
	84.5	86.8	86.5	78.6	87	**97.3**
	44.3	58.5	56.4	**96.9**	47.4	89
	93.4	94.5	93.4	94.2	90.1	**98.8**
	95	94.9	94.2	77.8	96.6	**99**
	35.1	35.1	40.4	**95.4**	63.3	73.3
	80.2	81.7	80.6	98.7	89	**99.2**
OA	74.4	77.6	78.2	81.1	82.2	**93.3**
AA	68.2	73.2	71.1	89.8	80.5	**90.6**
κ	70.5	74.2	74.9	78.6	79.6	**90.1**

TABLE 2.2
Classification Accuracies Computed over Pavia University Image Using State-of-the-art Methods and EWRF Method

Class	RF [5]	ERF [38]	BRF [40]	SVM 3DG [39]	CRF [41]	EWRF
	89.5	91.6	92	**97.4**	86.7	97.2
	100	**100**	99.7	97.3	99.8	99.4
	99.3	99.4	99.2	89.4	**99.6**	89.3
	98.1	93.5	96.1	97.3	96.2	95.3
	99.3	98	97.8	**99.6**	98.1	99.5
	53.3	59.2	63.3	**98.4**	72.8	76.4
	71.8	66.6	74.7	**98.2**	86.9	86.2
	82.5	84.3	83.9	84	86.6	**96.9**
	89.8	90	92.5	99.9	93.4	**100**
OA	72.3	74.8	77.7	**96.1**	82.2	95
AA	86.5	86.9	88.8	**95.7**	91.1	93.4
κ	65.7	68.2	71.9	**94.8**	76.6	93.2

TABLE 2.3
Classification Accuracies Computed over Kennedy Space Center Image Using State-of-the-art Methods and EWRF Method

Class	RF [5]	ERF [38]	BRF [40]	CRF [41]	EWRF
	94.2	94.3	94.2	94.3	**97**
	84.6	84.8	85	85.4	**91.5**
	88.4	88.3	88	88.1	**95.4**
	64.9	65.3	64.4	70.7	**80.7**
	51.8	53	54.3	**58.3**	54.8
	46.7	52.6	57.5	**58.1**	55.3
	80.9	82	79.2	84.1	**94.8**
	80.5	81.8	86.6	87.7	**94.1**
	91.5	92.8	93.7	96.1	**98**
	86.6	87.9	87.2	**91.1**	89.8
	96.1	96.1	96.7	**97.8**	97.3
	89	89.2	90.8	93.4	**95.8**
	99.4	99.5	99.7	99.7	**99.8**
OA	80.3	87.5	88.5	90.1	**92.4**
AA	86.1	82.1	82.9	84.9	**88**
κ	85.5	86.1	87.2	88.9	**92**

are encouraging since, in the SVM-3DG [43] the first features are extracted to perform classification, and the proposed approach performs better without extracting any features. Similarly, in the case of CRF [45] the feature selection approach is used. Still, the proposed EWRF outperforms. The corresponding original image, reference map, and classification map for both Indian Pines and Pavia University are shown in Figure 2.4.

The results for KSC images are shown in Table 2.3. We can observe from the table that the EWRF shows better class-specific accuracies for nine classes out of thirteen. The results obtained using the EWRF method for the values of OA, AA, and κ coefficients are significantly better than with the state-of-the-art methods.

2.4.2 Regression Application: Soil Moisture Prediction in Hyperspectral Dataset

2.4.2.1 Implementation Details and Performance Analysis

In this subsection, the performance of the proposed approach is investigated over the application of soil moisture prediction in hyper-spectral data. The details of the dataset can be found in Ref. [47]. The distribution of data points is shown in Figure 2.5. A total of 45 regressor trees are constructed under the EWRF approach. The dataset is divided into 50% training and 50% testing. The MSE and the coefficient of determination R^2 are computed to evaluate the performance of the EWRF method.

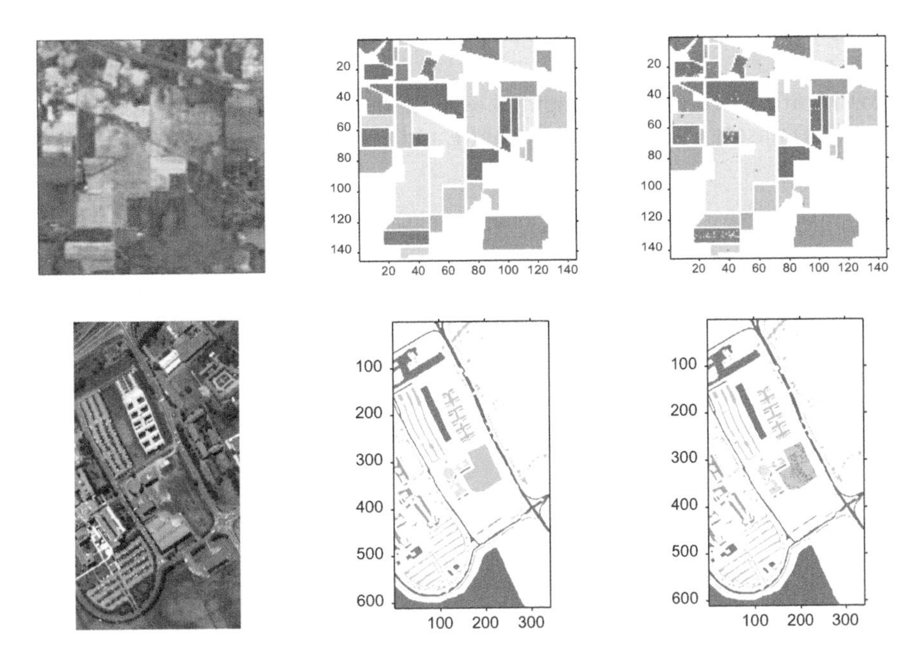

FIGURE 2.4 Classification map showing the original image, a referenced map to show ground truth, and predicted labels corresponding to each class labels for Indian Pines (Row 1) and Pavia University (Row 2) images.

TABLE 2.4
Mean Squared Error (MSE) and Coefficient of Determination (R^2) Values Computed for Soil Moisture Prediction over Hyperspectral Dataset

SN	Method	MSE	R^2
1	RF	1.35	0.89
2	EWRF	**1.18**	**0.91**

Table 2.4 shows the obtained results for the mean squared error and coefficient of determination. The proposed weighted scheme improves the regression capability by assigning more weight to the tree, showing more similarity with respect to the test sample. Therefore, the overall prediction performance improves.

2.4.3 Object and Digit Classification

2.4.3.1 Implementation Details and Performance Analysis

This subsection describes implementation details and performance measures over the Caltech-101, Caltech-256, and MNIST datasets. A total of 40 training images and 40 testing images chosen randomly and disjointed from each other are considered

TABLE 2.5
Classification Accuracies Computed over Caltech-101, Caltech-256, and MNIST Datasets Using Conventional RF and EWRF Approach

SN	Dataset	#Train images	#Test image	#Classes	Tree Depth	RF	EWRF
1	Caltech-101	40	40	26	12	56.3	**64.6**
2	Caltech-256	30	30	100	12	52.9	**59.8**
3	MNIST	40	40	10	12	50.6	**57.8**
4	Caltech-101	40	40	26	17	79.4	**85.2**
5	Caltech-256	30	30	100	17	58.9	**62**
6	MNIST	40	40	10	17	59.3	**69.4**
7	Caltech-101	40	40	26	20	87.1	**89.8**
8	Caltech-256	30	30	100	20	71.8	**74.1**
9	MNIST	40	40	10	20	64.8	**70.2**

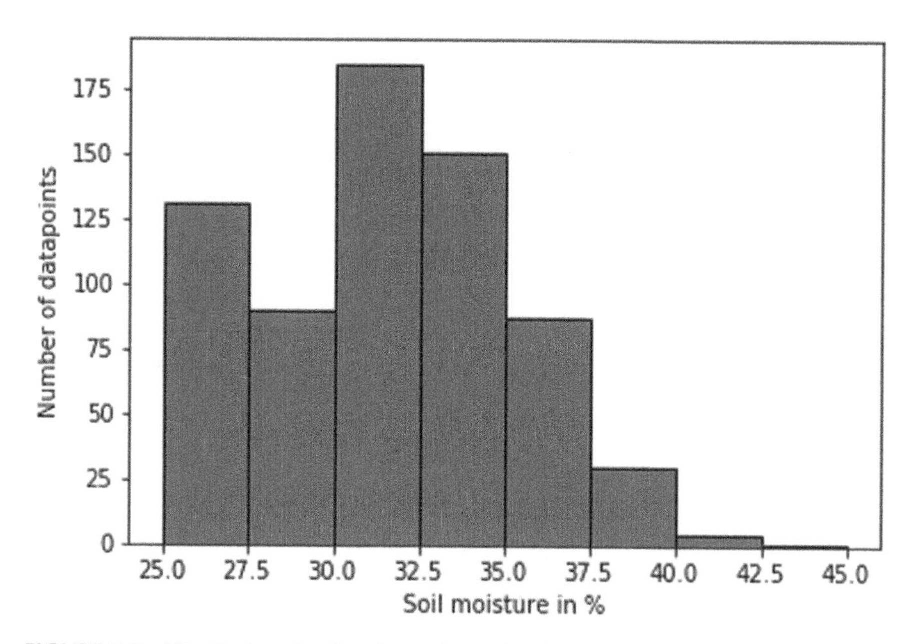

FIGURE 2.5 Distribution of soil moisture data points in hyperspectral data.

for the Caltech-101 [33] and MNIST [35] datasets. In the case of the Caltech-256 [34] dataset, a total of 30 images are considered for each training and testing set. Therefore, we have a total 26 classes for Caltech-101 and 100 classes for Caltech-256 datasets. To extract the features over these images, the Bag of Visual Words (BoVW) concept is used [48]. The K-means clustering is performed over all the training and testing images. A total of 45 trees are constructed under the EWRF approach.

Table 2.5 shows the obtained results over all the datasets using the RF and EWRF approach. We can observe that the EWRF shows significant improvement for all the

FIGURE 2.6 Examples of objects (a to f) which are miss-classified by the conventional RF and correctly classified by EWRF.

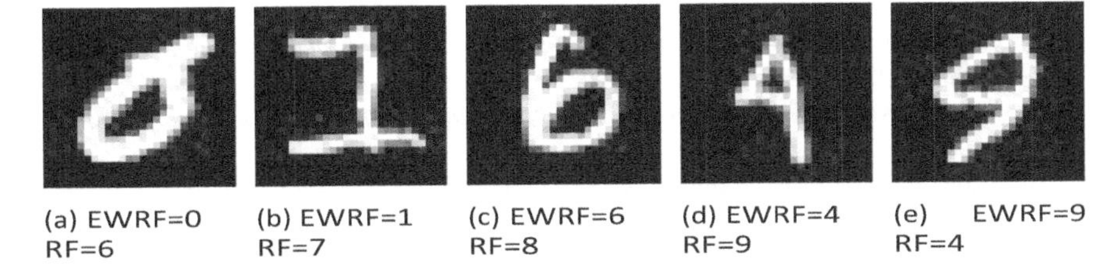

FIGURE 2.7 Examples of digits (a to e) which are miss-classified by the conventional RF and correctly classified by the EWRF.

datasets over the different values of the decision tree depth. We have also shown the examples that are misclassified by the conventional RF method and correctly classified by the EWRF approach (shown in Figure 2.6). Similarly, for the MNIST datasets, examples are shown in image form, which are misclassified by the conventional RF approach and correctly classified by the EWRF approach (shown in Figure 2.7).

2.5 CONCLUSION

The conventional RF predicts the class labels based on majority voting by considering equal weightage of every decision tree. While several weighting schemes have been proposed in the past, they assign the static weight to the decision tree. In this work, based on similarity between the test samples and decision trees, weight is assigned to the decision tree while performing majority voting. Thus, as the test sample changes, the computed weights also change. Hence, the derived weights are dynamic. The proposed EWRF method showed improvement over three hyper-spectral image datasets for classification. The results showed significant improvement in terms of average accuracy, overall accuracy, and kappa coefficient compared to current state-of-the-art methods. Furthermore, the efficacy of EWRF was tested over one more hyper-spectral dataset as a regression application. Finally, the object and digit classification tasks were performed using the EWRF approach. Overall, the obtained results confirmed the effectiveness of the proposed approach over the various domains.

CONFLICT OF INTEREST

On behalf of all authors, the corresponding author states that there is no conflict of interest.

NOTE

1 Referred to as conventional RF throughout the text.

REFERENCES

1. Christopher M Bishop. *Pattern recognition and machine learning*. springer, 2006.
2. Antonio Criminisi and Jamie Shotton. *Decision forests for computer vision and medical image analysis*. Springer Science & Business Media, 2013.
3. Cha Zhang and Yunqian Ma. *Ensemble machine learning: methods and applications*. Springer, 2012.
4. Shigeo Abe. *Support vector machines for pattern classification*, volume 2. Springer, 2005.
5. Chowdhury, M.M.H., Rahman, A., & Islam, M. R. (2018). Protecting data from malware threats using machine learning technique. In Proceedings of the 2017 12th IEEE Conference on Industrial Electronics and Applications (ICIEA) (pp. 1691–1694). IEEE, Institute of Electrical and Electronics Engineers. https://doi.org/10.1109/ICIEA.2017.8283111
6. Rahman, A. (2019). Statistics-based data preprocessing methods and machine learning algorithms for big data analysis. International Journal of Artificial Intelligence, 17(2): 44–65.
7. Uddin, M. G., Nash, S., Mahammad Diganta, M. T., Rahman, A., & Olbert, A. I. (2022). Robust machine learning algorithms for predicting coastal water quality index. Journal of Environmental Management, 321(11), 1–16. [115923]. https://doi.org/10.1016/j.jenvman.2022.115923
8. Uddin, M. G., Nash, S., Rahman, A., & Olbert, A. I. (2023). Performance analysis of the water quality index model for predicting water state using machine learning techniques. Process Safety and Environmental Protection, 1–30. https://doi.org/10.1016/j.psep.2022.11.073
9. Leo Breiman. Random forests. *Machine learning*, 45(1):5–32, 2001.

10. L Breiman, J Friedman, CJ Stone, and RA Olshen. Classification algorithms and regression trees, 1984.
11. J Ross Quinlan. C4. 5: Programming for machine learning. *Morgan Kauffmann*, 38:48, 1993.
12. Pierre Geurts, Damien Ernst, and Louis Wehenkel. Extremely randomized trees. *Machine learning*, 63(1):3–42, 2006.
13. Ashish Phophalia and Pradipta Maji. Multimodal brain tumor segmentation using ensem- ble of forest method. In *International MICCAI Brainlesion Workshop*, pages 159–168. Springer, 2017.
14. Hongliu Cao, Simon Bernard, Robert Sabourin, and Laurent Heutte. Random forest dissimilarity based multi-view learning for radiomics application. *Pattern Recognition*, 88:185–197, 2019.
15. AV Lebedev, Eric Westman, GJP Van Westen, MG Kramberger, Arvid Lundervold, Dag Aarsland, H Soininen, I K-loszewska, P Mecocci, M Tsolaki, et al. Random forest ensembles for detection and prediction of alzheimer's disease with a good between-cohort robustness. *NeuroImage: Clinical*, 6:115–125, 2014.
16. Chunyu Hu, Yiqiang Chen, Lisha Hu, and Xiaohui Peng. A novel random forests based class incremental learning method for activity recognition. *Pattern Recognition*, 78:277–290, 2018.
17. Hemant Ishwaran. The effect of splitting on random forests. *Machine Learning*, 99(1):75–118, 2015.
18. Vikas Jain, Ashish Phophalia, and Jignesh S Bhatt. Investigation of a joint splitting criteria for decision tree classifier use of information gain and gini index. In *TENCON 2018-2018 IEEE Region 10 Conference*, pages 2187–2192. IEEE, 2018.
19. Laura Elena Raileanu and Kilian Stoffel. Theoretical comparison between the gini index and information gain criteria. *Annals of Mathematics and Artificial Intelligence*, 41(1):77–93, 2004.
20. Lior Rokach and Oded Maimon. Top-down induction of decision trees classifiers-a sur- vey. *IEEE Transactions on Systems, Man, and Cybernetics, Part C (Applications and Reviews)*, 35(4):476–487, 2005.
21. Angshuman Paul, Dipti Prasad Mukherjee, Prasun Das, Abhinandan Gangopadhyay, Appa Rao Chintha, and Saurabh Kundu. Improved random forest for classification. *IEEE Transactions on Image Processing*, 2018.
22. GÃŠrard Biau, Luc Devroye, and GÃAbor Lugosi. Consistency of random forests and other averaging classifiers. *Journal of Machine Learning Research*, 9(Sep):2015–2033, 2008.
23. Misha Denil, David Matheson, and Nando De Freitas. Narrowing the gap: Random forests in theory and in practice. In *International conference on machine learning*, pages 665–673, 2014.
24. Yisen Wang, Shu-Tao Xia, Qingtao Tang, Jia Wu, and Xingquan Zhu. A novel consistent random forest framework: Bernoulli random forests. *IEEE transactions on neural networks and learning systems*, 2018.
25. GÃŠrard Biau. Analysis of a random forests model. *Journal of Machine Learning Research*, 13(Apr):1063–1095, 2012.
26. Stacey J Winham, Robert R Freimuth, and Joanna M Biernacka. A weighted random forests approach to improve predictive performance. *Statistical Analysis and Data Mining: The ASA Data Science Journal*, 6(6):496–505, 2013.
27. Angshuman Paul and Dipti Prasad Mukherjee. Enhanced random forest for mitosis de-tection. In *Proceedings of the 2014 Indian conference on computer vision graphics and image processing*, page 85. ACM, 2014.

28. Pritom Saha Akash, Md Eusha Kadir, Amin Ahsan Ali, Md Nurul Ahad Tawhid, and Mohammad Shoyaib. Introducing confidence as a weight in random forest. In *2019 Inter- national Conference on Robotics, Electrical and Signal Processing Techniques (ICREST)*, pages 611–616. IEEE, 2019.
29. Vikas Jain, Jaya Sharma, Kriti Singhal, and Ashish Phophalia. Exponentially weighted random forest. In *International Conference on Pattern Recognition and Machine Intelli- gence*, pages 170–178. Springer, 2019.
30. Yiyi Liu and Hongyu Zhao. Variable importance-weighted random forests. *Quantitative Biology*, 5(4):338–351, 2017.
31. Vikas Jain and Ashish Phophalia. Exponential weighted random forest for hyperspectral image classification. In *IGARSS 2019-2019 IEEE International Geoscience and Remote Sensing Symposium*, pages 3297–3300. IEEE, 2019.
32. L Breiman. Bagging predictors, technical report. *UC Berkeley*, 1994.
33. Li Fei-Fei, Rob Fergus, and Pietro Perona. Learning generative visual models from few training examples: An incremental bayesian approach tested on 101 object categories. In *2004 Conference on Computer Vision and Pattern Recognition Workshop*, pages 178–178. IEEE, 2004.
34. Gregory Griffin, Alex Holub, and Pietro Perona. Caltech-256 object category dataset. 2007.
35. Yann LeCun, L´eon Bottou, Yoshua Bengio, Patrick Haffner, et al. Gradient-based learning applied to document recognition. *Proceedings of the IEEE*, 86(11):2278–2324, 1998.
36. David A Landgrebe. *Signal theory methods in multispectral remote sensing*, volume 29. John Wiley & Sons, 2005.
37. Farid Melgani and Lorenzo Bruzzone. Classification of hyperspectral remote sensing im- ages with support vector machines. *IEEE Transactions on geoscience and remote sensing*, 42(8):1778–1790, 2004.
38. Rongrong Ji, Yue Gao, Richang Hong, Qiong Liu, Dacheng Tao, and Xuelong Li. Spectral- spatial constraint hyperspectral image classification. *IEEE Transactions on Geoscience and Remote Sensing*, 52(3):1811–1824, 2014.
39. Pedram Ghamisi, Javier Plaza, Yushi Chen, Jun Li, and Antonio J Plaza. Advanced spectral classifiers for hyperspectral images: A review. *IEEE Geoscience and Remote Sensing Magazine*, 5(1):8–32, 2017.
40. Zhi He, Lin Liu, Ruru Deng, and Yi Shen. Low-rank group inspired dictionary learning for hyperspectral image classification. *Signal Processing*, 120:209–221, 2016.
41. Hyperspectral dataset. http://lesun.weebly.com/hyperspectral-data-set.html. On- line Accessed: 25-September-2019.
42. Dhammika Amaratunga, Javier Cabrera, and Yung-Seop Lee. Enriched random forests. *Bioinformatics*, 24(18):2010–2014, 2008.
43. Xiangyong Cao, Lin Xu, Deyu Meng, Qian Zhao, and Zongben Xu. Integration of 3- dimensional discrete wavelet transform and markov random field for hyperspectral image classification. *Neurocomputing*, 226:90–100, 2017.
44. Yohei Mishina, Ryuei Murata, Yuji Yamauchi, Takayoshi Yamashita, and Hironobu Fu- jiyoshi. Boosted random forest. *IEICE Transactions on Information and systems*, 98(9):1630–1636, 2015.
45. Youqiang Zhang, Guo Cao, Xuesong Li, and Bisheng Wang. Cascaded random forest for hyperspectral image classification. *IEEE Journal of Selected Topics in Applied Earth Observations and Remote Sensing*, 11(4):1082–1094, 2018.
46. Liguo Wang and Chunhui Zhao. *Hyperspectral Image Processing*. Springer, 2016.

47. Hyperspectral benchmark dataset on soil moisture. https://zenodo.org/record/1227837#.XmYNw_fhVuQ.html. Online Accessed: 20-January-2020.
48. Jun Yang, Yu-Gang Jiang, Alexander G Hauptmann, and Chong-Wah Ngo. Evaluating bag-of-visual-words representations in scene classification. In *Proceedings of the interna- tional workshop on Workshop on multimedia information retrieval*, pages 197–206. ACM, 2007.

3 Study of Computational Statistical Methodologies for Modelling the Evolution of COVID-19 in India during the Second Wave

Harshit, Anubhuti Mittal and Priyanka Harjule
Department Of Mathematics, Malaviya National Institute of Technology Jaipur, India
2020pma5108.@mnit.ac.in
2020pma5104@mnit.ac.in
priyanka.maths@mnit.ac.in

CONTENTS

DOI: 10.1201/9781003253051-4

3.1 INTRODUCTION

The coronavirus disease or severe acute respiratory syndrome coronavirus-2 (SARS-CoV-2), in short called COVID-19, first emerged in the city of Wuhan located in China in late December 2019 [1]. This virus resembles Middle East Respiratory Syndrome coronavirus and Severe Acute Respiratory Syndrome coronavirus, and is said to be originated from bats [2]. COVID-19 has had catastrophic results and has resulted in more than 5 million deaths worldwide and 0.4 million deaths in India alone as on November 2021.

Symptoms of the COVID-19 include fever, shortness of breath, and sometimes loss of taste and smell [3]. Coronavirus has spread to more than 200 countries [4] and around 114 million cases have been reported as of February, 2021 worldwide [5]. Following the first cases of this respiratory tract infection in the city of Wuhan during late December 2019, COVID-19 quickly spread worldwide and ultimately the World Health Organization (WHO) declared it a global epidemic on March 11, 2020 [6].

India has reported the first case of Covid on January 27, 2020 in Kerala, when a 20-year-old female reported to General Hospital, Thrissur, Kerala, with some symptoms including dry cough and sore throat [7].

In the course of first wave, the Indian administration implemented a national lockdown on March 25, 2020. It was the first such lockdown in the country resulting in closure of schools, colleges, shops, buses, trains, etc. This resulted in a well-managed infection rate and the Indian policy was admired worldwide. In fact, many countries in Europe implemented lockdowns in different formats during the first wave of COVID-19 [8]. Daily Covid cases increased sharply by mid-September with more than 90,000 cases recorded each day, Covid cases declined to less than 15,000 in January 2021 [9].

Shockingly, during the second wave, it was observed that there was a lack of cooperation and lack of well-executed planning by the government of India. The second wave of COVID-19 starting in March 2021 was much more devastating and harmful than the first wave, with shortage of hospital beds, oxygen cylinders, medicines, and other medical equipment in different areas of the country. Vaccine shortage was also a key problem. By the end of April, India had the highest number of active cases of COVID-19 in the world. On April 30, 2021, over 400,000 new cases were reported in a single day [10].

Mathematical modelling is an important tool in understanding the nature and behaviour of disease dynamics in the population or the activities involved in the health system due to the disease [11–17]. SIR model of disease epidemiology is a powerful model to help us predict and understand the physical behaviour of the disease. It is an economic way of measurement and allows us to predict the future cases. COVID-19 is the deadliest disease of 21st century, and it is important to study its effects, behaviour, and impact on society. The SEIRD (Susceptible-Exposed-Infected-Recovered-Dead) and SEIR (Susceptible-Exposed-Infected-Recovered) models are being used here to study the effect of COVID-19 during the second wave (March 1 to October 10, 2021).

In order to model COVID-19 dynamics, we make use of computational statistical methodologies in the following study. Computational statistics is a combination of statistics and computer science. It refers to the methods of statistics that are implemented using computational methods. It uses numerical methods and statistics to solve numerous problems like estimation of parameters, testing of hypothesis, statistical modeling, etc. The main goal of computational statistics is to convert raw data into knowledge, but the main focal point here is computer-intensive statistical methods for the cases having very large sample size and non-homogeneous sets of data. It aims at designing an algorithm so that the statistical methods can be implemented on computers.

Since the field of data analytics and computational statistics is growing day by day, capturing the advances, current trends, and methodologies and evaluating their impact is of urgent importance. Advancements in simulation and graphical analysis also add to the pace of the statistical analytics field. Computational statistics play a very important role in various fields, be it financial applications like derivative pricing and risk management or biological applications like computational biology, bioinformatics, etc., that closely impact the lives of people.

The main objectives of this study include developing an understanding of computationally intensive modern methods for statistical learning, analyzation, inference, exploration of data, etc. In particular, here two computational methodologies for statistical learning will be introduced, first being the deterministic approach through stability analysis with the second being the stochastic approach through continuous time Markov chain and multi-type branching process. In addition, this study will also exemplify and demonstrate how to apply the above techniques in an effective manner for large realistic data sets like that of COVID-19.

3.2 RELATED WORK

In this section, we discuss the existing models on COVID-19 Epidemic.

Ian Cooper et al. [21] studied the novel coronavirus using the classical SIR (Susceptible-Infected-Removed) model for the communities of China, South Korea, India, Australia, Italy, and Texas in the United States. Sam Moore et al. [3] included the vaccination component in their research. They also highlighted the importance of co-morbidities in severity of the disease. They have shown how vaccination should be done in minimum time with maximum population covered. Brody H.Foy et al. [22] used the age-structured SEIR model with social contact matrices and compared the different COVID-19 vaccine allocation strategies. Mukesh Jakhar et al. [4] studied the COVID-19 outbreak in 24 states of India using the SIR model. They also calculated the basic reproduction number R_0 for different states of India. Ibrahim Halil Aslan et al. [23] analysed the local outbreak of COVID-19 in Hubei and Turkey using the SEIR model. They forecasted the peak of the outbreak and determined the total number of cases and deaths by means of social distancing, quarantine, etc. Shilei Zhao et al. [24] used the SUQC (Susceptible, Un-quarantined infected, Quarantined infected, Confirmed infected) model to predict the trend of COVID-19 in China.

Wenzhang Huang et al. [25] discussed the stability of disease-free equilibrium and studied the dynamics of STDs (Sexually Transmitted Diseases). Collin D. Funk et al. [26] focused on the development of COVID-19 vaccines, which is once again the most crucial factor to control the pandemic. Pierre Magal et al. [10] compared the deterministic SIR model and IBM (Individual-Based Model), which is stochastic in nature. Kunal Menda et al. [27] considered the extension of SEIRD model as R-SEIRD model, which has MSE (mean squared error) lower than that of the SEIRD model. They fitted the curves using CE-EM (certainty equivalent expectation maximization) method. Vishnu Vytla et al. [28] covered the Gaussian Model, which works on the central limit theorem and compared it with SIR, SIRD, and SEIRD models. Maher Ala'raj et al. [29] developed the SEIRD and ARIMA (Auto-Regressive Integrated Moving Average) model to analyse the US COVID-19 data and predicted the output for future with confidence intervals. Debashis Saikia et al. [30] did the modelling and analysis of COVID-19 using the SEIR model. They considered the time varying transmission rates, which gives more accurate results than time-independent rates. B. Malavika et al. [31] used the logistic growth model and standard SIR model for calculating the peak of COVID-19. Kelly R. Moran et al. [32] compared the epidemic forecasting with that of weather forecasting and concluded that it is more challenging to forecast epidemics than weather. Gang Xie [33] took the stochastic approach and studied the COVID-19 spread in the UK and Australia using Monte Carlo Simulation. Jose M. Carcione et al. [34] did the simulation using deterministic methods for the SEIR model. They took different R_0 values and calculated the peak of disease according to it.

Danane et al. [35] used an isolation strategy to investigate COVID-19 dynamics using stochastic model. They incorporated white noise and Levy jump perbutations in all model components. They also investigated dynamic properties of the stochastic model around equilibrium of the deterministic model. Hussain et al. [36] proposed a mathematical model including environmental white noise to study the COVID-19 spread. To analyse the model computationally, they developed a numerical scheme. They also predicted disease dynamics with the help of numerical simulations. Sher et al. [37] described the COVID-19 dynamics by considering a fractional order model using non-singular Kernel type. They used Laplace Adomian decomposition method (LADM) to solve the problem in a semi-analytical way. Peng et al. [38] applied the partical swarn optimization (PSO) algorithm to the system for parameter estimation. He introduced stochastic infection and seasonality and hence found the parameters and non-linear dynamics. Dorman et al. [39] surveyed and developed numerical methods for continuous time Markov branching process, which he applied to four models.

Shah Hussain et al. [40] provided a stochastic mathematical model to describe the spread of COVID-19. They built suitable Lyapunov functions and applied it to their model. They analysed the prevalence and extinction of the disease.

P. van den Driessche et al. [20] calculated the basic reproduction number R_0 (using the next-generation matrix technique). The authors present analysis of compartmental epidemiological models that possess a disease-free equilibrium. In the deterministic approach presented in this study, we used a similar technique to apply it on the SEIRD model and further used the obtained relation to calculate the transmission rate at the time of the second wave of COVID-19 in India (March

1 to 10 October, 2021). We further analysed the stability of basic reproduction number (R_0) for the SEIRD model.

Pakwan Riyapan et al. [18] discussed the positivity of the solution for the SEIQRD model. We took a similar approach to analyse the positivity of the solution for the SEIRD model.

Olabode and Culp et al. [42] proposed deterministic and stochastic models to study COVID-19 dynamics in Wuhan, China. They used the classical SEIR framework to formulate a system of ordinary differential equations and continuous time Markov chain to formulate the stochastic model, which is based on the Ordinary Differential Equation (ODE) model. They also used multi-type branching process approximation to predict the possibility of disease extinction. In the stochastic approach presented in this study, we applied similar methods on an SEIR model to analyse the spread of coronavirus in India and estimate the probability of it becoming extinct after the second wave.

Furthermore, Allen [43] used a numerical method called the Gillespie algorithm for numerical simulation of the SIR continuous time Markov chain model. We extend this approach on our SEIR model to numerically simulate the sample paths and estimate the probability of extinction of Covid after the second wave.

Vipin Tiwari et al. [44] calculated the recovery rate and death rate for the period of January 30 to July 10 2020 using the linear regression method. We extended a similar approach on our models (both deterministic and stochatic) to calculate the recovery rate and death rate for the second wave (i.e., 1st March to 10th October 2021 using linear regression method in MATLAB-2020. Further mathematical model prediction of COVID-19 peak was examined for the same period.

3.3 METHODOLOGY

In this section, we first introduce the SIR model, which is the most general form of compartmental models. We then discuss the SEIRD model for proceeding with the deterministic approach and SEIR model for modelling with the stochastic approach.

3.3.1 Preliminaries

The SIR model is a type of disease model that is used to study and analyse the growth of a particular disease during a particular time. In the SIR model the S stands for number of individuals who could potentially attract the disease, I is the infected people that currently have the disease, and R is the removed population.

The SIR model can be solved by deterministic methods such as solving differential equations numerically. In the SIR model the equations involved are as follows:

$$\frac{dS}{dt} = -aI(t)S(t) \tag{3.1}$$

$$\frac{dI}{dt} = aI(t)S(t) - rI(t) \tag{3.2}$$

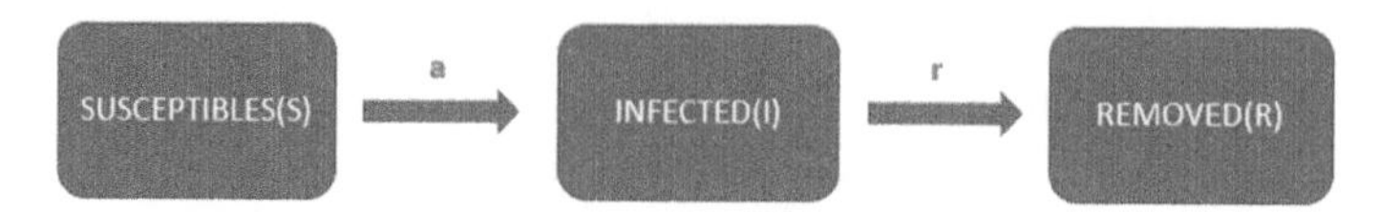

FIGURE 3.1 Schematic diagram of SIR model.

$$\frac{dR}{dt} = rI(t) \tag{3.3}$$

where:

a is the rate of contact or transmission rate.
r is the rate at which infected are getting removed as either recovered or dead.

The SIR model was first proposed by Ross and Hudson in 1916–1917 and further contributed by Kermack and McKendrick in 1927–1932 [10].

3.3.2 Deterministic Approach

3.3.2.1 Proposed Model 1

Here in this chapter the extended version of the SIR model (i.e., SEIRD model) is being used, which aims to study the exposed people (E) and dead people (D) as well.

In the extended model the whole population is categorised into five categories – SEIRD (susceptible-exposed-infectious-recovered-dead) – to analyse the progress of COVID-19 under its second wave in India.

Susceptible (S) in the model are the people who could potentially catch the disease, Exposed (E) are those who attract the disease but are not yet infectious, Infected (I) are those who currently have the disease and either are asymptomatic or have some symptoms, Recovered (R) are those who have recovered from the disease, and Dead (D) are those individuals who are deceased from the disease [18].

The SEIRD model can be expressed with the help of the following ordinary differential equations:

$$\frac{dS}{dt} = -aI(t)S(t) \tag{3.4}$$

$$\frac{dE}{dt} = aI(t)S(t) - pE(t) \tag{3.5}$$

$$\frac{dI}{dt} = pE(t) - wI(t) - oI(t) \tag{3.6}$$

$$\frac{dR}{dt} = oI(t) \tag{3.7}$$

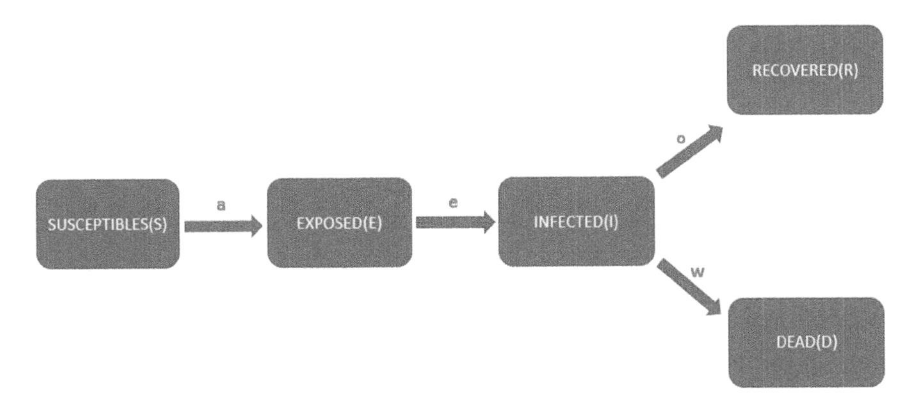

FIGURE 3.2 Schematic diagram of SEIRD model.

$$\frac{dD}{dt} = wI(t) \tag{3.8}$$

In the above equations 'a' is the transmission rate at which people are moving from susceptible to infected, and it was calculated with the help of R_0, transmission rate relation, 'p', is the rate at which exposed people are becoming infected (pre-infectious period assumed to be 5.2 days [19]); 'o' is the rate at which people are becoming recovered, and was calculated using curve fitting in MATLAB-2020; 'w' is the rate at which people are becoming deceased from the infection, and it was also calculated using curve fitting in MATLAB-2020.

Like every mathematical model, this SEIRD model works on the following assumptions:

1. Initially whole of the population is susceptible to infection.
2. People infected once cannot get reinfected again.
3. Second wave starts from March 1, 2021.
4. Data collected is accurate and reliable.

3.3.2.2 Calculation of Basic Reproduction Number (R_0)

The basic reproduction number is one of the most crucial parameters in studying the disease model. To determine the basic reproduction number (R_0) we use the next-generation matrix technique [27]. It is a technique that was developed by Diekmann et al. in 1990 who calculated R_0 as the largest eigenvalue of the next-generation matrix [48] consisting of disease classes.

To implement this, we take disease classes into consideration; i.e.,

$$\frac{dE}{dt} = aI(t)S(t) - pE(t) \tag{3.9}$$

$$\frac{dI}{dt} = pE(t) - wI(t) - oI(t) \quad (3.10)$$

where x=(E, I) are disease compartments.

Set

$$\mathcal{F} = \begin{bmatrix} aSI \\ 0 \end{bmatrix}$$

where entries of $\mathcal{F}$ are rate of arrival of new infection in compartment I (Infected).

$$\mathcal{V} = \begin{bmatrix} pE \\ -pE + wI + oI \end{bmatrix}$$

where entries of $\mathcal{V}$ are rate of moving of individuals out of compartment I by all other factors. The values of Jacobian matrices F(x) and V(x) are

$$\mathcal{F} = \begin{bmatrix} \frac{\partial F1}{\partial E} & \frac{\partial F1}{\partial I} \\ \frac{\partial F2}{\partial E} & \frac{\partial F2}{\partial I} \end{bmatrix}$$

$$= \begin{bmatrix} 0 & a \\ 0 & 0 \end{bmatrix}$$

∵ Initially the whole of the population is Susceptible so initial S is taken as 1.

$$V = \begin{bmatrix} \frac{\partial V1}{\partial E} & \frac{\partial V1}{\partial I} \\ \frac{\partial V2}{\partial E} & \frac{\partial V2}{\partial I} \end{bmatrix}$$

$$= \begin{bmatrix} p & 0 \\ -p & w+o \end{bmatrix}$$

The next-generation matrix is given by $K = FV^{-1}$ where

$$V^{-1} = \frac{1}{p(w+0)} \begin{bmatrix} w+o & 0 \\ p & p \end{bmatrix}$$

Spectral radius of matrix K will give us the basic reproduction number R_{0} :

$$K = FV^{-1} = \frac{1}{p(w+0)}\begin{bmatrix} 0 & a \\ 0 & 0 \end{bmatrix}\begin{bmatrix} w+o & 0 \\ p & p \end{bmatrix}$$

$$= \begin{bmatrix} \frac{a}{w+o} & \frac{a}{w+0} \\ 0 & 0 \end{bmatrix}$$

Eigen values of K are $\lambda_1 = 0, \lambda_2 = \frac{a}{w+o}$.

Spectral radius (i.e., magnitude of largest eigenvalue) is R_0. So, $R_0 = \frac{a}{w+o}$

where,

a is rate at which susceptible are getting exposed to the disease.
o is the rate of recovery of infected people.
w is the rate of death of infected people.

3.3.2.3 Stability Analysis

In this section we discuss the positivity of the solution and stability of disease-free equilibrium (DFE).

3.3.2.3.1 Positivity of the Solutions

As a differential equation can have both negative and positive solutions and here, we are calculating the number of people in a particular compartment, which cannot be negative so it is important to study the positivity of the solutions.

Theorem 1 *If* $S(0) > 0$, $E(0) > 0$, $I(0) > 0$, $E(0) > 0$, $D(0) > 0$, *then the solutions* $S(t) > 0$, $E(t) > 0$, $I(t) > 0$, $E(t) > 0$, $D(t) > 0$ *for all* t [18].

Proof. Considering the first equation of SEIRD model, we have that

$$\frac{dS}{dt} = -aI(t)S(t) - mS(t)$$

$$\Rightarrow \frac{dS}{dt} = (-aI(t) - m)S(t)$$

By using the variable separable method and integrating

$$\int \frac{dS}{S} = \int (-aI(t) - m)dt$$

$$\log S + \log S0' = \int (-aI - m)dt$$

$$\Rightarrow S = S_0 e^{\int (-(aI-m)dt}$$

Since the initial value S_0 and this exponential function is always positive $\Rightarrow S(t)$ is positive for all t > 0.

Using the same idea, we can claim that $S(t) > 0, E(t) > 0, I(t) > 0, R(t) > 0, D(t) > 0$. [11]

Lemma 1 *Assume q(K) to be the extreme real part of eigenvalues of K.*
$\rho(K)$ is the dominant eigenvalue of A.

A matrix $K = [kij]$ follows the Z pattern if $kij \leq 0 \ \forall \ i \neq j$, if K follows Z pattern and q(K) > 0 implies K will be a non-singular M-Matrix [13]

Lemma 2 *Assume S to be a non-singular M-Matrix and let T and TS^{-1} follow the Z pattern. Implies T is a non-singular M-Matrix iff TS^{-1} is a non-singular M-Matrix* .[13]

3.3.2.3.2 Stability of the Disease-Free Equilibrium

The DFE a0 is said to be locally asymptomatically stable if all the eigenvalues of Df(a0) have negative real part and is unstable when any eigenvalue of Df(a0) has a positive real part [13]

Theorem 2 *For the given disease transference model satisfying all the differential equations, if a0 is the DFE of model, then a0 is locally asymptomatically stable if reproduction number $R_0 < 1$ and unstable if $R_0 > 1$.* [13].

Proof. We have

$$F = \begin{bmatrix} 0 & a \\ 0 & 0 \end{bmatrix}$$

$$V = \begin{bmatrix} p & 0 \\ -p & w+o \end{bmatrix}$$

Let G1=F-V.

Now it can be observed that V is non-singular M-Matrix here and F is positive:

$$-G1 = V - F = \begin{bmatrix} p & 0 \\ -p & w+o \end{bmatrix}$$

$$-\begin{bmatrix} 0 & a \\ 0 & 0 \end{bmatrix}$$

$$= \begin{bmatrix} p & -a \\ -p & w+o \end{bmatrix}$$

which follows the Z pattern. Thus, q(G1) $< 0 \Rightarrow -G(1)$ is a non-singular M-Matrix where q(G1) is the extreme real part of eigenvalues of G1.

Now FV^{-} is non-negative:

$$FV^{-1} = \begin{bmatrix} \frac{a}{w+o} & \frac{a}{w+o} \\ 0 & 0 \end{bmatrix}$$

$$-G1V^{-1} = \begin{bmatrix} p & -a \\ -p & w+o \end{bmatrix} \begin{bmatrix} \frac{w+o}{p(w+o)} & 0 \\ \frac{p}{p(w+0} & \frac{p}{p(w+o)} \end{bmatrix}$$

$$= \begin{bmatrix} p & -a \\ -p & w+o \end{bmatrix} \begin{bmatrix} \frac{1}{p} & 0 \\ \frac{1}{w+0} & \frac{1}{w+o} \end{bmatrix}$$

$$FV^{-1} = \begin{bmatrix} \frac{a}{w+o} & \frac{a}{w+o} \\ 0 & 0 \end{bmatrix}$$

$$\begin{bmatrix} 1-\frac{a}{w+o} & \frac{-a}{w+o} \\ 0 & 1 \end{bmatrix}$$

$= I - FV^{-1}$ also follows the Z pattern.

Now using Lemma 1 by taking S=V and T=-G1=V-F
we have
$-G1$ is non-singular M-Matrix if and only if $I - FV^{-1}$ is a non-singular M-Matrix.

Due to FV^{-1} being non negative all eigenvalues of FV^{-1} have magnitude less than or equal to $\rho(FV^{-1})$.

So,
$I - FV^{-1}$ is a non-singular M-Matrix if and only if $\rho(FV^{-1}) < 1$:

$$I - FV^{-1} = \begin{bmatrix} 1 & 0 \\ 0 & 1 \end{bmatrix} \begin{bmatrix} \frac{a}{w+o} & \frac{a}{w+o} \\ 0 & 0 \end{bmatrix}$$

$$= \begin{bmatrix} 1 - \frac{a}{w+o} & \frac{-a}{w+o} \\ 0 & 1 \end{bmatrix}$$

Eigenvalues are $1 - \frac{a}{w+o}, 1$.

Since it is non-singular M-Matrix

$$1 - \frac{a}{w+o} > 0$$

$$1 > \frac{a}{w+o}$$

$$\text{i.e., } R0 < 1$$

Hence, $q(S1) < 0 \Leftrightarrow R0 < 1$.

Therefore, the DFE of model at $x0 = (1,0,0,0,0,0)$ is locally asymptomatically stable for $R0 < 1$. Similarly, it follows that

$q(G1) = 0 \Leftrightarrow$ -G1 is a singular M-Matrix
$\Leftrightarrow I - FV^{-1}$ is a singular M-Matrix
$\Leftrightarrow \rho(FV^{-1}) = 1$

Similarly, we can do it for non-singular case. Therefore,

$q(G1) = 0 \Leftrightarrow R_0 = 1$
implies

$q(G1) > 0 \Leftrightarrow R_0 > 1$ (i.e., DFE of model at a0) is locally unstable for $R_0 > 1$.

3.3.2.4 Data and Implementation

In this section we discuss the source and description of data used and how this data was implemented in the SEIRD model.

3.3.2.4.1 *Description of the Data Used*

Data of COVID-19 in India from March 1, 2021 to October 10, 2021 was collected from the open source website COVID19.org [48] and initial values of S,E,I,R,D were taken from worldometers.info/coronavirus/country/India website [46]. In this study, initial total susceptible population was determined by multiplying the total population of India (N) by a factor of 10^{-3} [47], initial exposed population is assumed to be nil, number of cases reported on February 28, 2021 were taken as initial value of I, and number of recoveries and number of deaths on February 28, 2021 were taken to be initial value of R and D, respectively.

3.3.2.4.2 *Implementation*

The five equations were solved using ODE45 in MATLAB-2020, which is a built-in numerical differential equations solver in MATLAB. R_0 value is taken to be 1.6 as assumption from previous data for the best fit. Transmission rate (a) was calculated using the relation $R_0 = \frac{a}{w+o}$. Recovery rate (o) and Deceased rate (w) were determined using curve-fitting method in MATLAB considering the number of confirmed cases and number of recovered cases, number of deceased cases, respectively [44].

3.3.3 Stochastic Approach

3.3.3.1 Proposed Model 2

In the following section, first we discuss the ordinary differential equations (ODE) model, which will serve as a basis for the actual stochastic model. We then introduce the stochastic model using continuous time Markov chain to investigate and predict the spread of COVID-19 in India during the second wave. Furthemore, we use a multitype branching process to estimate the probability of disease extinction after the second wave. We then compare the probabilities obtained from both methods and present the final conclusion.

3.3.3.1.1 *Underlying Deterministic Model*

The following SEIR ODE model [50]with parameter values that are constant will serve as a foundation to set up a continuous time Markov model.

The entire human population is subdivided into four sections: **Susceptibles** (S^*), **Exposed** (E^*), **Infected** (I^*), **and Recovered** (R^*) where susceptibles are individuals who are capable of catching the virus, exposed are the ones who have caught the disease but are not yet infectious (i.e., not yet capable of transmitting it to others),

TABLE 3.1
values and sources of parameters

Parameter	Parameter value	Source
a	0.5150	calculated using $R_0 = 1.6$ value
p	0.1923	fixed incubation period of 5.2 days
o	0.8642	estimated using curve fitting
w	0.009794	estimated using curve fitting

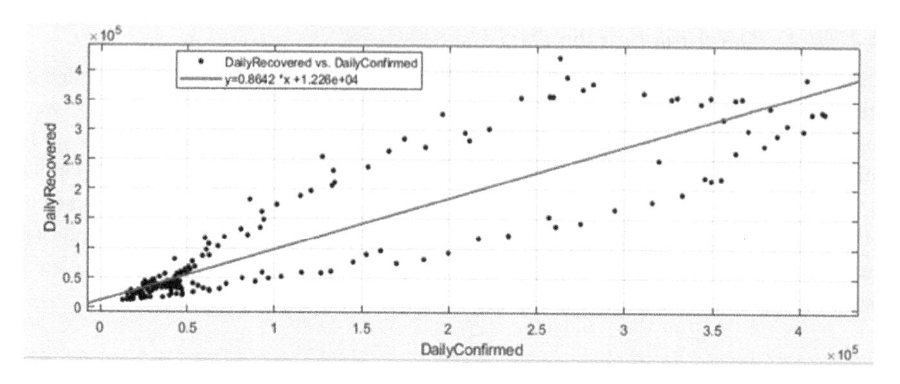

FIGURE 3.3 Calculation of recovery rate(o) in India by the method of linear regression with adjusted R-square: 0.7939 in MATLAB.

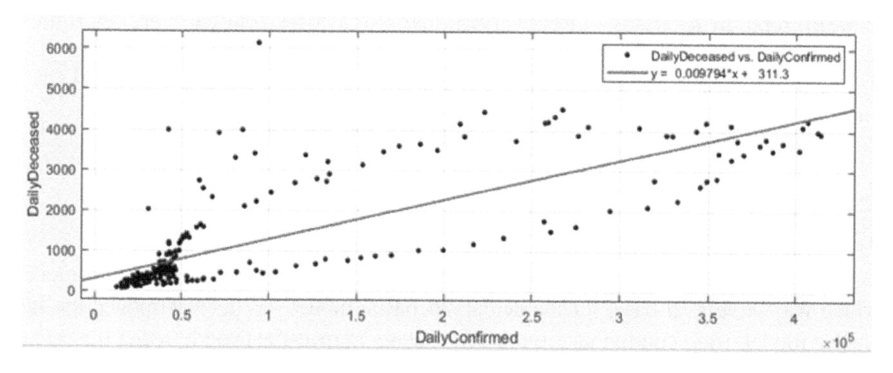

FIGURE 3.4 Determination of death rate in India by the method of linear regression with adjusted R-square: 0.6183 in MATLAB.

infected are those who are infectious and capable of transferring it to others, while recovered are those individuals who have recovered from the disease.

Corresponding ordinary differential equations:

$$\frac{dS^*}{d\tau} = b - \alpha I^* \frac{S^*}{N} - dS^*$$

$$\frac{dE^*}{d\tau} = \alpha I^* \frac{S^*}{N} - (d+\beta)E^*$$

$$\frac{dI^*}{d\tau} = \beta E^* - (d+\gamma+\lambda)I^*$$

$$\frac{dR^*}{d\tau} = \gamma I^* - dR^*$$

where

N is the **total size of the population** at some time τ
b and d are the **natural birth and death rates.**
N_0 is the **initial size of population.**
α is the **transmission rate** ie the rate at which the infection is transmitted.
β is the **latent period** (i.e., the period between the occurrence of the infection and the onset of infectiousness).
γ is the **infection period** (i.e., the period in which a person remains infected with the disease).
λ is the **mortality rate** induced by the disease.

- **Calculation of R_0 for SEIR model**
  R_0 is the number of people who can catch the infection from one infected host. The expression for this is calculated by the next-generation matrix technique as follows:
  First, we take the disease classes (E^* and I^*) into consideration; i.e.,

$$\frac{dE^*}{d\tau} = \frac{\alpha I^* S^*}{N} - dE^* - \beta R^*$$

$$\frac{dI^*}{d\tau} = \beta E^* - (d + \gamma + \lambda)I^*$$

Set

$$\mathcal{F} = \begin{bmatrix} \frac{\alpha S^* I^*}{N} \\ 0 \end{bmatrix}$$

where entries of $\mathcal{F}$ are rate of arrival of new infection in compartment I (Infected).

$$\mathcal{V} = \begin{bmatrix} (d+\beta)E^* \\ -\beta E^* + (d+\gamma+\lambda)I^* \\ \end{bmatrix}$$

where entries of $\mathcal{V}$ are rate of moving of individual out of compartment I by all other factors. The values of Jacobian matrices F(x) and V(x) are

$$F = \begin{bmatrix} \frac{\partial F1}{\partial E^*} & \frac{\partial F1}{\partial I^*} \\ \frac{\partial F2}{\partial E^*} & \frac{\partial F2}{\partial I^*} \end{bmatrix}$$

$$= \begin{bmatrix} 0 & \dfrac{\alpha S^*}{N} \\ 0 & 0 \end{bmatrix}$$

$\because$ Initially the whole of the population is Susceptible so $S^* = \mathrm{N}$.

$$V = \begin{bmatrix} \dfrac{\partial V1}{\partial E} & \dfrac{\partial V1}{\partial I} \\ \dfrac{\partial V2}{\partial E} & \dfrac{\partial V2}{\partial I} \end{bmatrix}$$

$$= \begin{bmatrix} d+\beta & 0 \\ -\beta & d+\gamma+\lambda \end{bmatrix}$$

The next-generation matrix is given by $K = FV^{-1}$ where

$$V^{-1} = \frac{1}{(d+\beta)(d+\gamma+\lambda)} \begin{bmatrix} d+\gamma+\lambda & 0 \\ \beta & d+\beta \end{bmatrix}$$

Spectral radius of matrix K will give us the basic reproduction number R_0:

$$K = FV^{-1} = \frac{1}{(d+\beta)(d+\gamma+\lambda)} \begin{bmatrix} 0 & \alpha \\ 0 & 0 \end{bmatrix} \begin{bmatrix} d+\gamma+\lambda & 0 \\ \beta & d+\beta \end{bmatrix}$$

$$= \begin{bmatrix} \dfrac{\alpha\beta}{(d+\beta)(d+\gamma+\lambda)} & \dfrac{\alpha}{(d+\gamma+\lambda)} \\ 0 & 0 \end{bmatrix}$$

Eigen values of K are $\lambda_1^* = 0, \lambda_2^* = \dfrac{\alpha\beta}{(d+\beta)(d+\gamma+\lambda)}$

Spectral radius (i.e., magnitude of largest eigenvalue) is R_0.
So,

$$R_0 = \frac{\alpha\beta}{(d+\beta)(d+\gamma+\lambda)}$$

3.3.3.1.2 Continuous-Time Markov Chain Model (CTMC)

Let $\mathrm{Y}(\tau) = (S^*(\tau), E^*(\tau), I^*(\tau), R^*(\tau))$ be a discrete random vector where $S^*(\tau)$, $E^*(\tau)$, $I^*(\tau)$, $R^*(\tau) \in \{0,1,\ldots,N\}$, $\tau \in [0,\infty]$.

$\Delta Y(\tau) = Y(\tau + \Delta\tau) - Y(\tau)$ represents the change $[\tau + \Delta\tau]$ during $[\tau, \Delta\tau]$.
s, e, i, r represent values of discrete random variables from the set {0,1,...,N}.
For all i, j, τ, the infinitesimal transition probabilities are defined as:

$$Pr(Y(\tau+\Delta\tau) = j / Y(\tau) = i) = \delta_{ij} + q_{ij}\Delta\tau + o(\Delta\tau)$$

where δ_{ij} is the kronecker delta.

$$\delta_{ij} = \begin{cases} 1 & i = j \\ 0 & i \neq j \end{cases}$$

i.e., δ_{ij} is 1 if we are transitioning from one state to that same state while it is 0 if we are transitioning from a state to some other state.

Therefore, the associated probabilities [36] of transitiong from state (s, e, i, r) to state (s+k,e+l,i+m,r+n) for $\Delta\tau > 0$; i.e.,

$$\begin{aligned} P_{(s,e,i,r),(s+k,e+l,i+m,r+n)}(\Delta\tau) &= P(S^*(\tau+\Delta\tau), E^*(\tau+\Delta\tau), I^*(\tau+\Delta\tau), R^*(\tau+\Delta\tau)) \\ &= (s+k, e+l, i+m, r+n) \mid P(S^*(\tau), E^*(\tau), I^*(\tau), R^*(\tau)) \\ &= (s,e,i,r) \end{aligned}$$

are given as Table 3.2:

TABLE 3.2
Infinitesimal transition probabilities of the continuous-time markov chain

Present State	Description	State Transition	Transition Probability
S*	Birth	[1,0,0,0]	$b\ \Delta\tau + o(\Delta\tau)$
S*	Death	[-1,0,0,0]	$dS^*(t)\ \Delta\tau + o(\Delta\tau)$
S*	Infection	[-1,1,0,0]	$\alpha\ \frac{S(\tau)}{N(\tau)} I^*(\tau)\ \Delta\tau + o(\Delta\tau)$
E*	Natural death	[0,-1,0,0]	$d\ E^*(\tau)\ \Delta\tau + o(\Delta\tau)$
E*	Loss of incubation	[0,-1,1,0]	$\beta\ E^*(\tau)\ \Delta\tau + o(\Delta\tau)$
I*	Natural death	[0,0,-1,0]	$d\ I^*(\tau)\ \Delta\tau + o(\Delta\tau)$
I*	Disease-induced death	[0,0,-1,0]	$\lambda\ I^*(\tau)\ \Delta\tau + o(\Delta\tau)$
I*	Recovery	[0,0,-1,1]	$\gamma\ I^*(\tau)\ \Delta\tau + o(\Delta\tau)$
R*	Natural death	[0,0,0,-1]	$d\ R^*(\tau)\ \Delta\tau + o(\Delta\tau)$

TABLE 3.3
Transition probabilities of disease classes for the branching process approximation

Present state	Description	State transition	Rate r_i
E^*	Infection	$[E^*, I^*] \rightarrow [E^*+1, I^*]$	$\alpha\ I^*(\tau)$
E^*	Natural death	$[E^*, I^*] \rightarrow [E^*-1, I^*]$	$d\ E^*(\tau)$
E^*	Loss of Incubation	$[E^*, I^*] \rightarrow [E^*-1, I^*+1]$	$\beta\ E^*(\tau)$
I^*	Loss of infection	$[E^*, I^*] \rightarrow [E^*, I^*-1]$	$(d+\gamma+\lambda) I^*(\tau)$

3.3.3.1.3 *Multitype Branching Process Approximation*

We make use of the multitype branching process theory for analysing continuous time Markov chain model dynamics near DFE $(S \approx N_0)$. We approximate the **birth** and **death** of exposed and infectious individuals near the origin since the extinction of the disease mainly depends on these two categories:

Assumption: $S = N_0$ and $R = 0$ where N_0 is the total initial population.

Transition probabilities for the branching process are $r_i\ \Delta\tau + o(\Delta\tau)$. Here, $\Delta Y(\tau) = Y(\tau+\Delta\tau) - Y(\tau)$ are given as:

- **Mathematical description of multitype branching processes**
  Let discrete state random variables be denoted by $Y = (Y_1, Y_2, \ldots, Y_n)$. Let Y_a correspond to the infected group l for l varying from 1 to n.
  Suppose $Y_a(0) = \delta_{ab}$; δ_{ab} is the kronecker delta

$$\delta_{ab} = \begin{cases} 1 & a = b \\ 0 & a \neq b \end{cases}$$

$$1 \leq a, b \leq n$$

Offspring probability generating function for the infectious individuals of type l is a function

$$f_a : [0,1]^n \rightarrow [0,1]$$

$$f_a(y_1, y_2, \ldots y_n) = \sum_{k_1=0}^{\infty} \sum_{k_2=0}^{\infty} \cdots \sum_{k_n=0}^{\infty} p_a(k_1, k_2, \ldots, k_n) y_1^{k1} y_2^{k2} \ldots y_n^{kn} \qquad [43]$$

where $p_a(k_1, k_2, \ldots, k_n)$ denotes the probability that there is **birth** of k_b individuals of type b, for b = 1,2,...,n from one individual of type l.

Applying the above theory on the exposed and infected population of the proposed model where $Y = (Y_1, Y_2) = (E^*, I^*)$ and n = 2, the offspring **probability generating function** [51]**for the exposed** can be represented as:

$$f_1(y_1, y_2) = p_0 + p_1 y_2; p_0 + p_1 = 1$$

where p_0 is the probability with which each exposed individual dies independently (death of an exposed individual can only be natural) and p_1 is the probability with which he/she survives (gives rise to an infected individual).

From the underlying determinsitic model,

$$p_0 = \frac{d}{d + \beta}$$

$$\Rightarrow p_1 = \frac{\beta}{d + \beta}$$

$$f_1(y_1, y_2) = \frac{d + \beta y_2}{d + \beta}$$

Likewise, the offspring **probability generating function for the infected** can be represented as:

$$f_2(y_1, y_2) = p_0 + p_2 y_1 y_2; p_0 + p_2 = 1$$

where p_0 is the probability with which each infected individual dies (death of an infected individual means he/she recovers or dies due to covid or dies naturally) and p_2 is the probability with which he/she survives and is replaced by two infected individuals.

Therefore,

$$p_0 = \frac{\gamma + \lambda + d}{\alpha + \gamma + \lambda + d}$$

$$\Rightarrow p_2 = \frac{\alpha}{\alpha + \gamma + \lambda + d}$$

$$f_2(y_1, y_2) = \frac{\alpha y_1 y_2 + \gamma + \lambda + d}{\alpha + \gamma + \lambda + d}$$

Now defining the expectation matrix [42] for the model:

$$A = \begin{vmatrix} \frac{\partial f_1}{\partial y_1} & \frac{\partial f_2}{\partial y_1} \\ \frac{\partial f_1}{\partial y_2} & \frac{\partial f_2}{\partial y_2} \end{vmatrix}$$

at $(y_1, y_2) = (1,1)$

$$A = \begin{vmatrix} 0 & \frac{\alpha}{\alpha+\gamma+\lambda+d} \\ \frac{\beta}{d+\beta} & \frac{\alpha}{\alpha+\gamma+\lambda+d} \end{vmatrix}$$

- The number of exposed and infected individuals that are expected to be produced by a single exposed host are represented by a_{11} and a_{21}, respectively. Likewise, the number of exposed and infected people produced by a single infected person are represented by a_{12} and a_{22}, respectively.
- A is **irreducible** and the offspring probability generating functions for E^* and I^* are **non-simple** (since $f_1(y_1, y_2)$ and $f_2(y_1, y_2)$ are not linear functions of y_1, y_2; i.e., each object can produce more than one offspring).
- As per the Jury conditions [48], $\rho(A) < 1$ if and only if $trace(A) < 1 + det(A) < 2$, where $trace(A) = \frac{\alpha}{\alpha+\gamma+\lambda+d}$ and $det(A) = \frac{-\alpha\beta}{\alpha+\gamma+\lambda+d}$

Now, in the inequality $trace(A) < 1 + det(A) < 2$, the second condition (i.e., $det(A) < 1$) is clearly satisfied (since all parameter values are positive) while the second condition is satisfied if and only if $R_0 < 1$; i.e., we imply that

$$R_0 < 1 \text{ iff } \rho(A) < 1$$

Now for the case when $R_0 > 1$, we find the fixed point of the offspring probability generating functions by equating $f_1(q_1, q_2)$ to q_1 and $f_2(q_1, q_2)$ to q_2. This **fixed point is useful in determining the possibility of extinction** of the disease.

- In conclusion, spectral radius of A (i.e., $\rho(A)$) decides whether a fixed point exists in $(0,1)X(0,1)$.
    a) If $\rho(A) < 1$ or $\rho(A) = 1$, then no other stable fixed point other than $(1,1)$ exists.
    b) If $\rho(A) > 1$, then another unique fixed point $(q_1^*, q_2^*) \in (0,1)^2$ exists.

Hence, on [0, 1], find the fixed point of offspring probability generating function.

$$f_1(q_1 + q_2) = q_1 \text{ on}[0,1]$$

$$\Rightarrow \frac{d + \beta q_2}{d + \beta} = q_1$$

$$\text{and } f_2(q_1, q_2) = q_2 \text{ on}[0,1]$$

$$\Rightarrow \frac{\alpha q_1 q_2 + \gamma + \lambda + d}{\alpha + \gamma + \lambda + d} = q_2$$

Combining these equations:

$$\frac{\alpha\left(\frac{d + \beta q_2}{d + \beta}\right)q_2 + \gamma + \lambda + d}{\alpha + \gamma + \lambda + d} = q_2$$

Solving further

$$\Rightarrow \frac{\alpha q_2(d + \beta q_2) + (d + \beta)(\gamma + \lambda + d}{(\alpha + \gamma + \lambda + d)(d + \beta)} = q_2$$

$$\Rightarrow q_2(d + \beta)[\alpha + (\gamma + \lambda + d)] - \alpha(d + \beta q_2)q_2 = (d + \beta)(\gamma + \lambda + d)$$

$$\Rightarrow q_2\alpha(d + \beta) + q_2(d + \beta)(\gamma + \lambda + d) - \alpha(d + \beta q_2)q_2 = (d + \beta)(\gamma + \lambda + d)$$

$$\Rightarrow (q_2 - 1)(d + \beta)(\gamma + \lambda + d) = (-\alpha(d + \beta) + \alpha(d + \beta q_2))q_2$$

$$\Rightarrow (q_2 - 1)(d + \beta)(\gamma + \lambda + d) = \alpha(-d - \beta + d + \beta q_2)q_2$$

$$\Rightarrow (q_2 - 1)(d + \beta)(\gamma + \lambda + d) = \beta\alpha(q_2 - 1)q_2$$

$$\Rightarrow (q_2 - 1)[\beta\alpha q_2 - (d + \beta)(\gamma + \lambda + d)] = 0$$

$$\Rightarrow (q_2 - 1)[c_1 q_2 + c_2] = 0$$

where c_1 and c_2 are the constants with $c_1 = \beta\alpha$ and $c_2 = -(d + \beta)(\gamma + \lambda + d)$.

On [0, 1], fixed points are:

$$(q_2 - 1)[c_1 q_2 + c_2] = 0$$

$$\Rightarrow q_2 = 1 \text{ or } c_1 q_2 + c_2 = 0$$

$$\Rightarrow q_2 = 1 \text{ or } q_2 = \frac{-c_2}{c_1} = \frac{1}{R_0}$$

Therefore, $q_2^* = 1$ or $q_2^* = \frac{1}{R_0}$

Clearly, $q_2^* < 1$ if and only if $R_0 > 1$, therefore minimal fixed point of q_2 is $q_2^* = min\left\{\frac{1}{R_0}, 1\right\}$.

Thus, the fixed point (q_1^*, q_2^*) is given by

$$q_2^* = \frac{1}{R_0}$$

$$q_1^* = \frac{d + \beta q_2}{d + \beta} = \frac{d}{d + \beta} + \frac{\beta q_2}{d + \beta} = \frac{d}{d + \beta} + \frac{\beta}{R_0(d + \beta)}$$

when $R_0 > 1$.

From the theory of multitype branching process approximation, the probability of extinction of disease is given as:

$$P_{extinction} = lim_{\tau \to \infty}[E^*(\tau) = I^*(\tau) = 0] = \begin{cases} 1 & R_0 \leq 1 \\ (q_1^*)^{e_0} (q_2^*)^{i_0} & R_0 > 1 \end{cases}$$

Here, $e_0 = E^*(0)$ and $i_0 = I^*(0)$.

q_1^* is probability of extinction of disease in exposed population.
q_2^* is the probability of extinction of disease in infected population.
Since sum of probabilities = 1 the probability of an outbreak = 1 – $P_{extinction}$
Hence,

$$P_{outbreak} = \begin{cases} 0 & R_0 \leq 1 \\ 1 - (q_1^*)^{e_0} (q_2^*)^{i_0} & R_0 > 1 \end{cases}$$

Clearly, in the stochastic model, $\rho(A)$ is the threshold for determining whether the disease will persist or become extinct. Unlike the deterministic model, the stochasticity here shows that even if $R_0 > 1$, there is still possibility of the disease becoming extinct.

3.3.3.2 Data and Implementation

In this section we discuss the source and description of data used and how this data was implemented in the SEIR model.

3.3.3.2.1 Estimation of Parameters

Data of COVID-19 in India from March 1, 2021 to October 10, 2021 was collected from the open source website COVID19.org [48] and initial values of S, E, I, R were taken from worldometers.info/coronavirus/country/India website [46]. In this study, initial total susceptible population was determined by multiplying the total population of India (N) by a factor of 10^{-3} [47], where initial infected population is assumed to be double the number of exposed people. The initial number of people recovered was taken to be nil.

3.3.3.2.2 Numerical Simulation

Generally, for multivariable processes, finding analytical solutions for transitions probabilities from backward and forward Kolmogorav differential equations is not easy. In probability theory, we have a straightforward numerical algorithm, which is similar to solving the master equation of a system involving chemical reactions. This is called the Gillespie algorithm [54] or stochastic simulation algorithm. It is used to generate a statistically correct trajectory or a possible solution of a system of stochastic equations for which the reaction rates are already known.

The crux of this algorithm is that two random numbers $v_1, v_2 \in U[0,1]$ are drawn at every step of time; one for selecting which reaction is going to take place next and the other to determine interevent time (i.e., after how much time this reaction will occur). From the Markovian property, it is inferred that the interevent time T follows exponential distribution $T \sim \mu e^{-\mu\tau}$ where parameter μ denotes summation of rates of all the events that are possible. For the SEIR CTMC model,

TABLE 3.4
Values and sources of parameters

Parameter	Parameter value	Source
α	0.46560	calculated using $R_0 = 1.6$ value
β	0.1923	fixed latent period of 5.2 days
γ	0.0714	fixed mean infectious period of 14 days[45]
d	0.007344	fixed death rate of 7.344 per 1000 people[46]
λ	0.009794	estimated using curve fitting

$$\mu = b + \alpha i \frac{s}{N} + ds + de + \beta e + di + \gamma i + \lambda i + dr$$

where (s,e,i,r) is the particular value for state of (S^*, E^*, I^*, R^*) at a given time τ. The interevent time is computed using uniform distribution U^* and the cumulative distribution F_k.

Using the underlying characteristics of U^*:

$$\text{s} = P(U^* \leq s) = P(1 - U^* \leq s), \quad s \in [0,1]$$

We derive an expression for T_k in terms of U^* -

$$F_k(\tau_1) = P(U^* \leq F_k(\tau_1))$$

But $P(U^* \leq F_k(\tau_1)) = P(F_k^-(U^*) \leq \tau_1)$.

Now using definition $F_k(\tau_1) = P(T_k \leq \tau_1)$ as well as the previous identity, $F_k^-(U^*)$ and T_k have the same cumulative distribution; i.e., $F_k(\tau_1)$.

Computing the inverse F_k^-:

Let $G(\tau_1) = 1 - e^{-\mu\tau_1} = \text{y}$

$\Rightarrow$ $1 - \text{y} = e^{-\mu\tau_1}$

$\Rightarrow$ $\ln(1-y) = -\mu\tau_1$

$\Rightarrow$ $\tau_1 = \dfrac{\ln(1-y)}{\mu}$

Therefore, $T_k = F_k^-(U^*) = \dfrac{\ln(1-U^*)}{\mu} = \dfrac{-\ln U^*}{\mu}$

Using the Gillespie algorithm, a uniformly distributed random number $v_1 \in U^*$ gives rise to a value $\tau_1 = -\dfrac{\ln v_1}{\propto}$ for the interevent time.

Now, which particular event will take place next is determined by the second random number v_2. Let us consider n events in general. The interval [0, 1] is subdivided according to the possibility of occurence of each event, $[0, P_1^*], [P_1^*, P_1^* + P_2^*], \ldots, (P_1^*, \ldots, P_{n-1}^*, 1]$, $\sum_{k=1}^{n} P_k^* = 1$.

If v_2 lies in the ith subinterval, then ith event occurs.

For the SEIR CTMC model, we have the following events with the corresponding probabilities:

$$P_1^* = \frac{b}{\mu}, P_2^* = \frac{\alpha i \frac{s}{N}}{\mu}, P_3^* = \frac{ds}{\mu}, P_4^* = \frac{de}{\mu}, P_5^* = \frac{\beta e}{\mu},$$
$$P_6^* = \frac{di}{\mu}, P_7^* = \frac{\gamma i}{\mu}, P_8^* = \frac{\lambda i}{\mu}, P_9^* = \frac{dr}{\mu}$$

If $i = e = 0$, an absorbing state is attained and termination of the process takes place.

We study the probability of extinction of the disease using a continuous time Markov chain in which 10,000 sample paths are simulated for the SEIR CTMC model. If the cumulative number of exposed and infected individuals reaches 'outbreak size' where outbreak size = $min\{25, 0.10N(0)\}$; N(0)=135 crores, then we count it as an outbreak, whereas if it hits zero, then we take it to be an extinction event. Then, we compare the probability of extinction of disease obtained from this CTMC with the estimate obtained from the theory of multitype branching process approximation.

3.3.3.3 Results and Discussion

In this section, we discuss the results obtained by both deterministic and stochastic approaches individually and then compare them.

3.4 RESULTS

3.4.1 Deterministic Approach

After analysing the data of COVID-19 for India during the second wave we saw that daily recovered and daily deceased data were close to trend line, which helped us predict the number of cases correctly for up to 50 days starting from March 1, 2021. It can be seen that the SEIRD model predicted the total number of COVID-19 cases correctly up to 50 days of the total 200 days. Further, it can be noticed that the peak of the cases was predicted at nearly 100 days (for $R_0 = 1.6$) since the start of the second wave (assumed to be started from March 1, 2021).

From the results we can say that Covid cases were almost seven times the actual reported number of cases during the peak period, which can be due to lots of cases being unreported and many infected people not getting tested. The second wave came to almost an end in October, which can be noticed in actual reported cases also. However, results would have been more accurate if we had considered vaccination as part of our model and different rates could have been time dependent for more accuracy. This is a study based on mathematical modelling, which works on some assumptions so it remains unclear whether this study can be useful in a real-world scenario.

3.4.2 Stochastic Approach

The 10,000 sample paths of the SEIR CTMC model are plotted in Figure 3.9 for N = 1350000000*0.001 for R_0 = 1.6. The graph represents the beginning, peak and

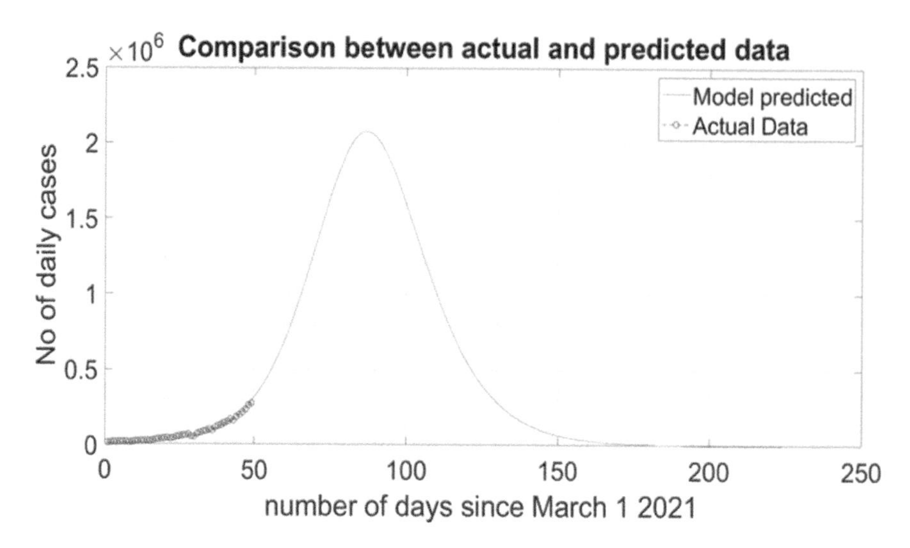

FIGURE 3.5 Comparison of actual and predicted number of cases and prediction of pandemic peak in India.

death of the second wave of Covid-19 in India. Close-up on the right side represents exponential growth at the initiation of the outbreak. The SEIR Model predicted that the peak of the outbreak would take place in nearly 90 days (for R0 = 1.6) since the start of 2nd wave (assumed to be started from March 1st 2021) that is maximum cases would be reported in the month of May. This is in line with the actual scenario of covid-19 in India during the second wave which can be observed from the available data[56]. The second wave almost came to an end in the month of October which can also be noticed in actual reported cases. Also from the sample paths of the CTMC model, we obtained the probability of extinction of covid after second wave to be 0.0829 which is nearly equal to the probability of disease extinction obtained using the theory of multi-type branching process approximation which is 0.0539.

However, results would have been more accurate if we had considered vaccination as a part of our model. This study is based on mathematical modelling which works on certain assumptions and hence does not always produce accurate results.

3.5 DISCUSSION

3.5.1 Deterministic approach

Basic reproduction number R_0 plays an important role in determining the peak of an epidemic. How the scenario is changing with change in R_0 value can be clearly seen in the graph of the trajectory of the SEIRD model with $R_0 = 1.2$ (Figure 3.7), $R_0 = 1.6$ (Figure 3.6), and $R_0 = 2.0$ (Figure 3.8). With $R_0 = 1.2$ (Figure 3.7) we can see that the peak of the disease comes after nearly 150 days with very few deaths due to COVID-19. When we took $R_0 = 1.6$ (Figure 3.6), and we can clearly observe the rise in cases and rise in deaths as well and peak comes nearly after 100 days. When

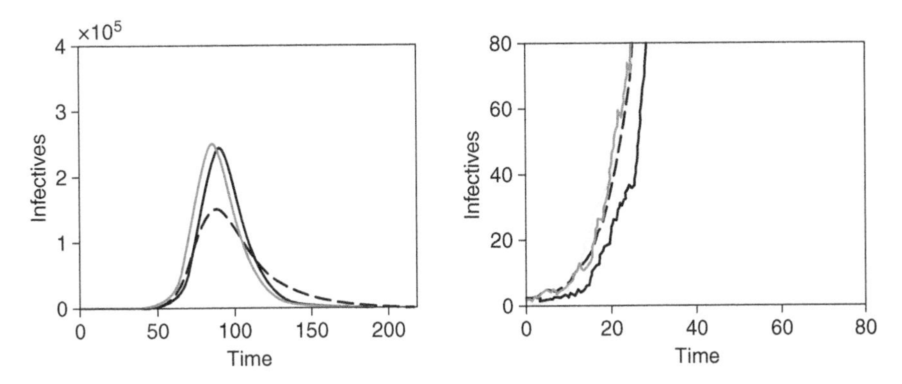

FIGURE 3.6 The graph is plotted for R_0 = 1.6 The black dashed curve represents the ODE solution of infected and other curves are sample paths of the SEIR CTMC model. Parameter values are b=17.377/1000, d=7.344/1000, alpha=0.46560, beta=0.1923, gam=(1/14), and lambda=0.009794. The initial conditions are i(0)=2, e(0)=1, r(0)=0, N = S(0)= 1350000000*0.001. The graph on the right of each figure is a close-up view on time interval [0, 80] of the graph on the left.

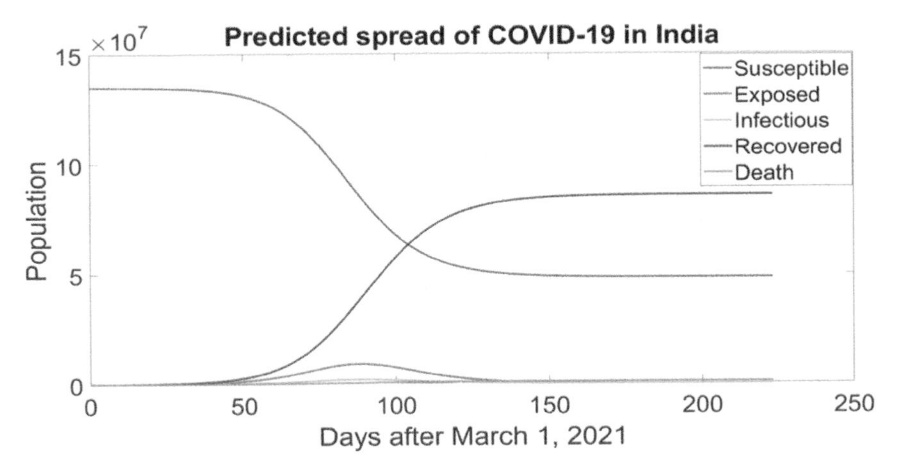

FIGURE 3.7 Trajectory of SEIRD model over time if R_0 =1.6

we took $R_0 = 2.0$ (Figure 3.8) the pandemic hits the peak just after 50 days, and there is increase in infected cases also.

3.5.2 Stochastic Approach

Basic reproduction number R0 plays a vital role in determining the dynamics of an outbreak. This can be clearly observed from the graphs of SEIR model with R_0 = 1.2, R_0 = 1.6 and R_0 = 2.0. After the second wave, if the basic reproduction

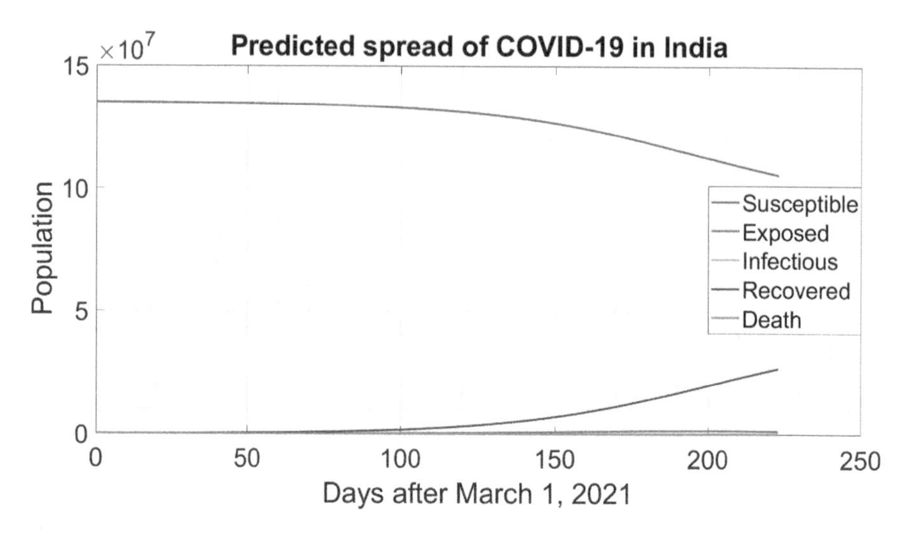

FIGURE 3.8 Trajectory of SEIRD model over time if R_0 was 1.2.

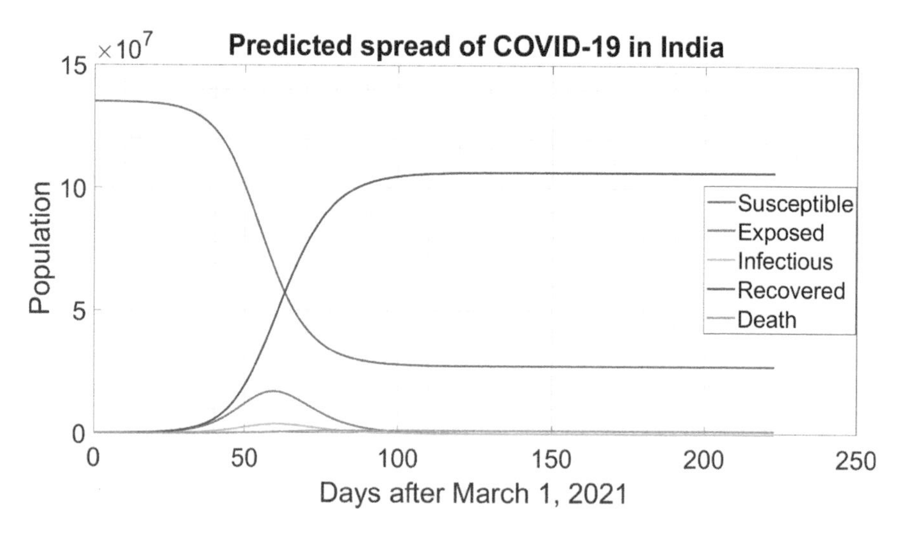

FIGURE 3.9 Trajectory of SEIRD model over time if R_0 was 2.0.

number remains the same as assumed (i.e., $R_0 = 1.6$), then we obtain the probability of extinction of disease to be 0.0829 from the sample paths of the continuous time Markov chain, which is nearly equal to the probability of disease extinction obtained from the multitype branching process approximation, which is 0.0539; while if the value of R_0 rises to around 2.0 (in the case of negligence in effective control strategies), then the probability of disease extinction is estimated to be 0.0613 and 0.0336 from CTMC and branching process, respectively. Whereas given proper control strategies being implemented and R_0 coming down to around 1.2, then the probabilities

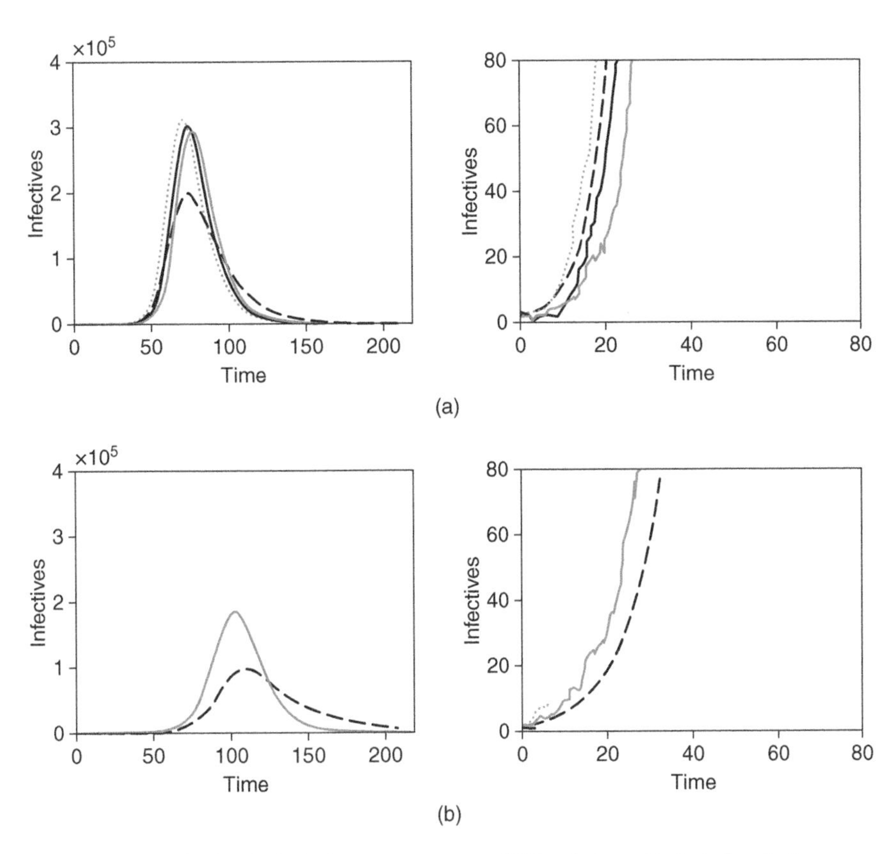

FIGURE 3.10 The black dashed curve represents ODE solution of infected and other curves are sample paths of the SEIR CTMC model. Parameter values are b=17.377/1000, d=7.344/1000, beta=0.1923, gam=(1/14), lambda=0.009794. The initial conditions are i(0)=2, e(0)=1, r(0)=0, N = S(0)= 1350000000*0.001. The graph on the right of each figure is a close-up view on time interval [0,80] of the graph on the left. The graphs are plotted for different values of R_0. (a) $R_0 = 2.0$ (i.e alpha = 0.5819937) and (b) $R_0 = 1.2$ (i.e alpha = 0.349196).

TABLE 3.5
Number of days taken for deterministic and stochastic approaches to hit the pandemic peak for different values of R0

Value of R_0	Deterministic approach	Stochastic approach
$R_0 = 1.2$	150 days	115 days
$R_0 = 1.6$	90 days	90 days
$R_0 = 2.0$	60 days	65 days

obtained are 0.1093 and 0.1003, respectively. The model is flexible enough to calculate probabilities of disease extinction for different values of R_0.

3.5.3 Comparison of both Approaches

We compare the number of days taken after the second wave to hit the peak of the pandemic using both the deterministic and stochastic approaches for different values of R0 in Table 5. From the table it is quite evident that both approaches provide equivalent results. For R0 = 1.6, both approaches suggest that the peak of covid cases will be attained 90 days after the beginning of second wave (which matches with the real life scenario). While if value of R0 falls down to 1.2, the peak of covid cases is estimated to be attained 150 days and 115 days after the second wave commencement using deterministic and stochastic approach respectively. Moreover, if R0 takes the value 2.0, we again obtain these values to be nearly equal i.e 60 days and 65 days respectively

3.6 CONCLUSION

In this paper, SEIRD model was studied using deterministic approach and an SEIR model using the stochastic approach in order to study, analyze and predict the transmission dynamics of COVID-19 in India during the second wave (assumed to have begun from 1st March 2021). The models and their stability studied in this paper contributes to the theory of epidemiological models. In this study authors investigate the applications of computational statistical techniques in mathematical modelling. After studying the COVID-19 cases in India from 1st March 2021 to 10th October 2021, using either of the approaches, it can be concluded that COVID-19 cases peaked in the end of May 2021 for $R_0 = 1.6$ which is quite high and could have been reduced with the help of effective control strategies and their proper implementation, by means of social distancing, use of face masks and regular sanitization of hands. Further it can be seen that COVID-19 cases dropped drastically and became almost negligible in the month of October which might be due to partial implementation of lockdown at regular intervals in different states of India and nationwide mass vaccination program. Further, the results obtained using the deterministic approach indicate that the value of basic reproduction number played as a sharp disease threshold: if $R_0 \leq 1$, the system is locally asymptomatically stable and hence the disease gets extinct while if $R_0 > 1$, the system is unstable and the disease continues to spread. However, results using the stochastic approach showed that even if $R_0 > 1$, there is still a possibility of the disease getting extinct. The limitations of the study involve lack of parameters to capture the dynamics of the Coronavirus due to lack of data. The future scope is to consider factors such as amount of medical help available in terms of oxygen supply and spatial parameters which can differentiate between the spread and evolution of pandemic in India as compared to that of other countries.

REFERENCES

[1] Krishna Mohan Agarwal, Swati Mohapatra, Prairit Sharma, Shreya Sharma, Dinesh Bhatia, Animesh Mishra, *Study and overview of the novel corona virus disease (COVID-19)*, Sensors International,Volume 1,2020,100037,ISSN 2666–3511.

[2] Olaimat AN, Aolymat I, Shahbaz HM, Holley RA (2020) *Knowledge and Information Sources About COVID-19 Among University Students in Jordan: A Cross-Sectional Study.*, Front. Public Health 8:254.

[3] Marco Cascella, Michael Rajnik, Abdul Aleem, Scott C. Dulebohn, Raffaela Di Napoli, *Features, Evaluation, and Treatment of Coronavirus (COVID-19)*, StatPearls Publishing; 2021 Jan.

[4] Wikipedia, *COVID-19 pandemic in India.* Accessed on 04/12/2021

[5] Sam Moore, Edward M. Hill, Louise Dyson, Michael J. Tildesley, Matt J. Keeling, *Modelling optimal vaccination strategy for SARS-CoV-2 in the UK*, PLoS Comput Biol 17(5): e1008849; 2021 May.

[6] Mukesh Jakhar, P. K. Ahluwalia, Ashok Kumar *COVID-19 Epidemic Forecast in Different States of India using SIR Model*,medRxiv Article ID 20101725; 2020 May

[7] worldometres.info/coronavirus/, *Worldometres* Accessed December 2021

[8] M.A. Andrews, Binu Areekal, K.R.Rajesh, Jijith Krishnan, R. Suryakala, Biju Krishnan, C.P. Muraly, and P.V. Santhosh, *First confirmed case of COVID-19 infection in India: A case report*, Indian J Med Res. 2020 May; 151(5): 490–492.

[9] Sujita Kumar Kar, Ramdas Ransing, S.M.Yasir Arafat, Vikas Menon, *Second wave of COVID-19 pandemic in India: Barriers to effective governmental response*Accessed on 27/03/2022

[10] Pierre Magal, Shigui Ruan, *Susceptible-Infectious-Recovered Models Revisited: From the Individual Level to the Population Level*, April 2014 Mathematical Biosciences 250(1).

[11] Rahman, A., & Kuddus, M. A. (2021). *Modelling the transmission dynamics of COVID-19 in six high-burden countries*. BioMed Research International, 2021, 1–17. [5089184]. https://doi.org/10.1155/2021/5089184

[12] Abdulla, F., Nain, Z., Karimuzzaman, M., Hossain, M. M., & Rahman, A. (2021). *A non-linear biostatistical graphical modeling of preventive actions and healthcare factors in controlling COVID-19 pandemic.* International Journal of Environmental Research and Public Health, 18(9), 4491. https://doi.org/10.3390/ijerph18094491

[13] Kuddus, M. A., & Rahman, A. (2021). Analysis of COVID-19 using a modified SLIR model with nonlinear incidence. *Results in Physics*, 27, 104478. https://doi.org/10.1016/j.rinp.2021.104478

[14] Rahman, A., & Kuddus, M. A. (2020). Cost-effective modeling of the transmission dynamics of malaria: A case study in Bangladesh. Communications in Statistics: Case Studies, Data Analysis and Applications, 6(2), 270–286. https://doi.org/10.1080/23737484.2020.1731724

[15] Kuddus, M. A., Mohiuddin, M., & Rahman, A. (2021). Mathematical analysis of a measles transmission dynamics model in Bangladesh with double dose vaccination. *Scientific Reports*, 11(1), 1–16. [16571]. https://doi.org/10.1038/s41598-021-95913-8

[16] Rahman, A., Kuddus, M. A., Ip, H. L., & Bewong, M. (2021). A review of COVID-19 modelling strategies in three countries to develop a research framework for regional areas. *Viruses*, 13(11), [2185]. http://10.3390/v13112185

[17] Sharif, O., Hasan, M. Z., & Rahman, A. (2022). Determining an effective short term COVID-19 prediction model in ASEAN countries. Scientific Reports, 12(2022), 1–11. [12:5083]. www.nature.com/articles/s41598-022-08486-5

[18] Pakwan Riyapan, Sherif Eneye Shuaib, Arthit Intarasit, *A Mathematical Model of COVID-19 Pandemic: A Case Study of Bangkok, Thailand*, Hindawi Computational and Mathematical Methods in Medicine Volume 2021, Article ID 6664483.

[19] Alene, M., Yismaw, L., Assemie, M.A. et al. *Serial interval and incubation period of COVID-19: a systematic review and meta-analysis.* BMC Infect Dis 21, 257 (2021).

[20] Van den Driessche, James Watmough, *Reproduction numbers and sub-threshold endemic equilibria for compartmental models of disease transmission*, Mathematical Biosciences 180 (2002).

[21] Ian Cooper, Argha Mondal, Chris G. Antonopoulos, *A SIR model assumption for the spread of COVID-19 in different communities*, Chaos, Solitons and Fractals,Volume 139,2020,110057,ISSN 0960-0779.

[22] Brody H. Foy, Brian Wahl, Kayur Mehta, Anita Shet, Gautam I. Menon, Carl Britto, *Comparing COVID-19 vaccine allocation strategies in India:A mathematical modelling study*, International Journal of Infectious Diseases 103 (2021) 431–438.

[23] Ibrahim Halil Aslan, Mahir Demir, Micheal Morgan Wise, Suzanne Lenhart, *Modeling COVID-19: Forecasting and analyzing the dynamics of the outbreak in Hubei and Turkey*, medRxiv preprint doi: https://doi.org/10.1101/2020.04.11.20061952

[24] Shilei Zhao, Hua Chen, *Modeling the epidemic dynamics and control of COVID-19 outbreak in China*, Higher Education Press and Springer-Verlag GmbH Germany, part of Springer Nature 2020.

[25] Wenzhang Huang, Kenneth L.Cooke, Carlos Castillo-Chavez, *stability and bifurcation for a multiple-group model for the dynamics of hiv/aids transmission*, Siam J. Appl. Math. Vol. 52, No. 3, pp. 835–854, June 1992.

[26] Funk CD, LaferriÃre C and Ardakani A (2020), /textitA Snapshot of the Global Race for Vaccines Targeting SARS-CoV-2 and the COVID-19 Pandemic., Front. Pharmacol. 11:937. doi: 10.3389/fphar.2020.00937

[27] Kunal Menda, Lucas Laird, Mykel J. Kochenderfer, Rajmonda S. Caceres, *Explaining COVID 19 outbreaks with reactive SEIRD models*, Scientific Reports | (2021) 11:17905 | https://doi.org/10.1038/s41598-021-97260-0.

[28] Vishnu Vytla, Sravanth Kumar Ramakuri, Anudeep Peddi, Kalyan Srinivas K, N. Nithish Ragav, *Mathematical Models for Predicting Covid-19 Pandemic: A Review*, IOCER 2020. Journal of Physics: Conference Series 1797 (2021) 012009 IOP Publishing, doi:10.1088/1742-6596/1797/1/012009.

[29] Maher Ala'raj, Munir Majdalawieh, Nishara Nizamuddin, *Modeling and forecasting of COVID-19 using a hybrid dynamic model based on SEIRD with ARIMA corrections*, Elsevier B.V. on behalf of KeAi Communications Co. Ltd,2020,https://doi.org/10.1016/j.idm.2020.11.007.

[30] Debashis Saikia, Kalpana Bora, Madhurjya P. Bora, *COVID-19 outbreak in India: an SEIR model-based analysis*, Nonlinear Dyn (2021) 104:4727-4751. https://doi.org/10.1007/s11071-021-06536-7.

[31] B. Malavika, S. Marimuthu, Melvin Joy, Ambily Nadaraj, Edwin Sam Asirvatham, L. Jeyaseelan, *Forecasting COVID-19 epidemic in India and high incidence states using SIR and logistic growth models*, Elsevier, a division of RELX India, Pvt. Ltd on behalf of INDIACLEN, https://doi.org/10.1016/j.cegh.2020.06.006.

[32] Kelly R. Moran, Geoffrey Fairchild, Nicholas Generous, Kyle Hickmann, Dave Osthus, Reid Priedhorsky, James Hyman, Sara Y. Del Valle1, *Epidemic Forecasting is Messier Than Weather Forecasting: The Role of Human Behavior and Internet Data Streams*

in Epidemic Forecast, The Journal of Infectious DiseasesÂ® 2016;214(S4):S404-8, DOI: 10.1093/infdis/jiw375.

[33] Gang Xie, *A novel Monte Carlo simulation procedure for modelling COVID 19 spread over time*, Scientific Reports | (2020) 10:13120 | https://doi.org/10.1038/s41598-020-70091-1.

[34] Jos M. Carcione, Juan E. Santos, Claudio Bagaini,Jing Ba, *A Simulation of a COVID-19 Epidemic Based on a Deterministic SEIR Model*, Front. Public Health 8:230. doi: 10.3389/fpubh.2020.00230.

[35] Jaouad Danane, Karam Allali, Zakia Hammouch, Kottakkaran Sooppy Nisar, *Mathematical analysis and simulation of a stochastic COVID-19 Levy jump model with isolation strategy*,Results in Physics, vol. 23, article 103994, 2021. https://doi.org/10.1016/j.rinp.2021.103994

[36] S. Hussain, E. N. Madi, H. Khan, *Investigation of the stochastic modeling of COVID-19 with environmental noise from the analytical and numerical point of view*, Mathematics, vol. 9, no. 23, p. 3122, 2021. https://doi.org/10.3390/math9233122

[37] M. Sher, K. Shah, Z. A. Khan, H. Khan, and A. Khan, *Computational and theoretical modeling of the transmission dynamics of novel COVID-19 under Mittag-Leffler power law*,Alexandria Engineering Journal, vol. 59, no. 5, pp. 3133–3147, 2020. https://doi.org/10.1016/j.aej.2020.07.014

[38] Shaobo He, Yuexi Peng, Kehui Sun, *Seir modeling of the COVID-19 and its dynamics, Nonlinear Dyn.*, Nonlinear Dynamics volume 101, pages 1667–1680 (2020). 10.1007/s11071-020-05743-y

[39] Karin S. Dorman, Janet S. Sinsheimer, Kenneth Lange, *In the garden of branching processes*, SIAM Rev., 46 (2004), 202–229. 10.1137/S0036144502417843

[40] Shah Hussain, *Advances in non-linear analysis and Applications*,Volume 2022 |Article ID 4320865. https://doi.org/10.1155/2022/4320865

[41] O. Diekmann, J. A. P. Heesterbeek, M. G. Roberts, *The construction of next-generation matrices for compartmental epidemic models*, J R Soc Interface. 2010 Jun 6; 7(47): 873–885. Published online 2009 Nov 5.s

[42] Damilola Olabode, Jordan Culp, Allison Fisher, Angela Tower, Dylan Hull-Nye, Xueying Wang, *Deterministic and stochastic models for the epidemic dynamics of COVID-19 in Wuhan, China*, Mathematical Biosciences and Engineering 2021, Volume 18, Issue 1: 950–967. doi: 10.3934/mbe.2021050

[43] Linda J.S. Allen, *A primer on stochastic epidemic models: Formulation, numerical simulation, and analysis*, 128–142. doi: 10.1016/j.idm.2017.03.001

[44] Vipin Tiwari, Namrata Deyal, Nandan S. Bisht, *Mathematical Modeling Based Study and Prediction of COVID-19 Epidemic Dissemination Under the Impact of Lockdown in India*, Front. Phys. 8:586899. doi: 10.3389/fphy.2020.586899,2020 November.

[45] *Covid19 India data availaible online at:* www.covid19india.org/. Accessed on 01/11/2021

[46] *Covid-19 data available online at:* www.worldometers.info/coronavirus/country/india/. Accessed on 01/11/2021

[47] Chatterjee S, Sarkar S, Chatterjee S, Karmakar K, Paul R.,*Studying the progress of COVID-19 outbreak in India using SIRD model*, Indian J Phys Soc Indian Assoc Cultiv Sci. (2020) 1–17. doi: 10.1007/s12648-020-01766-8

[48] *Covid19 India data availaible online at: www.covid19india.org/.* Accessed on 01/11/2021

[49] L. J. Allen, G. E. Lahodny Jr, *Extinction thresholds in deterministic and stochastic epidemic models*, Journal of Biological Dynamics, Vol. 6, No. 2, March 2012, 590–611. doi: 10.1080/17513758.2012.665502

[50] Damilola Olabode, Jordan Culp, *Deterministic and stochastic models for the epidemic dynamics of COVID-19 in Wuhan, China*, 2021, Vol 18, Issue 1: pages 950–967 10.3934/mbe.2021050

[51] Samuel Karlin, Howard Taylor, *A first course in stochastic Processes*, 1975, Academic Press, INC, page 400

[52] *World Health Organization* Accessed on 02/01/2022

[53] www.macrotrends.net Accessed on 02/01/2022

[54] Fernand Hayot and Ciriyam Jayaprakash, *A tutorial on cellular stochasticity and Gillespieâ's algorithm*, Department of Neurology, New York and Department of Physics, Columbus, 2016

[55] Linda J. S. Allen, *An introduction to mathematical biology*, Department of Mathematics and Statistics, Texas Tech University.

Theme 2

Machine Learning-adopted Models

4 Distracted Driver Detection Using Image Segmentation and Transfer Learning

Sanjit Kumar Dash, Aditya Prasad Tripathy, Sidhartha Bibekananda Dash, Debabrata Tripathy and Soumyajit Bal*
Odisha University of Technology & Research, Bhubaneswar, India
*Corresponding Author: sanjitkumar303@gmail.com

CONTENTS

4.1 INTRODUCTION

Research done by the World Health Organization (WHO) in 2020 suggests that around 1.35 million humans die from tragic traffic accidents worldwide every year [1]. Road accidents contribute approximately 3% of the gross domestic product (GDP) loss in almost every country. According to research, more than 90% of road collisions occur due to driver's fault, incorporating leading factors such as distraction, drunkenness, and exhaustion. Distracted driving has emerged as a fundamental cause of those collisions in recent years. According to the definition specified by the International Organization for Standardization (ISO), "distracted driving is the attention given to a non-driving related activity, typically to the detriment of driving performance" [2]. There are three major types of distracted driving:Manual – taking your hands off the wheel, Visual – looking away from the road, and Cognitive – being absentminded.

DOI: 10.1201/9781003253051-6

The popularity of onboard electronics like navigation systems and smartphoneshas introduced multiple elements that induce distracted driving action for drivers. Consequently, it has become essential to conduct an extensive analysis of distracted behaviours, find out their occurrence in drivers, and put forward corresponding solutions [3–8]. There is wide utilization of convolutional neural networks (CNNs) [9] for complex image processing tasks. Currently, several methods, like VGG19 [10], are proposed to deal with problems like image classification and recognition. The characteristics can be selected from the image dataset, and then the required classification can be performed using these methods. As a result, we will be able to detect distracted driving behaviours and improve driving safety.

Simonyan and Zisserman [11] introduced VGG architecture, which improves on the original LeNet networkput forward by Lecun et al. [12] by including additional convolutional layers. While LeNet had five convolutional layers, the proposed framework uses VGG architecture with 19 convolutional layers. The primary feature of VGG is that it uses small receptive fields that employ convolutions of 3x3 dimensions. Although it has a vast volume of parameters, high computational cost, and is time-consuming to train [13], on the other hand, it is considered a training model for the framework because it provides good performance in classification and localization tasks [11].

As we need to incorporate essential features into the feature descriptor, we extract features like the head and arm location of the driver along with its posture from the image, which will further improve the learning process. GrabCut [14], an iterative graph-based foreground segmentation algorithm, is used to extract the subject from the irrelevant background noise. It acts as an interactive tool that reduces the developer's effort and also helps in the configuration of the preprocessing section, which is helpful in a production platform. To adjust to changes within the system, the preprocessing module requires significant user effort or retraining by using other tools and techniques [15]. Furthermore, GrabCut outperforms other image segmentation techniques like Gaussian Mixture Model (GMM) implemented by Xing et al. [16] and Mask R-CNN [17] concerning speed and accuracy.

There are two globally accessible datasets:the State Farm Distracted Driver Detection (SFD3) dataset and the AUC Distracted Driver (AUCD2) dataset. The available images for training purposes in those datasets are classified into ten distinct classes of driving (one for safe driving and nine for unsafe driving):safe driving, operating the radio, texting (in right hand), texting (in left hand), talking on the phone (rightside), talking on the phone (leftside), reaching behind, drinking, application of makeup or hair adjustments, and talking with a passenger.

The main objective of this chapter is to (i) to build a CNN architecture for behaviour detection of the driver; (ii) to construct an efficient as well as a robust model by incorporating both image segmentation and image classification;(iii) to address the matter at hand, a robust machine learning model is proposed thatutilizes the advantages of image segmentation along with deep learning by using GrabCut;(iv) to use optimized image size for training as it provides a significant amount of the statistical information while maintaining a decent training time;and (v) to enhance the training process, we utilize the characteristics extracted from the posture of the driver on top of their face and hands position. The remainderof the chapteris structured as follows. Section 4.2 outlines the related work in this field. Section 4.3 focuses on the system model. Section 4.4 represents the datasets used in the experiment and its exploratory analysis. Section 4.5 discusses the results and offers observations and Section 4.6 finally concludes the chapter.

4.2 RELATED WORKS

This section discusses related works. Previous work in this area includes methods that rely on computer vision, deep learning, and other techniques based on the image of the position, the posture of the driver, and behaviourhave been done in this field to prevent these mishaps.

Ohn-bar et al. [18] suggested an approach to detect the position of the hand by using three classifiers on the given input images combining the three region-specific classifiers used on a second-stage classifier. It sees a small number of activities like the position of the hand, whether they are on the steering wheels, gearbox, or the instruction page (i.e., for using the radio or adjusting other manual functions), and this limits the detection of different kinds of distraction.

Zhao et al. [19] proposed a contourlet transformation to characterize the features and then compared the accuracy of four different classifiers: K Nearest Neighbour (KNN), Random Forest (RF), Linear Perceptron, and the Multilayer Perceptron (MLP). They used a broader dataset,the South East University (SEU) Driving Posture dataset. SEU has one-sided images of the drivers in classes like safe driving, lever operation, eating food, and talking over the phone. The mean classification accuracy of those above models was 85.09%, 87.16%, 37.21%, and 90.63%, respectively. After the comparison, they merged the features from the Pyramid Histogram of Oriented Gradients (PHOG) [20] and spatial feature extractors, which led to the improvement of the Multilayer Perceptron classifier by increasing the accuracy to 94.75%.

Yan et al. [21] fine-tuned the classification network by establishing a CNN architecture that uses sparse filtering, an unsupervised feature learning technique, and thereby attained a classification accuracy of 99.78% on the above SEU dataset.

Lately, several experiments have beendone with various deep learning frameworks for different tasks in the domain of natural language processing [22], computer vision [23], and audio processing [24]. Sathe et al. [25] utilized several data augmentation techniques like class-based, part-based data augmentation combined with the CNN, and key-point detection on theState Farm Distracted Driver Detection (SFD3) dataset [26] and achieved an accuracy of 96.72%. Abouelnaga et al. [27] built a dataset from scratch called AUC Distracted Driver (AUCD2). It also has ten classes similar to that of the SFD3 dataset. They implemented a weighted combination of AlexNet and InceptionV3 models to pull off an accuracy of 95.98% in testing. The input images for the architecture were a fusion of "raw," "hands-and-face," "face," "hand," and "skin-segmented" images.

Baheti et al. [28] introduced models like vgg-16, which is a regularized model, and a modified model with a similar number of parameters as the original and with 15M parameters, respectively. These models got an accuracy of 96.31% and 95.54%, respectively, with a standard of 4.44% on the actual VGG-16 model, on the AUC Distracted Driver dataset.

Koesdwiady et al. [29] studied the performance of both VGG-19 and Xgboost on their self-prepared dataset, which has ten classes identical to that of the State Farm dataset. Hssayeni et al. [30] studied the comparison of an SVM based on traditional handcrafted features with the transfer learning on ResNet, AlexNet, and VGG architectures and found that ResNet achieved the maximum accuracy of 85% on the SFD3 dataset.

Xing et al. [17] proposed a model that used an image segmentation technique that was based on Gaussian Mixture Modelling (MM) with the transfer learning frameworks. They attained an accuracy of approximately 94% on their dataset. Masood et al. [31] used CNN and studied the comparison of performance metrics of frameworks initialized with random weights and models, including the pre-trained weights.

Tran et al. [33] collected a dataset that contains images of various situations and states of the driver during normal and distracted driving and implemented four deep CNN models, which include AlexNet, residual network, VGG-16, and GoogleNet, as well as checked out those models on an embedded GPU platform. From this, it was determined that GoogleNet performed better than the remaining three models. N Moslemi et al. [34] used a 3D CNN and optical flow to enhance the detection of driver distraction. Their model used the Kinetics dataset for training posture and behaviour of driver was hyper-tuned to accomplish an accuracy of approximately 90% on the SFD3 dataset.

All the above-proposed models were based on deep learning and quiet time taking. Most of them were based on datasets having a smaller number of images. In contrast to those models, we propose an efficient and robust model that is trained and tested on a larger dataset: the SFD3 dataset having images with subjects of various ethnicities and age groups.

4.3 SYSTEM MODEL

The system model consists of two major components, the first one being image segmentation and the second one being image classification, as shown in Figure 4.1. The

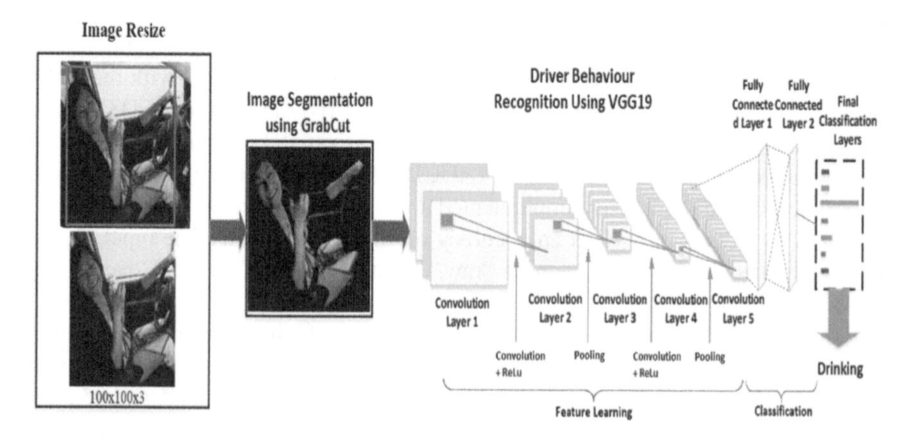

FIGURE 4.1 Overview of the system model.

images of drivers for behaviour detection were taken as input from the datasets. These dataset images are collected as static snapshots from an actual video feed. First, in the preprocessing section of the model, the unprocessed images are fed into GrabCut to remove the irrelevant background noise. This helps in separating the useful foreground features from these images. Second, the deep CNN model is trained to categorize the actions of the drivers into distinct classes. This model takes the images from the GrabCut as input and identifies those into distinct labels as output. Here, VGG19 is chosen as the deep CNN model.

4.3.1 Image Preprocessing

The raw images present in this dataset have a resolution of 640×640 and are preprocessed to minimize the hardware resources required for neural network training. Initially, the resolution of the raw static images is downscaled to 100×100, as shown in Figure 4.2. This particular resolution of the images is considered because images with higher resolution take more training time. On the other hand, the low-resolution images offer less statistical information due to their reduced size. The considered size is optimal for training as it provides a significant amount of statistical information while maintaining an optimized training time. After that, foreground extraction is performed with the help of GrabCut, an interactive foreground segmentation algorithm. It is used as it is an algorithm that allows changes to be carried out effortlessly in the preprocessing section with minimal input from the developer.

Further, we extracted the foreground region consisting of information about the driver's posture, a position as well as placement of the driver's head and arms by eliminating background noise. Elimination of such unwanted background not only significantly improves training time due to reduction of data size but also helps to improve the classification accuracy. This process is commonly known as image segmentation and implements the following steps:

1. *Import the required libraries, i.e., NumPy, cv2, os, glob.*
2. *Initialize the background and foreground model as an array of 0 with dimensions (1, 65).*
3. *Initialize Region of Interest (ROI) as a rectangle with coordinates (4, 0, 92, 98).*
4. *Load the image from the specified path.*
5. *Create a simple mask image similar to the loaded image.*
6. *Call the GrabCut method to mark the pixels with four flags, where 0 & 2 represent background elements while 1 & 3 represents foreground elements.*
7. *Create a mask to remove background elements, i.e., eliminate pixels marked with 0 & 2.*
8. *Apply the created mask to the image.*
9. *Save the image in the desired path.*

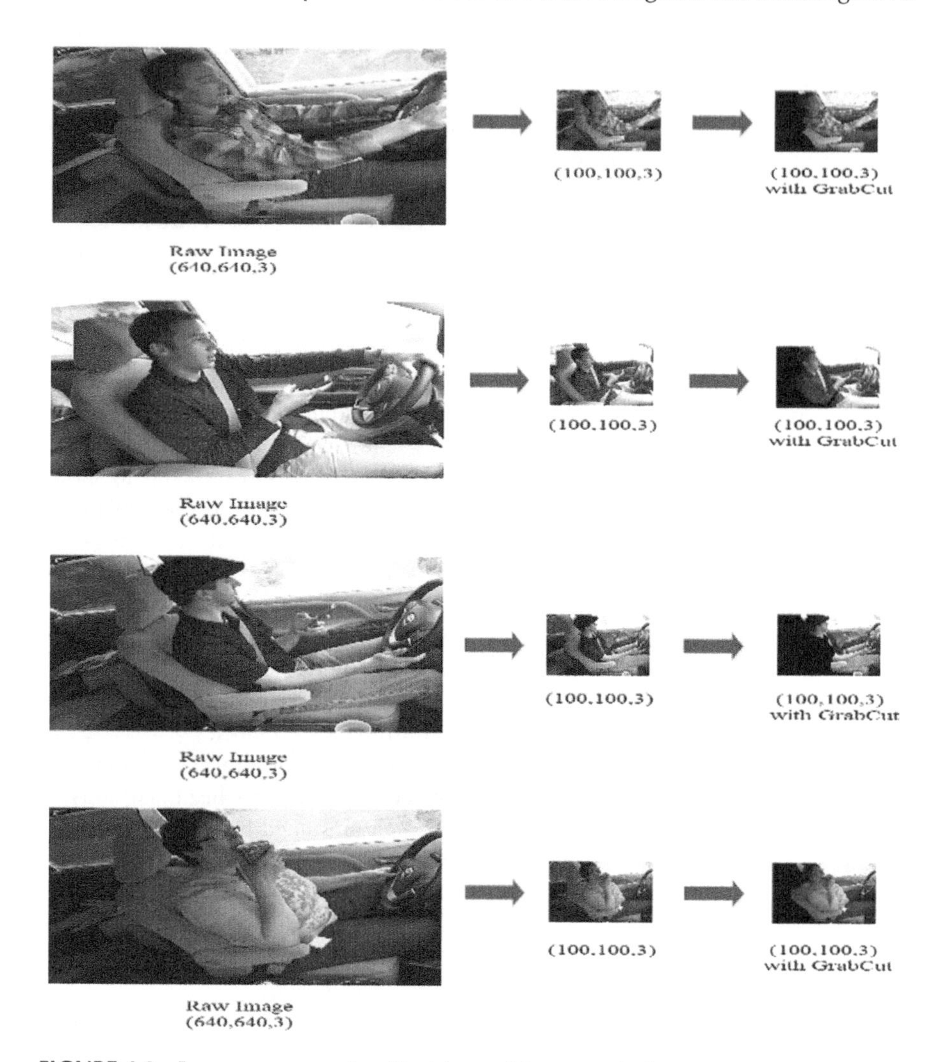

FIGURE 4.2 Image pre-processing (Resizing and Segmentation).

4.3.2 Classification Function

VGG19 architecture has 19 deep layers and uses small receptive fields that employ convolutions of 3x3 dimensions. This model is initialized with the default weights provided by the ImageNet dataset, which is a very large dataset that contains 15 million labelled images of more than 22,000 categories and these categorized images are of high resolution. Hence this model is mentioned as a pretrained model. Due to its massive number of labelled images, it helps in improving the accuracy of the training models.

The first two dense layers with output dimensions 1024 and 512 use "ReLU" as an activation function, also known as Rectified Linear Unit, that uses A(x) = max (0, x) and provides an output x if x is positive and 0 otherwise. This function is used

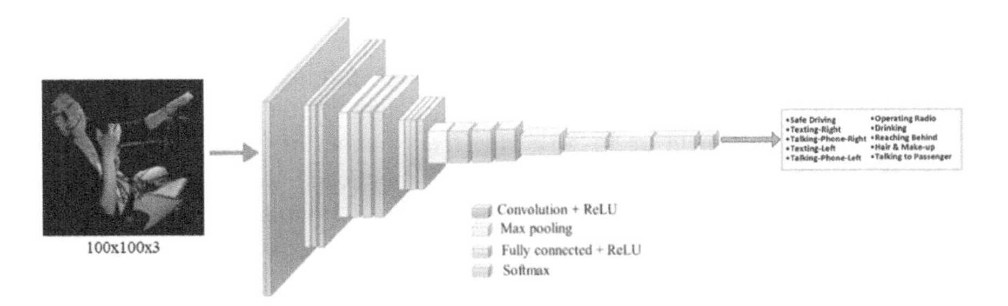

FIGURE 4.3 VGG19 model visualizing low to high-level features.

because it is significantly lessexpensive than other activation functions with respect to computational requirements like sigmoid and tanh. After all, ReLU comprises undemanding mathematical operations. Only a few neurons are triggered, making the network sparse efficient and simple for computation. The last dense layer with output dimension 10 (as there are ten distinct classes) uses "SoftMax" as an activation function as it is ideally used in the output layer of the classifier where we are trying to attain the likelihood to characterize the class of each image.

To avoid overfitting in the neural network model, dropout is used at regular intervals with a probability of 0.3 and 0.2, respectively, to make a 30% and 20% chance that the output of a given neuron will be forced to 0. In the end, an "Adam" optimization, which is a stochastic gradient descent method, is used with a learning rate value of 0.001. This optimization is based on an adaptive approximation of first- and second-order moments. This model is used as a feature extractor to extract statistically valuable features from the processed images and then used to categorize the drivers'behaviours from those images into ten distinct classes, as shown in Figure 4.3.

So, classification function carries out the following steps:

1. *initialize 'conv' = VGG19 (weights = ImageNet).*
2. *initialize variable 'input' with the default shape of the image as (100, 100, 3).*
3. *output = conv(input).*
4. *flatten(output).*
5. *add a dense layer with output dimension 1024 with'ReLU' as activation function.*
6. *add dropout (0.3).*
7. *add a dense layer with output dimension 512 with'ReLU'as activation function.*
8. *add dropout (0.2).*
9. *add a dense layer with output dimension 10 with'SoftMax'as activation function.*
10. *apply 'input' to model.*
11. *implement optimiser = 'Adam (learning rate = 1e-3)'.*
12. *compile and save model.*

4.3.3 Training Algorithm

We initialize x_train and y_train as empty lists and start appending the processed images and their corresponding class id. Then, we initialize the current fold number as 0, which will increment up to 5(i.e., the total number of folds). Then, we initialize the model with the classification function mentioned above in Section 4.3.2 and set the weights of neural networks. We then set early stopping patience at ten and monitor the model checkpoint with validation loss, which acts as a stopping condition for epochs. The stopping condition is satisfied when the validation loss does not improve continuously for ten epochs in a fold. Then, the training variables are fitted to the model, and the process is repeated in five times. The final model is saved as a JSON file.

The CNN architecture is trained by implementing the following steps:

1. *initializex_train and y_train as empty lists i.e., [].*
2. *for i in range (10): #as there are 10 classes*
   a. *x_train.append(img) #img is image file of* c_i
   b. *y_train.append(i)*
3. *end for*
4. *initialize current_fold = 0, total_folds = 5, log_list = []*
5. *initialize kf with KFold method*
6. *for train, test in kf:*
   a. *model = train_model()*
   b. *foldn += 1*
   c. *set weights_path*
   d. *set callbacks object with EarlyStopping at patience = 10 and ModelCheckpoint at monitor = 'val_loss'*
   e. *model.fit(x_train, y_train, batch_size = 32, epochs = 40)*
   f. *load this to log*
   g. *plot log.history of 'loss', 'val_loss', 'accuracy', 'val_accuracy'*
7. *end for*
8. *save model in 'json' format*

The trained model in JSON format is later used in Streamlight, which is an open-source app framework for machine learning along with the updated weights. This is used to view the actual real-time simulation of distracted driver detection. Keras and tensor flow libraries are used to create a CNN. The model is built using the Google Colab research platform, which provides a high-performing setup to train and test the model.

4.4 DATASET AND EXPLORATORY ANALYSIS

The dataset utilized in the training of this model is obtained from State Farm from the Distracted Driver detection Competition in Kaggle. The original dataset has images of 26 subjects consisting of both men and women of various ages and ethnicity displaying safe driving and nine other distracted driving conditions. These ten situations, also referred to as classes, are safe driving, operating the radio, texting (in right hand), texting (in left hand), talking on the phone (rightside), talking on the phone (leftside), reaching behind, drinking, application of makeup or hair adjustments, and talking with a passenger. The meta-data, such as creation dates, etc., are removed from the images of the dataset to ensure that such meta-data does not affect the training model.

This dataset is designed for frame-by-frame classification. Images are extracted from a live video feed, and behaviours and posture of the driver are categorized into distinct labels. Similar video samples of various sizes and durations are used for creating raw images. There are 102,150 images in this dataset, from which 22.424 are considered as training images and the remaining as testing images. Images are initially 640×640 pixels in size. The class distribution of the State Farm dataset is represented in Figure 4.4.

The classes were developed with various images of subject drivers. This distribution of drivers over diverse classes can be visually seen in Figure 4.5. Despite all standard features among images of each subject, it is worth mentioning that classifying these images is a difficult task since the number of pixels that permit class separation is minuscule in comparison to the total size of the image. In images in which the subjects are talking on the phone, i.e., in classes two and four, the phone appears partly concealed and nearer to the face. Finally, in images that involve texting on the phone, i.e., classes one as well as 3, it seems that the face and the wheel are partially hiding the phone.

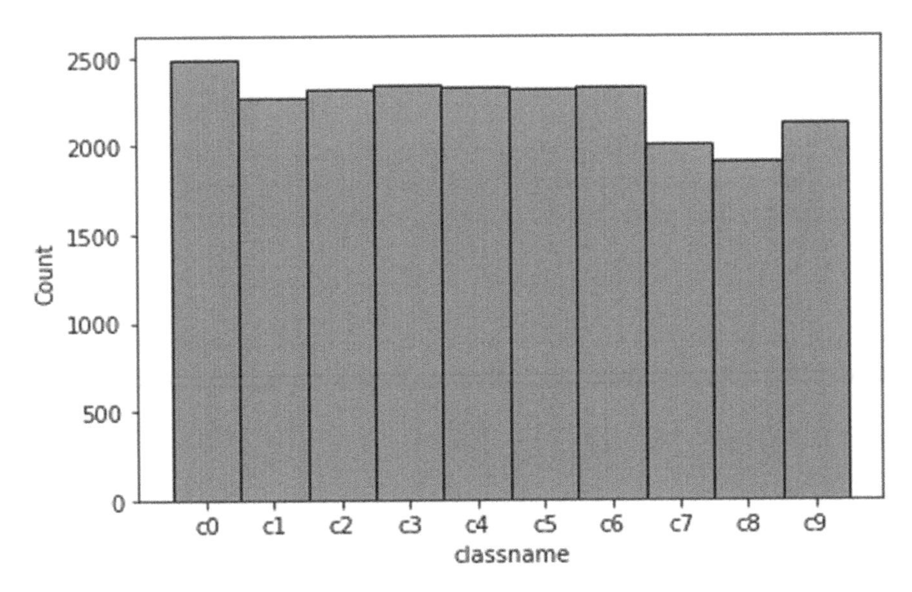

FIGURE 4.4 Class distribution histogram of State Farm Dataset.

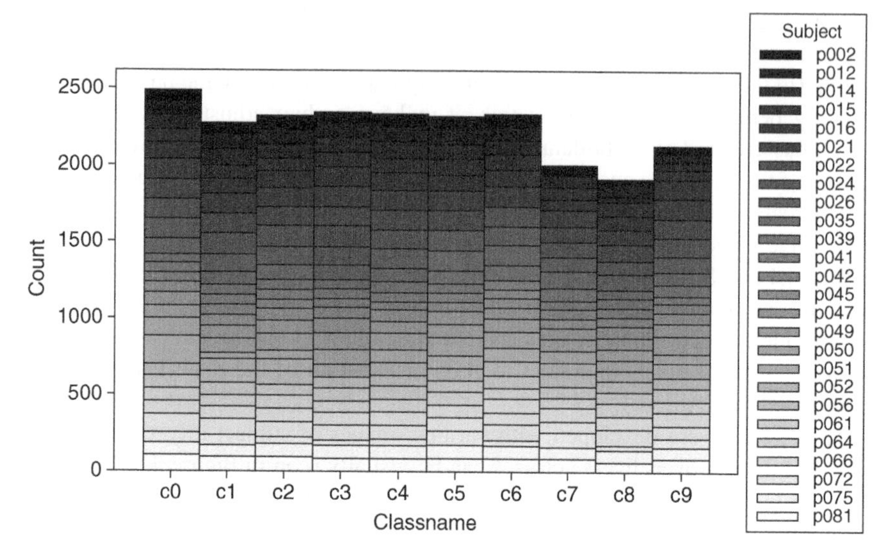

FIGURE 4.5 State farm dataset class histogram visualizing image distribution of the subject.

4.5 RESULT AND DISCUSSION

In this section, we discuss the accuracies achieved in various training folds of the model. Drivers head and hand position and body postures are considered as feature descriptor for the training of the model to acknowledge distracted driver images. The new descriptor preserves the foreground relevant features and eliminates the irrelevant background noise from the 100x100 resolution image of the driver. This supports the model in reducing the irrelevant information that did not get eliminated in the simple preprocessing method of the convolution filters of the generalized VGG-19 architecture.

The final training accuracy achieved by this model after five times is 96.87%, with a cross-validation accuracy of 92.79%, as seen in Figure 4.10, which is an improvement over the initial training accuracy after one-fold, which was 81.60%. The validation loss is 11.96%, which is less than the training loss, which is 12.19%, proving that our model is appropriately fitted. Initially, the model pulled off anaccuracy of 81.60% in training and 79.14% in testing, in the firstfold, as represented in Figure 4.6.

Then, the model improved the training accuracies up to 85.65%, 89.76%, and 93.87%, as well as improved the testing accuracies up to 79.63%, 85.38%, and 91.76%, respectively, in 2nd, 3rd, and 4th fold, which are shown in Figures 4.7, 4.8, and 4.9, respectively. Finally, in the 5th fold, the model managed to achieve an impressive accuracy of 96.87% in training and 92.79% in testing, as shown in Figure 4.10. The 5-folds were processed through 25, 25, 20, 14, and 35 epochs out of 40 epochs, respectively, in each fold, by reaching the set patience point of tenepochs as the validation loss did not improve further.

Figure 4.11 depicts the confusion matrix for ten classes of distracted driving behaviours applying the proposed VGG19 model. The blue diagonal in the matrix

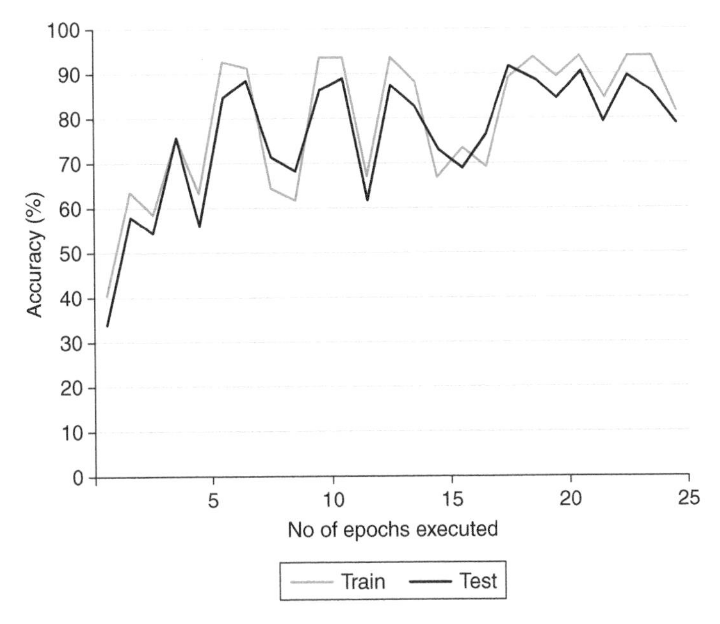

FIGURE 4.6 Graph showing training and validation accuracy w.r.t number of epochs in 1st fold.

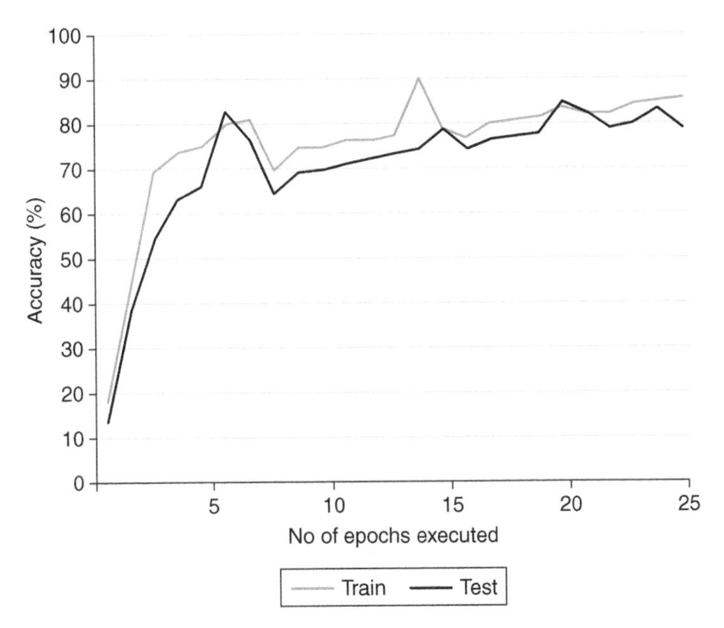

FIGURE 4.7 Graph showing training and validation accuracy w.r.t number of epochs in 2nd fold.

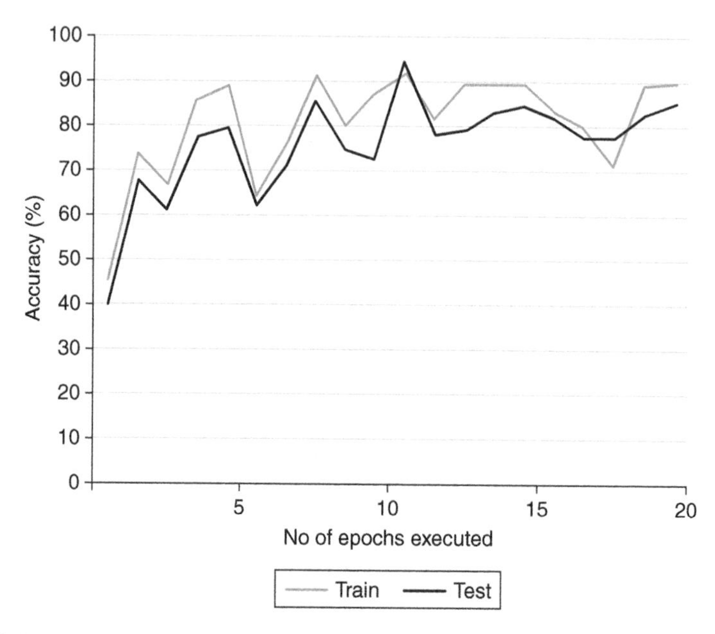

FIGURE 4.8 Graph showing training and validation accuracy w.r.t number of epochs in 3rd fold.

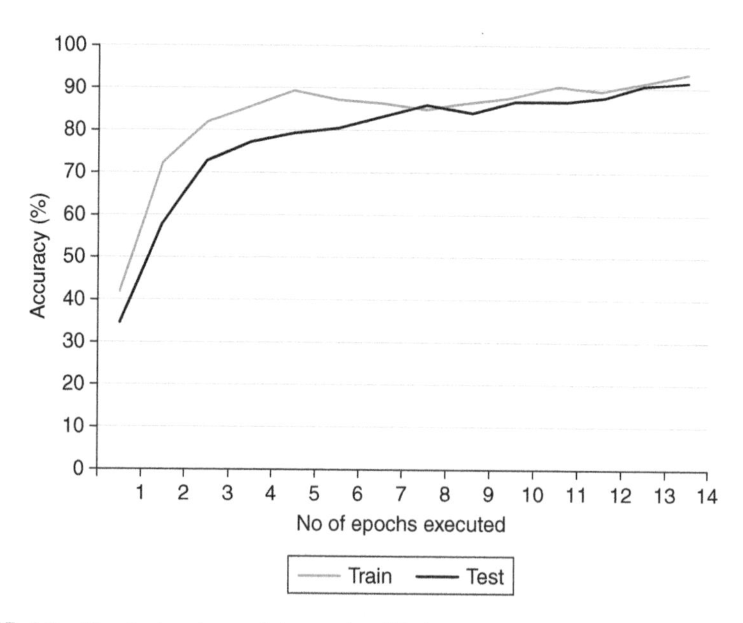

FIGURE 4.9 Graph showing training and validation accuracy w.r.t number of epochs in 4th fold.

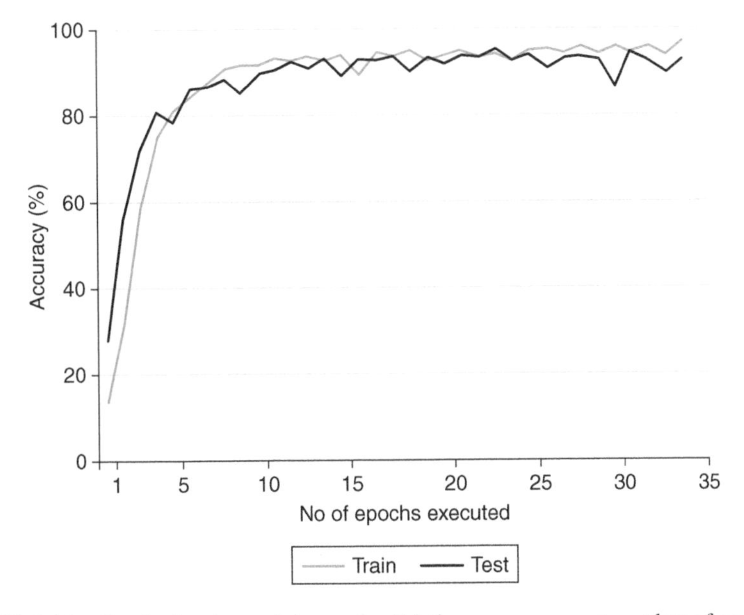

FIGURE 4.10 Graph showing training and validation accuracy w.r.t number of epochs in 5th fold.

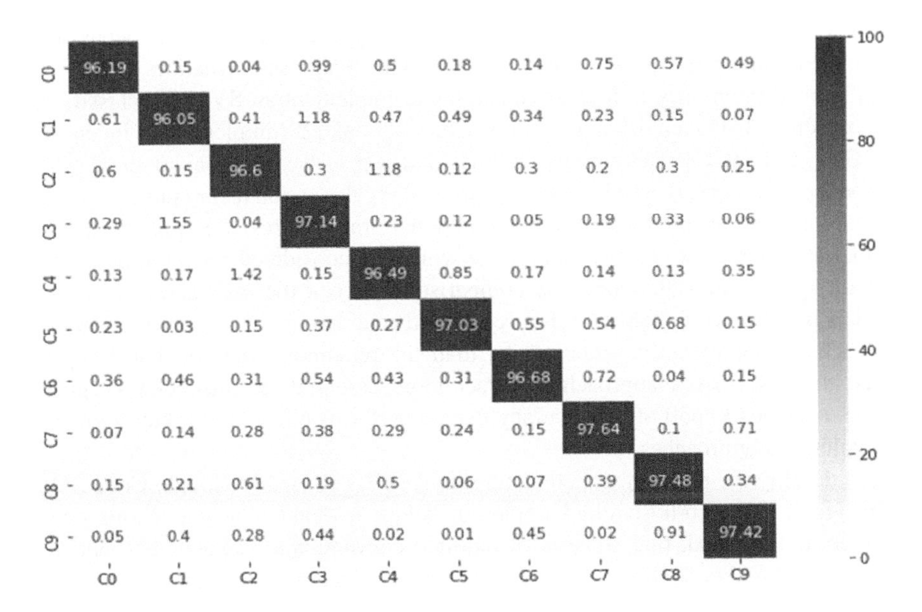

FIGURE 4.11 Confusion Matrix of the proposed model.

TABLE 4.1
Comparison of the Various Models along with Their Accuracy

Approach	Accuracy (%)
Handcrafted Features with SVM [33]	27.70
AlexNet [33]	72.60
VGG-16 [33]	82.50
ResNet152 [33]	85.00
VGG-19 [34]	77.00
Inception V3 [34]	73.00
AlexNet and InceptionV3 combined [28]	95.98
CNN along with Data Augmentation [25]	96.72
Proposed VGG-19 with Image Segmentation	**96.87**

shows the percentage of true positives(i.e., accurately detected classes). The left column displays the original label of the ten classes, and the bottom row indicates the predicted labels. The remaining yellow boxes in the confusion matrix are false positives with respect to the predicted labels but are false negatives with respect to the original labels of the mentioned ten classes. The row-wise cumulative percentage is always equalling 100. In contrast, the blue diagonal line of numbers has a mean of 96.87%, which is also the final training accuracy of the model after the 5th fold.

We chose various frameworks for recognizing distracted driving behaviours compared with the proposed VGG-19 model with image segmentation. Tran et al. [27] applied previously trained architectures to implement an SVM for classifying by extracting distracted driving features. Chawan et al. [28] implemented Inception, VGG16, and VGG19 to classify distracted driver behaviors. Baheti et al. [22] introduced an alternate VGG model that applied regularization techniques.

To compare our model with other traditional architectures, we split the dataset in the ratio of 8:2 per tenimages, eight of which are considered for training and the remaining two for validation. The comparison between the accuracies of various models is presented in Table 4.1.The results indicate that the CNN-based techniques can achieve significantly better results than the remaining traditional techniques, since the CNN-based approaches extract more essential features. Our proposed model managed to pull off an accuracy that outperforms other similar but promising convolutional approaches.

To develop the mentioned architecture, Python 3.7 is used in Google Colab IDE, which is an IDE web-hosted by Google for Python, to enable machine learning with storage on the cloud, that consists of required conventional packages like Numpy, Pandas, Matplotlib, Seaborn, Scikit-Learn, OpenCV, and deep learning modules like Keras and TensorFlow. The saved .json file is later used in Streamlit, along with the weights of the 5th fold of the neural network. The dashboard, designed with the help of Streamlit, helps to simulate the distraction detection technique of the driver, as shown in Figure 4.12. There are two ways to select an image for the classification: a. by clicking the browse files button or b. by dragging and dropping the image in the box, as shown in Figure 4.13.

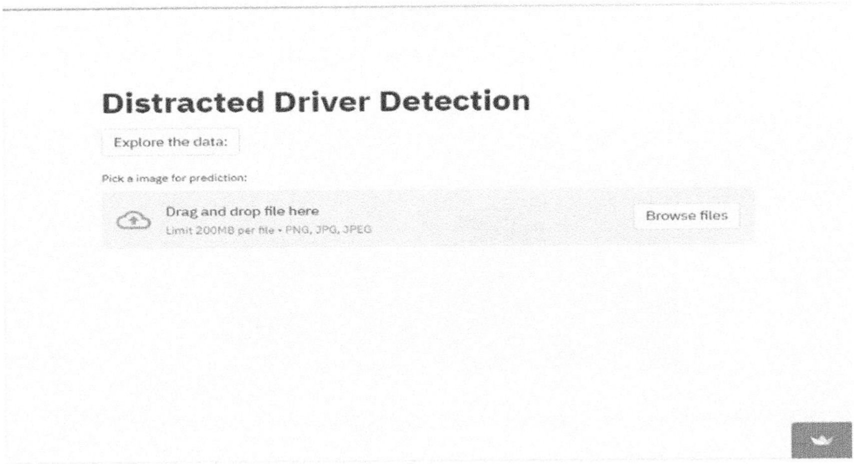

Distracted Driver Detection
Explore the data:
Pick a image for prediction:
Drag and drop file here
Limit 200MB per file • PNG, JPG, JPEG
Browse files

FIGURE 4.12 Dashboard of the model.

Distracted Driver Detection
Explore the data:
Pick a image for prediction:
Drag and drop file here
Limit 200MB per file • PNG, JPG, JPEG
+ Copy
Browse files

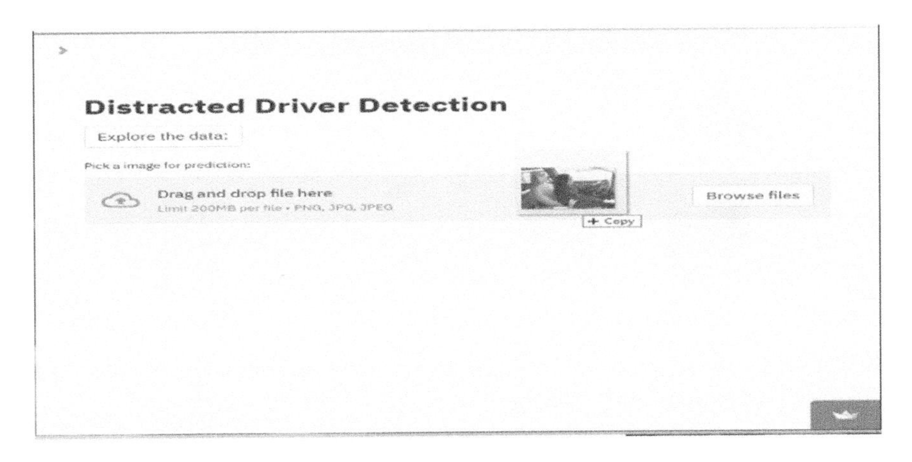

FIGURE 4.13 Uploading an image using dragging and dropping the image into the input area.

After selecting and uploading the desired image, the dashboard updates itself with the original image as shown in Figure 4.14, along with the images of all preprocessing steps: a. the resized image and b. the GrabCut segmented image as shown in Figure 4.15. At the end of the page, the predicted result along with the image gets updated, as shown in Figure 4.16.

4.6 CONCLUSION

In this chapter, a much more efficient and accurate model for classifying different images of drivers based on their postures and positions of the other body parts was introduced. The discussed method uses an image-segmentation process that

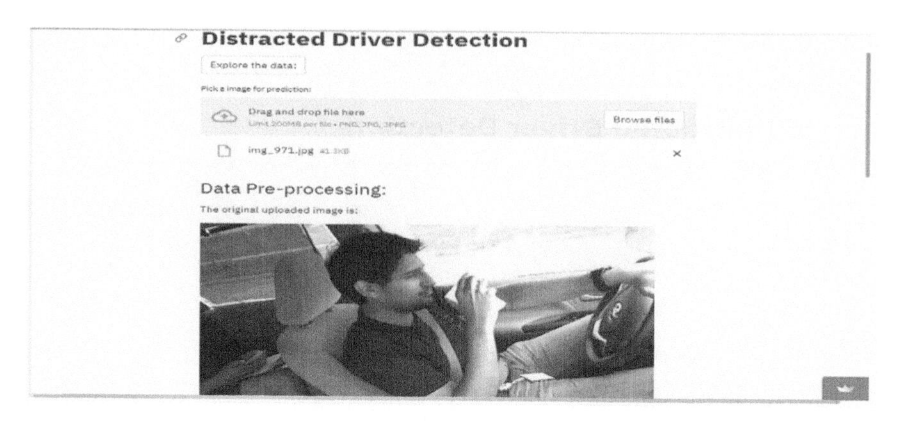

FIGURE 4.14 Initially, the uploaded image is gets updated into the dashboard.

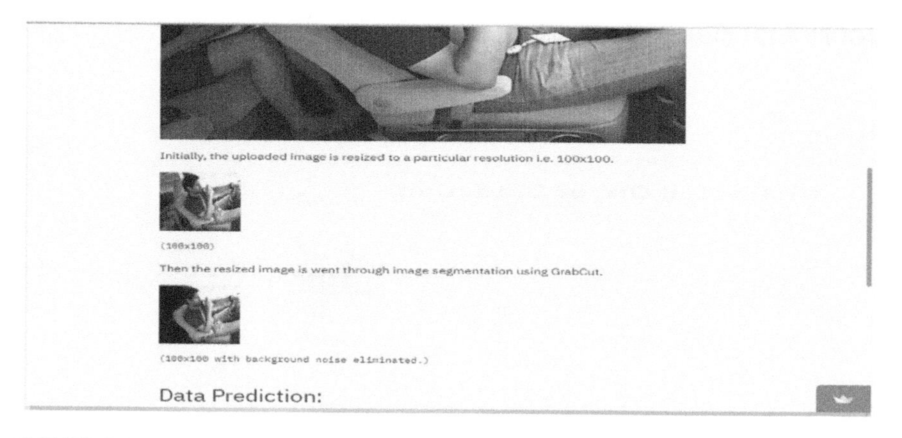

FIGURE 4.15 The pre-processed images, both resized and segmented, are updated as a part of the prediction process in the dashboard.

eliminates the irrelevant background noise pixels from images. It maintains the suitable foreground pixels, making the image-classification model more efficient than others. The proposed method has positively produced high accuracy even with less image resolution in all ten classes of driver distractions. With an increase in resolution, the algorithm's runtime will increase significantly without affecting the detection accuracy much. This model can be used as a base architecture for the future development of the real-time detection implementation model.

As this chapterfocused only on the one-sided image with a particular lighting condition, future works lie in analyzing the driving behaviour from different camera angles other than the right-hand side, which is provided by State Farm Dataset, along with images with different lighting conditions like low-light, extreme sunlight, etc., and more focus should be given towards decreasing the required time for computation as well as extending the desired number of features. Furthermore, the objective is not

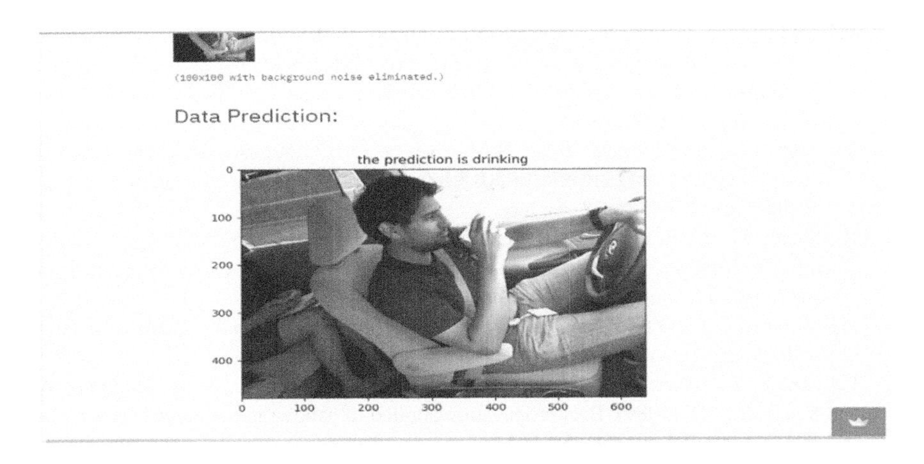

FIGURE 4.16 The predicted class of distracted driving appears automatically along with the input image in the dashboard.

just the detection of distracted driver but also to reduce the frequency of distracted driving.

REFERENCES

[1] Nogayeva, S., Gooch, J., &Frascione, N. (2020). The Forensic Investigation of Vehicle-Pedestrian Collisions: A Review. Science & Justice.

[2] Young, K. L., &Regan, M. A. (2013). Defining the relationship between behavioral adaptation and driver distraction. Behavioural Adaptation and Road Safety Theory, Evidence and Action. Boca Raton: CRC Press, Taylor and Francis Group, 2013, 227–43.

[3] Rahman, A. (2020). Statistics for data science and policy analysis. (1st ed.) Springer. https://doi.org/10.1007/978-981-15-1735-8

[4] Chowdhury, M.M.H., Rahman, A., & Islam, M. R. (2018). Protecting data from malware threats using machine learning technique. In Proceedings of the 2017 12th IEEE Conference on Industrial Electronics and Applications (ICIEA) (pp. 1691–1694). IEEE, Institute of Electrical and Electronics Engineers. https://doi.org/10.1109/ICIEA.2017.8283111

[5] Uddin, M. G., Nash, S., Mahammad Diganta, M. T., Rahman, A., & Olbert, A. I. (2022). Robust machine learning algorithms for predicting coastal water quality index. Journal of Environmental Management, 321(11), 1–16. [115923]. https://doi.org/10.1016/j.jenvman.2022.115923

[6] Rahman, A. (2019). Statistics-based data preprocessing methods and machine learning algorithms for big data analysis. International Journal of Artificial Intelligence, 17(2): 44–65.

[7] Rahman, A., Nimmy, S. F., & Sarowar, G. (2019). Developing an automated machine learning approach to test discontinuity in DNA for detecting tuberculosis. In J. P. Davim (Ed.), Proceedings of the Twelfth International Conference on Management Science and Engineering Management (pp. 277–286). Springer. https://doi.org/10.1007/978-3-319-93351-1_23

[8] Uddin, M. G., Nash, S., Rahman, A., & Olbert, A. I. (2023). Performance analysis of the water quality index model for predicting water state using machine learning techniques. Process Safety and Environmental Protection, 1–30. https://doi.org/10.1016/j.psep.2022.11.073

[9] Albawi, S., Mohammed, T. A., &Al-Zawi, S. (2017, August). Understanding of a convolutional neural network. In 2017 International Conference on Engineering and Technology (ICET) (pp. 1–6). Ieee.

[10] Shaha, M., &Pawar, M. (2018, March). IEEE. Transfer learning for image classification. In 2018 Second International Conference on Electronics, Communication and Aerospace Technology (ICECA) (pp. 656–660).

[11] Simonyan, K.,&Zisserman, A. (2014). Very deep convolutional networks for large-scale image recognition. arXiv preprint arXiv:1409.1556.

[12] LeCun, Y., Boser, B., Denker, J. S., Henderson, D., Howard, R. E., Hubbard, W., &Jackel, L. D. (1989). Backpropagation applied to handwritten zip code recognition. Neural computation, 1(4), 541–551.

[13] Szegedy, C., Vanhoucke, V., Ioffe, S., Shlens, J., &Wojna, Z. (2016). Rethinking the inception architecture for computer vision. In Proceedings of the IEEE conference on computer vision and pattern recognition (pp. 2818–2826).

[14] Rother, C., Kolmogorov, V., & Blake, A. (2004). "GrabCut" interactive foreground extraction using iterated graph cuts. ACM transactions on graphics (TOG), 23(3), 309–314.

[15] Kaur, D., &Kaur, Y. (2014). Various image segmentation techniques: a review. International Journal of Computer Science and Mobile Computing, 3(5), 809–814.

[16] He, K., Gkioxari, G., &Doll, P. (2017)´.,ar, and Ross, Girshick. Mask r-cnn. In, ICCV, 1(6).

[17] Xing, Y., Tang, J., Liu, H., Lv, C., Cao, D., Velenis, E., &Wang, F. Y. (2018, June). End-to-end driving activities and secondary tasks recognition using deep convolutional neural network and transfer learning. In 2018 IEEE Intelligent Vehicles Symposium (IV) (pp. 1626-1631). IEEE.

[18] Ohn-Bar, E., Martin, S., &Trivedi, M. (2013). Driver hand activity analysis in naturalistic driving studies: challenges, algorithms, and experimental studies. Journal of Electronic Imaging, 22(4), 041119.

[19] Baheti, B., Gajre, S., &Talbar, S. (2018). Detection of distracted driver using convolutional neural network. In Proceedings of the IEEE conference on computer vision and pattern recognition workshops (pp. 1032–1038).

[20] Zhao, C. H., Zhang, B. L., Zhang, X. Z., Zhao, S. Q., &Li, H. X. (2013). Recognition of driving postures by combined features and random subspace ensemble of multilayer perceptron classifiers. Neural Computing and Applications, 22(1), 175–184.

[21] Yan, C., Coenen, F., &Zhang, B. (2016). Driving posture recognition by convolutional neural networks. IET Computer Vision, 10(2), 103–114.

[22] Shah, R. R., Yu, Y., Verma, A., Tang, S., Shaikh, A. D., &Zimmermann, R. (2016). Leveraging multimodal information for event summarization and concept-level sentiment analysis. Knowledge-Based Systems, 108, 102–109.

[23] Kumar, Y., Jain, R., Salik, M., ratn Shah, R., Zimmermann, R., &Yin, Y. (2018, December). Mylipper: A personalized system for speech reconstruction using multi-view visual feeds. In 2018 IEEE International Symposium on Multimedia (ISM) (pp. 159–166). IEEE.

[24] Yu, Y., Tang, S., Raposo, F., &Chen, L. (2019). Deep cross-modal correlation learning for audio and lyrics in music retrieval. ACM Transactions on Multimedia Computing, Communications, and Applications (TOMM), 15(1), 1–16.

[25] Sathe, V., Prabhune, N., &Humane, A. (2018). Distracted driver detection using cnn and data augmentation techniques. International Journal of Advanced Research in Computer and Communication Engineering, 7(4).

[26] Torres, R., Ohashi, O., Carvalho, E., &Pessin, G. (2017, September). A deep learning approach to detect distracted drivers using a mobile phone. In International Conference on Artificial Neural Networks (pp. 72–79). Springer, Cham.

[27] Abouelnaga, Y., Eraqi, H. M., &Moustafa, M. N. (2017). Real-time distracted driver posture classification. arXiv preprint arXiv:1706.09498.

[28] Baheti, B., Gajre, S., &Talbar, S. (2018). Detection of distracted driver using convolutional neural network. In Proceedings of the IEEE conference on computer vision and pattern recognition workshops (pp. 1032–1038).

[29] Karray, F., Campilho, A., &Cheriet, F. (Eds.). (2017). Image Analysis and Recognition: 14th International Conference, ICIAR 2017, Montreal, QC, Canada, July 5–7, 2017, Proceedings (Vol. 10317). Springer.

[30] Hssayeni, M. D., Saxena, S., Ptucha, R., &Savakis, A. (2017). Distracted driver detection: Deep learning vs handcrafted features. Electronic Imaging, 2017(10), 20–26.

[31] Masood, S., Rai, A., Aggarwal, A., Doja, M. N., &Ahmad, M. (2018). Detecting distraction of drivers using convolutional neural network. Pattern Recognition Letters.

[32] da Silva Oliveira, F. R., &Farias, F. C. (2018, November). Comparing transfer learning approaches applied to distracted driver detection. In 2018 IEEE Latin American Conference on Computational Intelligence (LA-CCI) (pp. 1–6). IEEE.

[33] Tran, D., Do, H. M., Sheng, W., Bai, H., &Chowdhary, G. (2018). Real-time detection of distracted driving based on deep learning. IET Intelligent Transport Systems, 12(10), 1210–1219.

[34] Moslemi, N., Azmi, R., & Soryani, M. (2019, March). Driver distraction recognition using 3D convolutional neural networks. In 2019 4th International Conference on Pattern Recognition and Image Analysis (IPRIA) (pp. 145–151). IEEE.

5 Review Analysis of Ride-Sharing Applications Using Machine Learning Approaches

Bangladesh Perspective

*Taminul Islam[1], Arindom Kundu[1], Rishalatun Jannat Lima[1], Most Hasna Hena[1], Omar Sharif[*2], Azizur Rahman[3] and Md Zobaer Hasan[4]*

[1]Department of Computer Science & Engineering, Daffodil International University, Ashulia, Bangladesh

[2]Universal College Bangladesh (Monash College), Dhaka, Bangladesh

[3]School of Computing, and Mathematics and Engineering, Charles Sturt University, Wagga Wagga, Australia

[4]School of Science, Monash University Malaysia, Selangor D. E., Malaysia

*Corresponding Author: omar.sharif@monashcollege.edu.au

CONTENTS

DOI: 10.1201/9781003253051-7

5.1 INTRODUCTION

There are several ride-sharing applications available in Bangladesh like Uber, Patho, Obhai, Grab, etc. The quality of each company depends on their providing services. Users are able to submit their reviews of services using number of stars as well as comments. The ranking and popularity of a ride-sharing app are determined by the reviews left by its users. Reviews help future passengers better evalute ride-sharing apps, while the authenticity of those evaluations is unknown and could have varied effects. The use of machine learning techniques can make it easier to spot fake ride-sharing app reviews. Web mining techniques (Sharma et al., 2022) employ a variety of machine learning algorithms to locate and collect specific data from the internet. Content mining is used to gather reviews (Lai et al., 2021), and involves using machine learning to train a classifier to assess review attributes and user sentiments for determining the user overall experience (positive or negative) (Agarwal, 2022c; Agarwal et al., 2016). Fake reviews are usually detected by looking at specific factors that are not directly related to the content of ride-sharing, such as the category in which the review appears. Some people manipulate reviews to disseminate false information (Agarwal et al., 2022a). False information can be used to boost or degrade a company or application, depending on the intent. Fake reviews, review spams, and opinion spams are all terms used to describe this type of activity. In accordance with Rausch et al. (2022), creating a false review is a form of opinion spamming. Instead of expressing their true ideas or experiences, reviewers attempt to deceive readers or automated opinion mining and sentiment analysis algorithms (Hossain et al., 2021; Kumari et al., 2021), which is considered as an unlawful behaviour (Rausch et al., 2022).

A fake review is one in which the reviewer knowingly provides untruthful or irrelevant information regarding the review item, whether it is partly false or completely false. Aside from being dubbed fake reviews, other terms for them include bogus, scamming, misleading, and spam (Chowdhury et al., 2018a). Spammers may intend to build excitement for a product or service by generating good reviews in large numbers. This is the fundamental concern with review spams (Chowdhury et al., 2018b). False reviews now have a significant impact on how customers perceive a brand (Krishna Rao et al., 2022). For organizations, positive reviews can result in large financial gains; on the other hand, poor evaluations can quickly destroy a company's good name. Automated systems or paid reviewers can create reviews. Fake positive evaluations for a company's products or services can be written by people or third-party groups hired by companies or merchants. Since anybody can simply create and submit a review on the internet, the practice of spamming ride-sharing applications with fake reviews has become more common. Therefore, we utilize some machine learning approaches such as Decision Tree, Random Forest,

TABLE 5.1
Research Question Criteria

Criteria	Details
Population	Bangladeshi ride-sharing application users
Mediation	Machine learning and deep learning approaches to prediction
Outcome	Important attributes, accuracy, and classification
Context	Ride-sharing application's review section

Gradient Boosting, AdaBoost and Bidirectional Long Short-Term Memory (Bi-LSTM) (Rahman, 2019), and Bi-LSTM to achieving an optimal level of accuracy of reviews. A suitable research question is vital for uncovering related works in machine learning and approaches for ride-sharing applications data analysis. Kitchenham et al. (2010) outlines the steps necessary to answer the appropriate research questions, such as population, intervention, outcomes, and context. Table 5.1 shows the research topic criteria.

According to the review findings, the following research questions should be pursued:

Q1: How can we fetch reviews from ride-sharing applications?
Q2: What are the approaches to finding real reviews?
Q3: What are the approaches to pre-processing data?
Q4: What is the market value of ride-sharing applications?
Q5: What machine learning approaches are used for review analysis?
Q6: What is the performance of the present proposed models?

This chapter is organized as follows. A background analysis is found under the "Related Work" section. In the methodology section, the essential modelling idea is thoroughly explained. Figures and tables are used in the process to illustrate the suggested thought. All the models' performances are evaluated in great depth in the Results section. This study's Discussion and Conclusion sections discuss the results of this study and where the research is headed.

5.2 RELATED WORK

The research community in the domain of natural language processing with machine learning and deep learning has grown significantly (Agarwal, 2022c; Agarwal, 2019). More than 15 million evaluations from more than 3.5 million users from three major travel sites were included in a study by Minnich et al. (2015). There were three main goals in their work. They developed brand-new tools for detecting disparities across many sites. They also carried out the first comprehensive research of cross-site variations using real data and produced a data-science-based technique with 93% accuracy. The TrueView score was then presented, and 20% of hotels appeared to have a low trustworthiness score, based on the results. Moreover, the study by

Heydari et al. (2015) analyzes various modelling tools to categorize them according to models that mostly identify spam in reviews. It is important to note that each sort of detection method has various strengths and drawbacks. Although the above study method has continued to improve in terms of accuracy and output, it's main limitation is that they can't achieve more precise results without going through the process of systematic analysis (Crawford et al., 2016). Most of these studies have one thing in common: they turn reviews into word vectors, which can provide tens of thousands of unique characteristics. However, little research has been done on how to appropriately reduce the size of the feature subset to a tolerable quantity. Filter-based element rankers and term feature selection were applied by researchers to lower the size of a feature subset. These approaches are used in the review spam domain.

These results illustrate that there is no one-size fits-all method to feature selection. Also, the optimum technique to minimize the size of the feature subset depends on the classifier employed and the intended size of the feature subset. Researchers have used Decision tree, Logistic Regression, Naïve Bayes, SVM, and Multinomial Naïve Bayes to determine the accuracy of their proposed model (Shiraz et al., 2017; Thevaraja & Rahman, 2019), and many found decision trees to be most accurate at about 83%. Review spam detection is no different in that finding labelled datasets is always a difficulty for machine learning researchers. Using Amazon Mechanical Turk (AMT) to produce fake reviews for their dataset and combining them with "true" TripAdvisor ratings, Ott et al. (2011) developed a unique technique. To come up with their final dataset, they gathered a total of 400 false and 400 true reviews. These classifiers were tested on a variety of different datasets, including unigrams, bigrams, and trigrams. There was no statistical analysis done to see if the difference between SVM and bigrams in terms of performance was significant since the dataset was rather small.

Some published fake news detecting works are summarized in Table 5.2.

5.3 METHODOLOGY

This section introduces the main methodology of this study. An outline of this research process is shown in Figure 5.1 This section explores the dataset's source and features. In addition, contextual aspects are addressed here. Some classification models and evaluation procedures are briefly explored in the later portion of this chapter. The steps in our study process are as follows:

Dataset Creation: The dataset was created from the review sections of Bangladeshi ride-sharing applications: Uber, Pathao, Obhai, and Shohoz.

Pre-processing: To deal with noisy and inconsistent data, pre-processing techniques are employed (Rahman et al., 2013; Rahman & Harding, 2016; Rahman, 2017; Rahman, 2019). Many different pre-processing procedures are used to improve the quality of the final product of the data. Main techniques that were applied include tokenization, lemmatization, Punctuation removal, Stopwards removal and others.

TABLE 5.2
Summary of Related Research Work

References	Year	Contribution	Dataset	Models	Accuracy
(Vachane and D. 2021)	2021	Incorporated spam identification architecture that uses display review illuminating lists as metadata structures.	HIN Resource	Naïve Bayes, Decision Tree	Decision Tree achieved the best 92.06% accuracy
(Manaskasemsak et al., 2021)	2021	Developed graph model to detect fake reviews and fake reviewers.	Yelp	CNN	75%
(Yao et al., 2021)	2021	Developed model to detect fake reviews from hotel and restaurant sections.	Crowdsourcing. Amazon Mechanical Turk. TripAdvisor	RF, AdaBoost, SVM, CNN, LSTM	Random Forest achieved the best 90% accuracy
(Budhi et al., 2021)	2021	Proposed a data sampling technique that improves the accuracy of the fake review class.	Yelp	LR, SVM, Multilayer Perceptron (MLP), Bagging Predictor (BP), RF, AdaBoost	SVM achieved the best 85.74% accuracy
(Wang et al., 2020)	2020	Proposed a method to identify false reviews using multiple feature fusion and collaborative rolling training.	Yelp	Rf, LR, Latent Dirichlet Allocation, KNN, DT, Naïve Bayes, SVM	SVM achieved the best 84.45% accuracy
(Kumar, 2020)	2018	Developed model according to behaviour feature of reviewer to detect fake and true reviews.	Yelp	RF, SVM	Random Forest achieved the best 91.396% accuracy

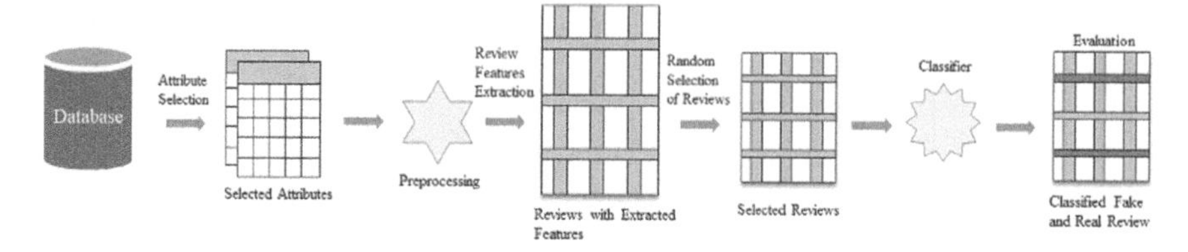

FIGURE 5.1 A step-by-step guide to detecting fake and real reviews.

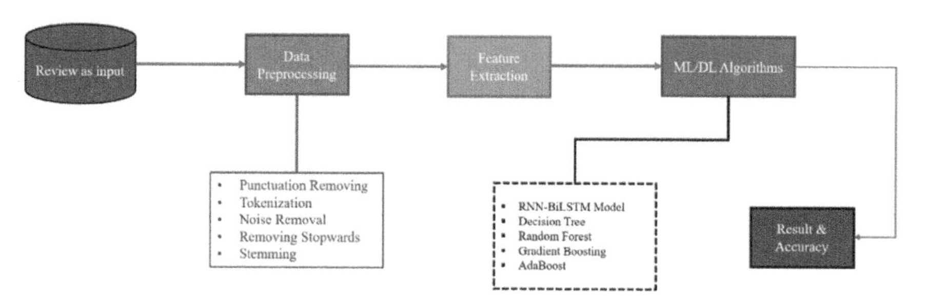

FIGURE 5.2 Proposed model workflow.

Feature Extraction: A feature set for the classification model was built using attributes that were retrieved after pre-processing the data in the review database such as review of quality, service, experience, satisfaction, etc. of hotels.

Training the Model: Several classification algorithms were then trained for experiments associated with our study concerning the accuracy of ride-sharing applications reviews.

Figure 5.2 shows the workflow of our proposed research methodology.

5.3.1 Data Description

Data collection is always a challenging part of a research study. It was a tough task to collect all the fresh data within a short time from all the authorized sites. Data (online user reviews) was collected from four Bangladeshi ride-sharing applications review sections: Uber, Pathao, Obhai, and Shohoz. This data was manually collected from individual websites and social platforms and GooglePlay Store reviews section, resulting in a total of 3315 online reviews from individuals from four platforms. Table 5.3 illustrates the amount of collected data from the individual apps.

Table 5.4 lists the amount of total True reviews, Partially False reviews, and False reviews. The exact number of this statistics is – True (1365), Partially False (930), and False (1020). From Table 5.3 and Table 5.4 it is clearly seen that the data is balanced . This is an important part of this research.

These data were classified into three categories: – True, False, and Partially False. Table 5.5 gives the descriptions of the categories as classified by Balouchzahi et al. (2021).

TABLE 5.3
Amount of Collected Individual Data

Applications	Data Size
Uber	1000
Pathao	985
Obhai	815
Shohoz	515
Total	3315

TABLE 5.4
Amount of Categorical Data

Category	Amount of data
True	1365
False	1020
Partially False	930

TABLE 5.5
Descriptive Category

Category	Description
True	The given text includes contents that are clearly apparent or capable of being logically proven.
Partially False	Main claim in given text might be true but also contain false or misleading information information, not surely true and not certainly false.
False	The main content of given text is fake.

5.3.2 Data Pre-Processing

Pre-processing of data is the initial step in doing research and is the first stage in data mining (Rahman, 2020; Agarwal et al., 2022b). There are several ride-sharing apps platforms from which we obtained our reviews datasets that require preprocessing. These datasets were split down into a wide range of numerical values and processed one at a time since machine learning and deep learning can handle numerical data only. For the text and rating dimensions, we used the following data preparation approaches. We focused on the data pre-processing steps followed by data cleaning, data integration, data transformation, data reduction, and finally data discretization.

Reduce of Dimension: To reduce the size of the data, it is necessary to convert it from a high-dimensional space to a low-dimensional one, while still retaining as many of the original data's attributes as possible (Burges, 2010). Unnecessary features in the data are responsible for increasing the length of time it takes to complete an operation. The public id and title characteristics were deleted from the dataset prior to the data being entered into the model. The text column in the dataset contains the input data, while the target column has the rating.

Punctuation Removal: There are many punctuation marks, links, numbers, and other special characters used in reviews, none of which have any influence on whether the review is true or incorrect in the vast majority of cases. In addition, punctuation appears often and has a substantial influence on the measurements for punctuation, but it has no effect on the classification of the text, which is a mixed bag (Pradha et al., 2019). Figure 5.3 shows an example of punctuation removal.

Noise Removal: Noise removal refers to the process of removing letters, numbers, and fragments of text that might obstruct text analysis. It is a vital step in the preparation of data (Tang et al., 2022). All data must be clear and free of noise. Words that are unnecessary to tokenize and vectorize must be removed from the input sequence. Tokenization is improved by converting uppercase characters to lowercase ones.

Tokenization: Tokenization is the process of separating review material into words (tokens). To calculate the Reviewer Content Similarity (RCS) and capital variety, tokenization is a critical step since it allows each word in the review to be separated (Sockin et al., 2022). For word tokenization, we made use of the NLTK library. For example, for the word "greatest." we used character tokens: g-r-e-a-t-e-s-t and subword tokens: great-est.

Removing Stopwards: Due to the fact that stopwards are widespread in natural language and do not convey any unique meaning, they are not significant in a phrase (Gerlach et al., 2019). Stopwards such as 'is', 'an', 'the', etc., may increase the

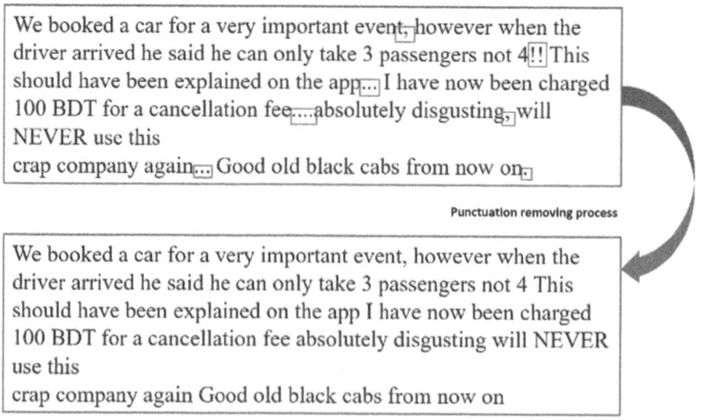

We booked a car for a very important event, however when the driver arrived he said he can only take 3 passengers not 4!! This should have been explained on the app... I have now been charged 100 BDT for a cancellation fee.... absolutely disgusting, will NEVER use this
crap company again... Good old black cabs from now on.

Punctuation removing process

We booked a car for a very important event, however when the driver arrived he said he can only take 3 passengers not 4 This should have been explained on the app I have now been charged 100 BDT for a cancellation fee absolutely disgusting will NEVER use this
crap company again Good old black cabs from now on

FIGURE 5.3 Example of punctuation removing from the text.

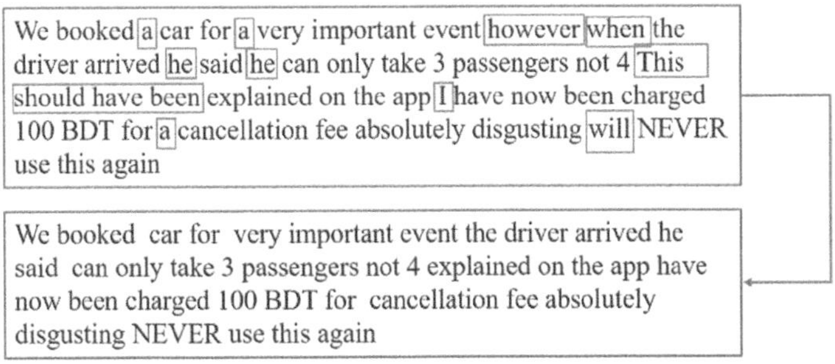

We booked a car for a very important event however when the driver arrived he said he can only take 3 passengers not 4 This should have been explained on the app I have now been charged 100 BDT for a cancellation fee absolutely disgusting will NEVER use this again

We booked car for very important event the driver arrived he said can only take 3 passengers not 4 explained on the app have now been charged 100 BDT for cancellation fee absolutely disgusting NEVER use this again

FIGURE 5.4 Example of stopwards removing from the text.

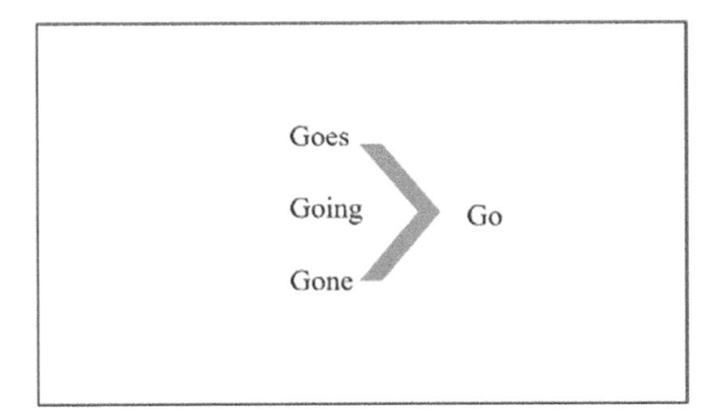

FIGURE 5.5 Example of stemming removal from the text.

amount of time it takes to process data in data analysis. Because of this, it is important to remove stopwords from phrases. We employed the NLTK library for this purpose. We must remove all of the unnecessary text (e.g., stopwards) and strings from the data in order to make it trainable. Because of this, we convert all of our text into numbers so that it may be utilized as a teaching aid. Figure 5.4 is an example of stopwords being removed from a text.

Capitalization: In a computational model, it is ideal to use the same register level regardless of whether upper- or lowercase characters are used (Păiş et al., 2022). It does not matter what kind of register level you use when it comes to digits. Lowercase letters were used in this study.

Stemming Removal: Eliminating suffixes and prefixes from a word is known as stemming. Stemming is a technique that is frequently employed in information retrieval activities. Numerous researchers have demonstrated that stemming increases information retrieval system performance (Atchadé et al., 2022). Using the stemming method, we can get a word back to its root structure. Figure 5.5 shows how stemming is done.

5.3.3 Proposed Model Working Procedure

In this research data was collected from four different ride-sharing applications. Our main goal was to find the best machine learning model to detect fake, true, or partially fake reviews smoothly. Five machine learning algorithms were applied to find the best accuracy on this dataset: Decision tree, Random Forest, Gradient Boosting, AdaBoost, and Bi-LSTM. We got the best accuracy (85%) from the Bi-LSTM model.

To begin the modeling procedure, data were divided into two parts:

- Dataset for Training
- Dataset for Testing

About 80% of the data was utilized for training, and 20% for testing. And this is also what we expected to observe in our model.

Feature Selection and Extraction

The Keras library was used to create our Bi-LSTM model, which was then tested. It is possible to make a model using a glove embedding of 100d. The sequential model was used as the foundation for this experiment's analysis. A number of different techniques are used in the model for accurate feature extraction, including embedding, dropout layers, and a layer with 256 neurons that is totally connected to the rest of the network. This dataset has many classes, which is why soft-max activation was used to apply the output layer to the final layer. It is consistent with other algorithms, such as Random Forest, Gradient Boosting, and AdaBoost. N-gram features such as unigram, bigram, and trigram are employed in all machine learning techniques for improved model outcomes. The model was trained using 20 epochs and 128 batch sizes of training data to achieve optimal performance. The accuracy of this model was determined to be 85%, while the F_1 Score was found to be 89%.

5.3.3.1 Machine Learning Models

In this study, the major aim was to create the best machine learning model to recognize fake, real, or partially fake reviews. On this dataset, we used five different machine learning methods to determine the one with the greatest accuracy: Decision Tree, Random Forest, Gradient Boosting, AdaBoost, and Bi-LSTM models. These models are discussed in this section.

Decision Tree

Classification and regression models may be built at regular intervals using a decision tree. In terms of categorization and predictions, this is the most effective and widely used technology available today. There are many different types of decision trees; the most common is the flowchart-like tree structure, in which each internal node symbolizes a test on a certain characteristic, and each branch reflects a conclusion of the test (Pappalardo et al., 2021). The last word is a node in a tree having nodes for decisions and nodes for leaves. Other nodes are either a few or many branches in the decision tree. Decisions or classifications are represented by a leaf node. The root node of a passing tree, which corresponds to the highest successful predictor, is the simplest decision node in the tree. Decision trees can deal with any type of data, whether it is numerical or categorical (Fletcher et al., 2019).

Random Forest

The choice tree is the basic component of random forest classifications. The choice tree is littered with living trees including a variety of elements at each node. The entropy of a specified collection of characteristics is supported by the nodes. In the random forest, a collection of decision trees is linked to a collection of bootstrap samples derived from the source dataset. Trees are the building blocks of a forest, and the more trees there are, the more stable it will be (Khan et al., 2022). By creating call trees out of data samples, the random forest algorithm receives the forecast for every one of these trees, then votes on which is best. Breiman (2001) include extensive information on random forest classifiers. At times while using the quality random forest strategy, the bootstrapping technique is used to help create an appropriate random forest with the requisite number of decision trees thus boosting classification accuracy using the notion of overlap dilution as described. To train and optimize the process, random forests are often used such as growing trees, making it easier to achieve a decision in each level. As a result, random forest is a good method for numerous packets (Magidi et al., 2021).

Gradient Boosting

Many machine learning methods are combined into Gradient Boosting Classifiers (GBCs) in order to create a strong predictive model. Gradient boosting is a technique in which decision trees are occasionally employed. Gradient boosting models have lately been used to win multiple Kaggle informatics challenges due to their success in categorizing large datasets. The main goal is to lower the amount of error in the next model by aligning the desired outcomes (Bahad et al., 2020). There are many different methods to build gradient boosting classifiers in the Python machine learning, Scikit-Learn. This reference examines the theory underlying gradient boosting models and looks at two distinct techniques to construct gradient boosting models in Scikit-Learn (Chakrabarty et al., 2019).

AdaBoost

Multi-learner approaches to problem-solving are referred to as "ensemble learning" (Liu et al., 2022). When it comes to learning, ensemble techniques are a popular choice because of their superior capacity to generalize. Due to its strong theoretical foundation, precise prediction, tremendous simplicity (Wang noted it required only "only 10 lines" of code), and extensive and successful use cases, the AdaBoost algorithm created by Wang et al. was among the most significant ensemble techniques. Because AdaBoost is the most widely used ensemble algorithms, its huge influence is not surprising. The theoretical and practical aspects of these two topics are briefly discussed in this reference (Wang et al., 2021). Because of AdaBoost, there has been an abundance of theoretical research on ensemble approaches, which is readily available in the machine learning and statistical literature.

Bi-LSTM

In comparison to Long Short-Term Memory, Bi-Directional Long Short-Term Memory (Bi-LSTM) excels at categorizing sequences (LSTM). It is the process of creating a neural network that can process information in both forward and reverse orientations. The Bi-LSTM is composed of two LSTMS, one for forward and one for reverse input. It is feasible to communicate data in both directions using disguised states. Each time step, the outputs of two LSTMs are merged to generate one (Liu

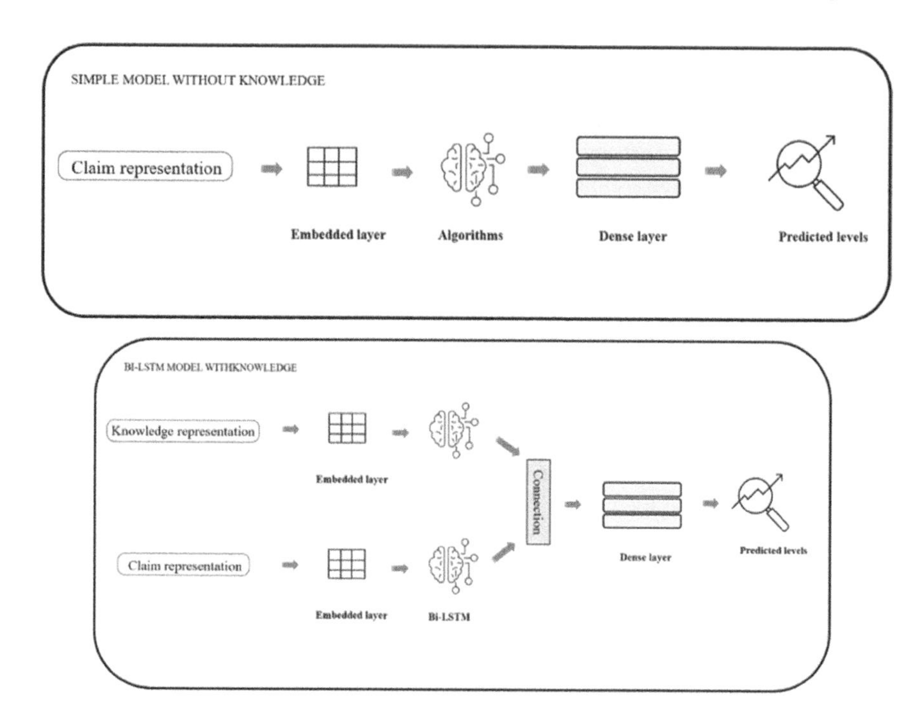

FIGURE 5.6 Working sketch between Simple and Bi-SLTM models.

et al., 2022). The Bi-LSTM technique contributes to the reduction of the restrictions associated with traditional RNNs. The context is more easily comprehended as a result of Bi-LSTM's high degree of accuracy. However, with bi-directional input, we can ensure that both the future and the past are preserve (Braşoveanu et al., 2019). Based on the previous validated reviews, a natural language inference (NLI) model is built using Bi-LSTM neural networks in this phase to assess the validity of each individual claim. We started by training a simple machine learning model with simply the assertions (hypotheses) as input. The NLI-based model is then trained to infer the accuracy of the claim based on previous information (premises). We test the suggested NLI-based strategy to identifying false reviews by comparing the outcomes of these two models. Figure 5.6 shows the process of this method.

A key feature of Bi-LSTM is the ability to learn the forward and backward information contained within the input words. A given input phrase X has N words and is represented as a vector $(x_1, x_2 \ldots, x_n)$. After taking into account prior hidden states h_{t-1} and cell states c_{t-1} the Eq. 5.1 is used to determine the present state:

$$i_t = \sigma\left(W_i w_t + U_i h_{t-1} + b_i\right) f_t = \sigma\left(W_f w_t + U_f h_{t-1} + b_f\right) 9 \quad 99v$$

$$= tanh\left(W_c w_t + U_c h_{t-1} + b_c\right) C_t = f_t C_{t-1} + i_t \tilde{c}_t h_t = o_t tanh\left(c_t\right)$$

$$o_t = \sigma\left(W_o w_t + U_o h_{t-1} + b_o\right) \tag{5.1}$$

The sigmoid function denoted by σ and the hyperbolic tangent function tanh are both used in the equation proposed by Zheng and Chen (2021). The authors provide an algorithmic process for matrix multiplication that stands for elementwise multiplication. There are two sets of weight matrixes: W represents the current input vector w_t and U represents the prior hidden state vector h_{t-1}; both sets of weight matrixes are shown in the above equation. The Bi-LSTM method's pseudocode is given in the following example.

Input: Character embeddings that have been pre-trained *X*.

Output: The probability distribution *P* of the input sequence is returned:

(1) The forward LSTM layer receives character vectors from *X*
(2) **for** *i length*(X) do
(3) send X_i to Bi-LSTM layer
(4) end **for**
(5) Set 2: The current LSTM network's cell state was updated.
(6) $f_t = \sigma\left(W_f\left[h_{t-1}, x_t\right] + b_f\right)$
(7) $i_t = \sigma\left(W_i\left[h_{t-1}, x_t\right] + b_i\right)$
(8) $\tilde{c}_t = tan\ tanh\left(W_f\left[h_{t-1}, x_t\right] + b_c\right)$
(9) $c_t = f_t * c_{t-1} + i_t * c_t$
(10) $o_t = \sigma\left(W_o\left[h_{t-1}, x_t\right] + b_o\right)$
(11) $h_t = o_t * tanh\left(c_t\right)$
(12) Step 3: Send the *X* character vectors to the reverse LSTM layer and repeat the previous two steps.
(13) Step 4: The hidden layers' forward and backward sequencing are spliced together to produce a sentence-level hidden unit sequence *C* that is rich in context.
(14) Step 5: The prediction matrix *P* is obtained once *C* was delivered via a complete connection layer.
(15) **Return** *P*.

5.3.3.2 Performance Measurement Unit

Various writers utilized a number of criteria to judge the effectiveness of their models. Despite the fact that the bulk of the research utilized many indicators to measure their efficiency, a low amount is also used a single statistic. In this study Accuracy, Precision, Recall, and F_1- Score is examined for evaluating this research effort. Text data analysis benefits greatly from using these four measurement units.

5.3.3.3 Accuracy

The ratio of correctly predicted items to all possible predictions indicates the accuracy of a model outcomes. Eq. 5.2 defines the accuracy:

$$Accuracy = \frac{TP + TN}{TP + FP + TN + FN} \tag{5.2}$$

5.3.3.4 Precision

The precision of a machine learning model's prediction is an essential performance parameter. Divide the number of correct forecasts by the number of correct positives. Eq. 5.3 defines the precision:

$$Precision = \frac{TP}{TP + FP} \tag{5.3}$$

5.3.3.5 Recall

Recognizing all possible real values is the ability of a detector to reliably discover and identify them. It is defined as the ratio of TP to the sum of TP and FN in Eq. 5.4:

$$Recall = \frac{TP}{TP + FN} \tag{5.4}$$

5.3.3.6 F_1-Score

This is known as the harmonic mean since it relies on both accuracy and memory. Mathematical formulation of a memory retrieval is given in Eq. 5.5:

$$F_1 - Score = 2\left(\frac{Precision \times Recall}{Precision + Recall}\right) \tag{5.5}$$

5.4 RESULT

Here five machine learning algorithms were applied on this fresh dataset. There is a tight comparison between the algorithms. However, the Bi-LSTM achieved the best accuracy, precision, recall, and F_1 score. Bi-LSTM achieved the best 85% accuracy but Random Forest and AdaBoost work well and achieved 83% accuracy. We found 80% accuracy in Gradient Boosting and 79% accuracy on Decision Tree algorithm. Table 5.6 gives the results of the five machine learning algorithms.

In the following we look at the results of the five different algorithms, then evaluate the results.

TABLE 5.6
Result Comparison between Five Machine Learning Algorithms

Algorithms	Accuracy	Precision	Recall	F_1 Score
Decision Tree	0.798	0.691	0.723	0.691
Random Forest	0.834	0.690	0.723	0.693
Gradient Boosting	0.803	0.703	0.718	0.629
AdaBoost	0.830	0.706	0.731	0.702
Bi-LSTM	0.850	0.862	0.910	0.893

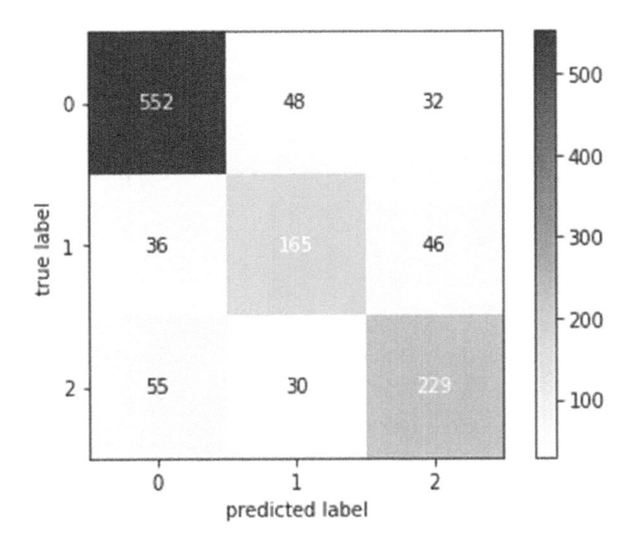

FIGURE 5.7 Confusion matrix of Decision Tree.

Decision Tree

After implementing the decision tree approach to train our model we acquired an accuracy of 79% and a F_1score of 69% where precision was 0.69 and recall 0.72. Figure 5.7 shows the confusion matrix of this algorithm.

Random Forest

After using Random Forest approach to train our model we acquired an accuracy of 83%, which is better than decision tree and a F_1 score of 69%, which is similar with DT, and precision and recall were the same as decision tree. However, the difference on average accuracy was 4%. Figure 5.8 shows the confusion matrix of this algorithm.

Gradient Boosting

In the Gradient Boosting approach after training this model achieved accuracy of 80%, performing relatively better than decision tree and worse than random forest; its F_1score of 62% was not good. The result of precision and recall was 0.70 and 0.71. Figure 5.9 shows the confusion matrix of this algorithm.

AdaBoost

In the AdaBoost approach after training our model we achieved accuracy of 83%, which performs relatively better than DT and GB and it scores F_1score of 70% that performs good. The result of precision and recall was 0.70 and 0.73. Figure 5.10 shows the confusion matrix of this algorithm.

Bi-LSTM

The proposed model's output identifies the review item presented. The review is either true, false, or partially false. The RNN model can't handle text, that's why true

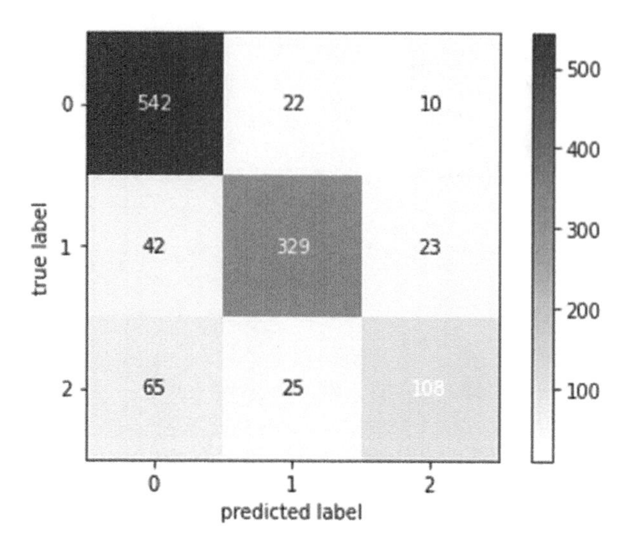

FIGURE 5.8 Confusion matrix of Random Forest.

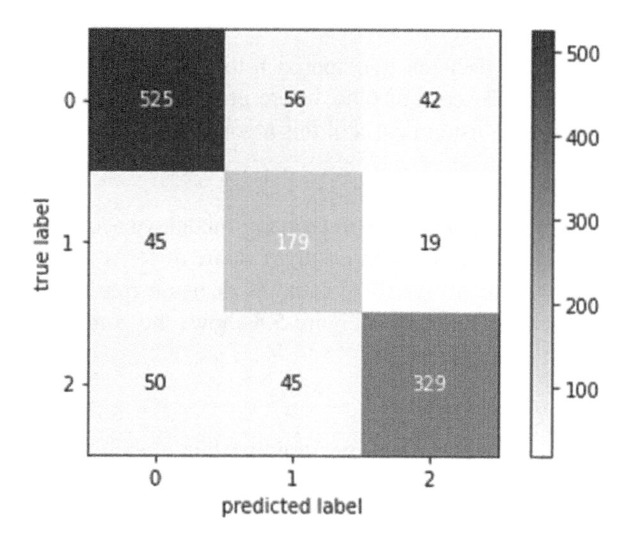

FIGURE 5.9 Confusion matrix of Gradient Boosting.

is considered to be 2, false is 0, and partially false is 1. We applied 20 epochs to train our model and got the best accuracy of 85% and the best F_1score of 89%, where precision and recall was 0.86 and 0.91. We found the best performance using the Bi-LSTM method. Table 5.7 shows the classification report of Bi-LSTM. Figure 5.11 shows the confusion matrix of Bi-LSTM.

Figure 5.12 demonstrates the connection between our proposed model's accuracy and evaluation accuracy and loss and evaluation loss, both of which are achieved using the Bi-LSTM model. These graphics demonstrate that our suggested model

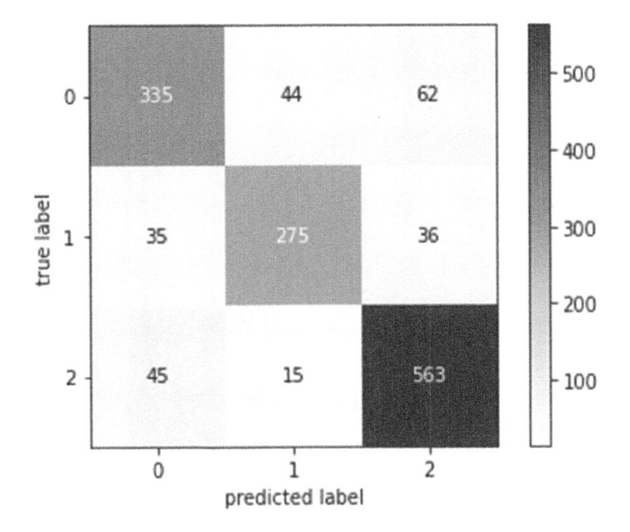

FIGURE 5.10 Confusion matrix of AdaBoost.

TABLE 5.7
Classification Report of Bi-LSTM

Class	Precision	Recall	F_1-Score
True	0.86	0.91	0.89
False	0.91	0.89	0.90
Partially False	0.49	0.37	0.42

accumulates knowledge from its predecessors. From this figure we can see that there are some overfitting issues in the loss vs. evaluation loss graph. While it is not clear the actual reason behind this overfitting, the difference between the learning loss and the evaluation loss increased in later epochs, which may be the reason for this issue, and is a significant limitation of this work.

5.4.1 Cross-Validation

Cross-validation is a term that refers to a variety of similar model validation approaches used to determine the generalizability of the results of a statistical analysis to an independent dataset (Wieczorek et al., 2022). In this work, K-fold cross-validation was used to determine the result. The dataset was divided into k subsets using k-fold cross-validation. The cross-validation process is then used to validate each subset, while the remaining k-1 subsets are merged to utilize as training samples. According to statistical concepts, the best k relies on the number of variables and the type of predictor. K refers to how many groups each data sample should be split into, and it is the only parameter in the technique. k-fold cross-validation is the common

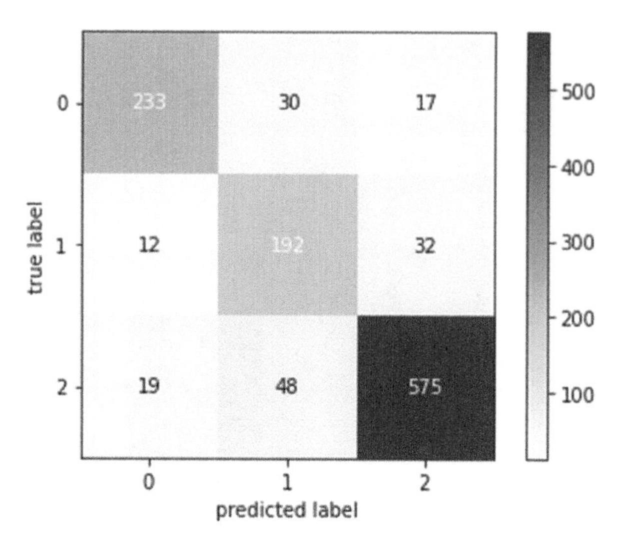

FIGURE 5.11 Confusion Matrix of Bi-LSTM.

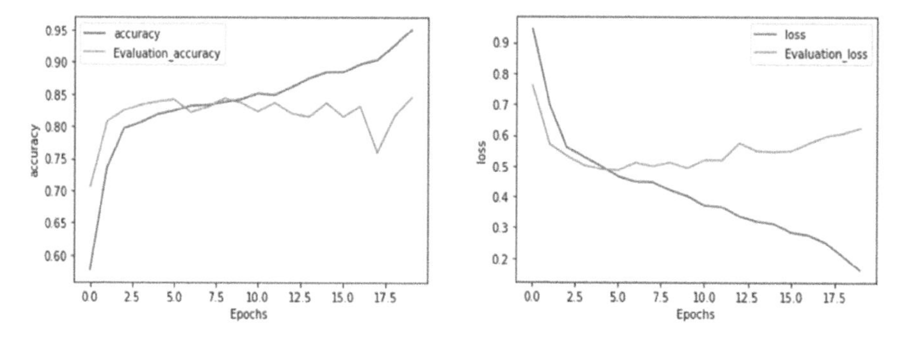

FIGURE 5.12 Graphical representation of accuracy vs evaluation accuracy and loss vs evaluation loss.

name given to this technique (AVUÇLU and E, 2022). It is possible to use a specified value for k in the reference to the model, such as k=10 as 10-fold cross-validation. Table 5.8 clearly shows the k-fold cross-validation of accuracy for each and every model of this work.

In this work, five machine learning algorithms were implemented to find the best model performance. As noted, we got the best model accuracy of 85% from the Bi-LSTM method. To determine this performance, k-fold cross-validation method was used; the detailed results are given in Table 5.8. The five algorithms – Decision Tree, Random Forest, Gradient Boosting, AdaBoost, and Bi-LSTM – performed well in the accuracy cross-validation. We found an 80% 10-fold cross-validation mean score for accuracy from the decision tree whereas we found 84% from the random rorest. On the other hand, gradient boosting performed little bit less accurately than other

TABLE 5.8
K-fold Cross-Validation (for Accuracy) of Five Applied Algorithms

Algorithm	cv=10	cv_score	cv_score (mean)
Decision Tree	1	0.802999	0.803640
	2	0.807806	
	3	0.795142	
	4	0.806100	
	5	0.796545	
	6	0.812104	
	7	0.804949	
	8	0.795977	
	9	0.805268	
	10	0.809513	
Random Forest	1	0.845032	0.847256
	2	0.843038	
	3	0.851093	
	4	0.840588	
	5	0.862186	
	6	0.835555	
	7	0.854597	
	8	0.842178	
	9	0.861001	
	10	0.837284	
Gradient Boosting	1	0.794761	0.794529
	2	0.800492	
	3	0.797103	
	4	0.800222	
	5	0.796060	
	6	0.793273	
	7	0.793011	
	8	0.795018	
	9	0.791043	
	10	0.784308	
AdaBoost	1	0.859676	0.842638
	2	0.839634	
	3	0.864538	
	4	0.828949	
	5	0.854451	
	6	0.823677	
	7	0.827245	
	8	0.864300	
	9	0.823963	
	10	0.839943	

(continued)

TABLE 5.8 (Continued)
K-fold Cross-Validation (for Accuracy) of Five Applied Algorithms

Algorithm	cv=10	cv_score	cv_score (mean)
Bi-LSTM	1	0.858471	0.855861
	2	0.857227	
	3	0.854545	
	4	0.856899	
	5	0.860044	
	6	0.859404	
	7	0.854067	
	8	0.850897	
	9	0.855279	
	10	0.851772	

algorithms. We found a 79% 10-fold cross-validation mean score from this algorithm, whereas AdaBoost performed at 84%. Finally, Bi-LSTM performed better than all algorithms and achieved 85% 10-fold cross-validation mean score, which was the highest from all the algorithms. It is clear that an 10-fold cross-validation score is good and shows that the quality of this model for this esearch is good.

5.5 DISCUSSION

Ride-sharing applications are largely influenced by customer and passenger evaluations found online. Passengers who are considering using a service might benefit from reading online reviews before making their final choice of provider. Users' purchasing decisions might be influenced by fake online reviews. On ride-sharing apps, fake reviews are used to promote or degrade services, and can tarnish a good service's reputation, resulting in financial loss for a well-known business. Customers and companies alike are harmed by fake reviews, which are detrimental to both parties. Since 2007, researchers have been focusing on the identification of bogus reviews. Fake reviews, individual spammers, and spammer groups are the focus of the majority of the current research work being done in these fields. The purpose of this research was to determine whether or not some machine learning algorithms could be used to detect bogus reviews. Four Bangladeshi ride-sharing apps were analyzed by our researchers. The classification methods used in this study performed well. Our research has effectively revealed a previously unknown aspect (i.e., machine learning based cross-valiadation of the fake review) by accurately detecting through appropriate classifier.

While all methods behaved similarly, each classifier is notably different from the others, with Bi-LSTM, AdaBoost, and Random Forest being the best and Decision tree and Gradient Boosting being the worst. Among all other algorithms, Bi-LSTM produced the best figure. Figure 5.13 illustrates the sharp difference more clearly. It is critical to notice the confidence interval for the mean accuracy score across all subset sizes, indicating that random forst and AdaBoost are not mutually exclusive.

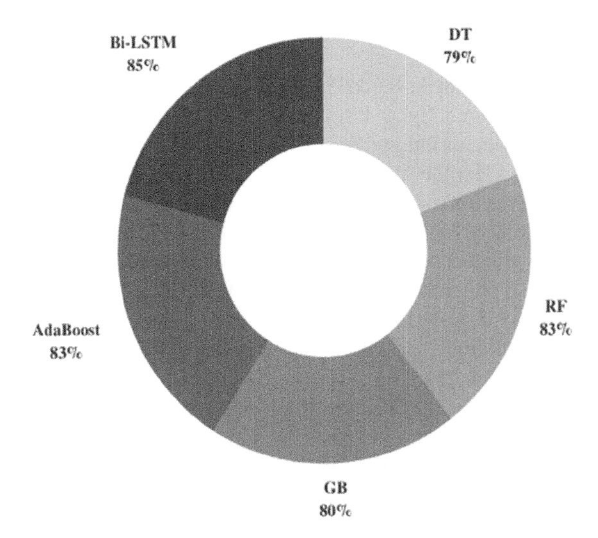

FIGURE 5.13 Accuracy comparison between classifiers.

In this work, data was collected manually by researchers. It was a challenging task to collect quality fresh reviews from different platforms. The result satisfies us in terms of the new dataset. If we work hard to collect the best possible data, then our model should achieve more accuracy in every sector.

5.6 CONCLUSIONS

Today, ride-sharing applications are an important factor of Bangladesh's infrastructure. Fake reviews significantly impair consumers' ability to obtain authentic information. The focus of this research was on identifying fraudulent reviews through the use of well-known machine learning techniques. Data was collected from Uber, Pathao, Shohoz, and Obhai, four of Bangladesh's most popular ride-sharing applications. Following feature extraction and model construction, Bi-LSTM attained the highest accuracy of 85%. Additionally, Random Forest and AdaBoost performed well, with an accuracy of 83%. The model was tested using newly collected data. There are some limitations of this work. Data overfitting is the primary issue of this research. This problem can be solved by using a more appropriate dataset. We could work with a more comprehensive dataset (e.g., big data); this is another limitation of this work. Collecting high-quality data can help enhance accuracy. Performance could be better too. In the future, we hope to test our proposed technique on a bigger, more varied dataset in order to get past existing limitations.

ACKNOWLEDGMENTS

The authors are thankful to the anonymous reviewers for their help in improving this manuscript. We would also like to acknowledge the editing support provided by Charles Sturt University.

REFERENCES

Agarwal, B., Rahman, A., Patnaik, S., & Poonia, R. C. (2022a). Proceedings of International Conference on Intelligent Cyber-Physical Systems. (Algorithms for Intelligent Systems). Springer. www.springer.com/gp/book/9789811671357.

Agarwal, B., Agarwal, A., Harjule, P., & Rahman, A. (2022b). Understanding the intent behind sharing misinformation on social media. Journal of Experimental and Theoretical Artificial Intelligence, 1–15. https://doi.org/10.1080/0952813X.2021.1960637.

Agarwal, B. (2022c) Financial sentiment analysis model utilizing knowledge-base and domain-specific representation. Multimed Tools Appl (2022). https://doi.org/10.1007/s11042-022-12181-y.

Agarwal B., Nayak R., Mittal N., Patnaik S., "Deep Learning Based Approaches for Sentiment Analysis," In Springer. 2015.

Agarwal B., Mittal N., "Prominent Feature Extraction for Sentiment Analysis," In Springer Book Series: Socio-Affective Computing series, Springer International Publishing, ISBN: 978-3-319-25343-5, DOI: 10.1007/978-3-319-25343-5, 2016.

Atchadé, M. N., & Sokadjo, Y. M. (2022). Overview and cross-validation of COVID-19 forecasting univariate models. Alexandria Engineering Journal, 61(4), 3021–3036.

Avuclu, E. (2022). A Statistical Evaluation Between Hybrid, Random, and K-fold Cross Validation to Medical Diagnosis Using Cancerous Breast Tissues.

Bahad, P., & Saxena, P. (2020). Study of adaboost and gradient boosting algorithms for predictive analytics. In International Conference on Intelligent Computing and Smart Communication 2019 (pp. 235–244). Springer, Singapore.

Balouchzahi, F., Shashirekha, H., & Sidorov, G. (2021). MUCIC at CheckThat! 2021: FaDo-fake news detection and domain identification using transformers ensembling. CEUR Workshop Proceedings (CEUR-WS.org). pp. 1–10.

Braşoveanu, A. M., & Andonie, R. (2019). Semantic fake news detection: a machine learning perspective. In International Work-Conference on Artificial Neural Networks (pp. 656–667). Springer, Cham.

Breiman, L. (2001). Random Forests. Machine Learning, 45, 5–32. https://doi.org/10.1023/A:1010933404324

Budhi, G. S., Chiong, R., & Wang, Z. (2021). Resampling imbalanced data to detect fake reviews using machine learning classifiers and textual-based features. Multimedia Tools and Applications, 80(9), 13079–13097.

Burges, C. J. (2010). Dimension reduction: A guided tour. Now Publishers Inc.

Chakrabarty, N., Kundu, T., Dandapat, S., Sarkar, A., & Kole, D. K. (2019). Flight arrival delay prediction using gradient boosting classifier. In Emerging technologies in data mining and information security (pp. 651–659). Springer, Singapore.

Chowdhury, M. M. H., Rahman, A., & Islam, M. R. (2018a). Protecting data from malware threats using machine learning technique. In Proceedings of the 2017 12th IEEE Conference on Industrial Electronics and Applications (ICIEA) (pp. 1691–1694). IEEE, Institute of Electrical and Electronics Engineers. https://doi.org/10.1109/ICIEA.2017.8283111.

Chowdhury, M. M. H., Rahman, A. & Islam, M. R. (2018b) Malware analysis and detection using data mining and machine learning classification. *Advances in Intelligent Systems and Computing*, 580, pp. 266–274. Springer. https://doi.org/10.1007/978-3-319-67071-3_33.

Crawford, M., Khoshgoftaar, T. M. & Prusa, J. D. (2016). "Reducing Feature SetExplosion to Faciliate Real-World Review Sapm Detection," In Proceeding of 29th International Florida Artificial Intelligence Research Society Conference, Association for the Advancement of Artificial Intelligence (www.aaai.org).

Fletcher, S., & Islam, M. Z. (2019). Decision tree classification with differential privacy: A survey. ACM Computing Surveys (CSUR), 52(4), 1–33.

Gerlach, M., Shi, H., & Amaral, L. A. N. (2019). A universal information theoretic approach to the identification of stopwords. Nature Machine Intelligence, 1(12), 606–612.

Heydari, A., ali Tavakoli, M., Salim, N., & Heydari, Z. (2015). Detection of review spam: A survey. Expert Systems with Applications, 42(7), 3634–3642.

Hossain, A., Karimuzzaman, M., Hossain, M. M., & Rahman, A. (2021). Text mining and sentiment analysis of newspaper headlines. Information, 12(10), 1–15. [414]. https://doi.org/10.3390/info12100414

Khan, M. A., Shah, M. I., Javed, M. F., Khan, M. I., Rasheed, S., El-Shorbagy, M. A., ... & Malik, M. Y. (2022). Application of random forest for modelling of surface water salinity. Ain Shams Engineering Journal, 13(4), 101635.

Kitchenham, B., Pretorius, R., Budgen, D., Brereton, O. P., Turner, M., Niazi, M., & Linkman, S. (2010). Systematic literature reviews in software engineering–a tertiary study. Information and software technology, 52(8), 792–805.

Krishna Rao, L. V., Sai Vamsi, S., Naazneen, S., Gowthami, S., & Bindu Bharati, G. (2022). Product Review Analysis. In ICT Systems and Sustainability (pp. 659–669). Springer, Singapore.

Kumar, J. (2020). Fake Review Detection Using Behavioral and Contextual Features. arXiv preprint arXiv:2003.00807.

Kumari S., Agarwal B., Mittal M., "A Deep Neural Network Model for Cross-Domain Sentiment Analysis," International Journal of Information System Modeling and Design (IJISMD), 12(2), 1–16, 2021.

Lai, C. H., & Hsu, C. Y. (2021). Rating prediction based on combination of review mining and user preference analysis. Information Systems, 99, 101742.

Liu, B., Liu, C., Xiao, Y., Liu, L., Li, W., & Chen, X. (2022). AdaBoost-based transfer learning method for positive and unlabelled learning problem. Knowledge-Based Systems, 108162.

Liu, Y., Wang, L., Shi, T., & Li, J. (2022). Detection of spam reviews through a hierarchical attention architecture with N-gram CNN and Bi-LSTM. Information Systems, 103, 101865.

Magidi, J., Nhamo, L., Mpandeli, S., & Mabhaudhi, T. (2021). Application of the random forest classifier to map irrigated areas using google earth engine. Remote Sensing, 13(5), 876.

Manaskasemsak, B., Tantisuwankul, J., & Rungsawang, A. (2021). Fake review and reviewer detection through behavioral graph partitioning integrating deep neural network. Neural Computing and Applications, 1–14.

Minnich, A. J., Chavoshi, N., Mueen, A., Luan, S., & Faloutsos, M. (2015, May). Trueview: Harnessing the power of multiple review sites. In Proceedings of the 24th International Conference on World Wide Web (pp. 787–797).

Ott, M., Choi, Y., Cardie, C., & Hancock, J. T. (2011). Finding deceptive opinion spam by any stretch of the imagination. arXiv preprint arXiv:1107.4557.

Păiş, V., & Tufiş, D. (2022). Capitalization and punctuation restoration: a survey. Artificial Intelligence Review, 55(3), 1681–1722.

Pappalardo, G., Cafiso, S., Di Graziano, A., & Severino, A. (2021). Decision tree method to analyze the performance of lane support systems. Sustainability, 13(2), 846.

Pradha, S., Halgamuge, M. N., & Vinh, N. T. Q. (2019, October). Effective text data preprocessing technique for sentiment analysis in social media data. In 2019 11th international conference on knowledge and systems engineering (KSE) (pp. 1–8). IEEE.

Rahman, A. (2019). Statistics-based data preprocessing methods and machine learning algorithms for big data analysis. International Journal of Artificial Intelligence, 17(2): 44--65.

Rahman, A. (2020). *Statistics for Data Science and Policy Analysis*. Springer.

Rahman, A. (2017). Small area housing stress estimation in Australia: Calculating confidence intervals for a spatial microsimulation model. Communications in Statistics Part B: Simulation and Computation, 46(9), 7466–7484. https://doi.org/10.1080/03610918.2016.1241406

Rahman, A., & Harding, A. (2016). Small area estimation and microsimulation modeling. Chapman and Hall/CRC.

Rahman, A., Harding, A., Tanton, R., & Liu, S. (2013). Simulating the characteristics of populations at the small area level: new validation techniques for a spatial microsimulation model in Australia. Computational Statistics and Data Analysis, 57(1), 149–165. https://doi.org/10.1016/j.csda.2012.06.018

Rausch, M., & Zehetleitner, M. (2022). Evaluating false positive rates of standard and hierarchical measures of metacognitive accuracy.

Sharma, P. S., Yadav, D., & Thakur, R. N. (2022). Web Page Ranking using Web Mining Techniques: A comprehensive survey. Mobile Information Systems (article ID 7519573), https://doi.org/10.1155/2022/7519573

Shiraz, M., Gani, A., Khokhar, R., Rahman, A., Iftikhar, M., & Chilamkurti, N. (2017). A distributed and elastic application processing model for mobile cloud computing. Wireless Personal Communications, 95(4), 4403–4423. https://doi.org/10.1007/s11277-017-4086-6.

Sockin, M., & Xiong, W. (2022). Decentralization through tokenization (No. w29720). National Bureau of Economic Research.

Tang, X., Dong, S., Luo, K., Guo, J., Li, L., & Sun, B. (2022). Noise Removal and Feature Extraction in Airborne Radar Sounding Data of Ice Sheets. Remote Sensing, 14(2), 395.

Thevaraja, M., & Rahman, A. (2019). Assessing robustness of regularized regression models with applications. In J. Xu, S. E. Ahmed, F. L. Cooke, & G. Duca (Eds.), Proceedings of the Thirteenth International Conference on Management Science and Engineering Management (1 ed., Vol. 1, pp. 401–415). (Advances in Intelligent Systems and Computing; Vol. 1001). Springer. https://doi.org/10.1007/978-3-030-21248-3_30.

Vachane, D. (2021). Online Products Fake Reviews Detection System Using Machine Learning. Turkish Journal of Computer and Mathematics Education (TURCOMAT), 12(1S), 29–35.

Wang, J., Kan, H., Meng, F., Mu, Q., Shi, G., & Xiao, X. (2020). Fake review detection based on multiple feature fusion and rolling collaborative training. IEEE Access, 8, 182625-182635.

Wang, W., & Sun, D. (2021). The improved adaboost algorithms for imbalanced data classification. Information Sciences, 563, 358–374.

Wieczorek, J., Guerin, C., & McMahon, T. K-fold cross-validation for complex sample surveys. Stat, e454.

Yao, J., Zheng, Y., & Jiang, H. (2021). An Ensemble Model for Fake Online Review Detection Based on Data Resampling, Feature Pruning, and Parameter Optimization. IEEE Access, 9, 16914–16927.

Zheng, X., & Chen, W. (2021). An attention-based bi-LSTM method for visual object classification via EEG. Biomedical Signal Processing and Control, 63, 102174.

6 Nowcasting of Selected Imports and Exports of Bangladesh

Comparison among Traditional Time Series Model and Machine Learning Models

Md. Moyazzem Hossain[*1], *Faruq Abdulla*[2] *and Azizur Rahman*[3]

[1]Department of Statistics, Jahangirnagar University, Savar, Dhaka, Bangladesh; School of Mathematics, Statistics and Physics, Newcastle University, Newcastle upon Tyne, UK

[2]Department of Applied Health and Nutrition, RTM Al-Kabir Technical University, Sylhet, Bangladesh

[3]School of Computing, Mathematics and Engineering, Charles Sturt University, Wagga Wagga, Australia

* Corresponding Author: hossainmm@juniv.edu

CONTENTS

DOI: 10.1201/9781003253051-8

6.1 INTRODUCTION

Nowcasting was originally defined as "the description of the current state of the weather in detail and the prediction of changes that can be expected on a timescale of a few hours" [1]. Nowcasting, a combination of "now" and "forecast", is the estimation of a target variable's current state, or a close approximation of it, either forwards or backwards in time, utilizing information that is available in a more timely manner [2]. It is frequently applied in this research of meteorological phenomena [3]. Before being accepted into the economic literature in the 2000s, the concept and terminology stayed in the meteorological realm for years. Mariano and Murasawa (2003) show that the concept of real-time macroeconomic estimates predates the use of the nowcasting terminology [4]. The availability of a diverse set of innovative, timely indicators should ostensibly have resulted in significant advancement in the development of economic nowcasting, in which real-time macroeconomic variables with a significant lag are obtained using a set of more timely indicators [5,6]. Previous research focused on the methodology and applications of nowcasting included dynamic factor models (DFM) [7,8], mixed frequency data sampling (MIDAS) [9,10] and mixed-frequency vector autoregression (VAR) [9], GDP [11,12], and trade [7,13]. Moreover, common nowcasting has relevance in the context of the 2030 Agenda for Sustainable Development [14].

In the era of globalization, we now live in an open economy, where international trade plays a critical role in the production and consumer decision-making processes [15]. The performance of a country's exports and imports is crucial to its economic progress [16]. Export competitiveness results in economies of scale and a faster rate of technological advancement [17]. The proper utilization of our natural resources is impossible due to a lack of capital and technology. As a result, for importing industrial items and exporting raw materials, we must rely on international trade [18]. Intermediate products, capital machinery, and industrial raw materials imports have increased in recent years [19]. Several studies have been undertaken in Bangladesh on various aspects of import, export, and economic growth. Love and Chandra (2005) examined the hypothesis associated with export-led growth for Bangladesh using annual data on GDP, exports, and imports in a multivariate framework, finding short- as well as long-run unidirectional causality from income to exports [20]. Utilizing annual data over the period 1960 to 2003, Clarke and Ralhan (2005) show support for a causal relationship between export and growth for Bangladesh and reported that ancillary variables reveal a causal relationship between export and GDP when causation is studied across a time horizon [21]. Shirazi and Manap (2005) use co-integration and multivariate Granger Causality analyses to investigate the hypothesis of export-led growth for five South Asian countries, including Bangladesh. Their findings demonstrated the impact of exports on GDP and imports on GDP in Bangladesh [22]. In a study of trade policy and economic growth in Bangladesh, the authors explored short-run feedback effects between exports and output growth, as well as between imports and output growth [23]. Several previous studies applied the widely used auto-regressive moving average (ARIMA) model for prediction and forecasting purposes [24–34]. Moreover, researchers compared the performance of ARIMA with ANN model and reported that ANN performed better than ARIMA model [35–37].

Because of its nonlinear mapping capabilities as well as data processing properties, artificial neural networks (ANN) have recently gotten a lot of interest in financial forecasting. The application of Markov switching models to gross national product (GNP) of the United States [38] accelerated the interest in using nonlinear models, leading to the development of large techniques such as logistic regression, Holts exponential smoothing, ANN, Support Vector Machine (SVM), and machine learning techniques [39–42]. The SVM is a supervised learning technique that was first introduced in the early 1960s. It belongs to the generalized linear classifiers family, and it employs classification as well as regression machine learning theory to maximize predicted accuracy while avoiding overfitting data at the same time [43]. The SVM models, which were introduced by Vapnik in the 1990s [44,45], have proven to be effective as well as promising techniques for handling both linear and nonlinear classification and regression [46]. Moreover, SVMs, in contrast to other nonparametric approaches like neural networks, have evolved into powerful tools for solving issues and overcoming certain classic challenges like over-fitting [43]. A previous study utilized several machine learning algorithms to nowcast GDP growth [47]. Although machine learning algorithms were originally designed for prediction, they are increasingly being used for nowcasting as well [47]. Previous research has found that machine learning methods, such as boosted trees, random forests, support vector machines, and neural networks, outperform conventional models for nowcasting [48–50]. A previous study pointed out that the weighted machine learning model demonstrates to be the top performer, outperforming the ARIMA benchmark model in nowcasting the volume of both exports and imports [51]. A group of researchers examined the most effective approaches for nowcasting global GDP and trade growth [52] and another group of researchers presented nowcast results for 2019Q4 and 2020Q1 GDP for Kenya, Uganda, Nigeria, Ghana, and South Africa [53]. Nowcasting models are estimated to have captured roughly 67% of the drop in services exports and 60% of the fall in imports due to the COVID-19 shock across G7 economies [54].

However, in the context of Bangladesh, the majority of past research relied on econometric and/or time series models. In Bangladesh, there is a dearth of research based on machine learning models in the literature. As a result, the authors aimed to use machine learning models for nowcasting, such as the ANN and SVM models, and compare the performance of these models to a traditional ARIMA model using selected Bangladesh imports and exports.

6.2 METHODOLOGY

6.2.1 Data and Variables

This study is based on the secondary dataset extracted from the website of Humanitarian Data Exchange (HDX) [55]. This study considered four imports, such as Communications, computers, etc. (% of service imports), Insurance and financial services (% of service imports), Transport services (% of service imports), and Goods and services (% of GDP) and four exports variables including Communications, computers, etc. (% of service exports), Insurance and financial services (% of service exports), Transport services (% of service exports), and Goods and services (% of GDP). This study considered annual data covering the period 1976 to 2020 for seven variables; however, for the variable Insurance and financial services (% of service exports) data covered from 1983 to 2020. The study period depends on the availability of data.

6.2.2 Methods

6.2.2.1 ARIMA Model

Herein, the target variable can be defined as a discrete time series as follows (equation (6.1)):

$$Y_t = f(t); t = 1, 2, \ldots, T \tag{6.1}$$

where t represents time. Now, an ARIMA(p, d, q) can be represented as

$$\Phi(B)(1-B)^d\ y_{tij} = \mu + \Theta(B)e_{tij} \tag{6.2}$$

where $\Phi(B)$ and $\Theta(B)$ are polynomial operators in B of degrees p and q, respectively, and $By_t = y_{t-1}, B^2 y_t = y_{t-2}, \ldots$ [56].

We use the maximum likelihood method for estimating the parameters of the model presented in equation (6.2). Assume that $Y_n = (Y_1, \ldots, Y_n)'$ and let $\hat{Y}_n = (\hat{Y}_1, \ldots, \hat{Y}_n)'$, where $\hat{Y}_1 = 0$ and $\hat{Y}_j = E(Y_j \mid Y_1, \ldots, Y_{j-1}) = P_{j-1}X_j$, $j \geq 2$.

Let Γ_n denote the covariance matrix $\Gamma_n = E(Y_n Y_n')$ with non-singularity assumption. The likelihood of Y_n is

$$L(\Gamma_n) = (2\pi)^{-n/2}\ (\det \Gamma_n)^{-1/2} \exp\left(-\frac{1}{2} Y_n' \Gamma_n^{-1} Y_n\right) \tag{6.3}$$

The likelihood given in equation (6.3) of the vector Y_n can be reduced to

$$\mathrm{L}(\Gamma_n) = \frac{1}{\sqrt{(2\pi)^n\, v_0 \cdots v_{n-1}}} \exp\left(-\frac{1}{2}\sum_{j=1}^{n}\left(Y_j - \hat{Y}_j\right)^2 / v_{j-1}\right) \tag{6.4}$$

The likelihood given in equation (6.4) for data from an ARMA(p, q) process is simply calculated from the innovations form of the likelihood by estimating the one-step predictors $\hat{Y}_{i+1}$ as well as the corresponding mean squared errors v_i [57].

In this study, the Box-Jenkins ARIMA modelling approach was used to model building and nowcasting. The underlying principle of the Box-Jenkins ARIMA approach is that it assumes that the current observed value of the variable of interest, Y_t depends on its past observed values $Y_{(t-k)}$ and an unobservable random process called random white noise e_t; where e_t are assumed to be normally distributed with mean zero along with constant variance. Therefore, the current value of the variable of interest is defined as a forecast function of the explanatory variables: past values of the variable of interest and/or the random error term. The Box-Jenkins ARIMA modelling approach is an iterative procedure including four steps: "model identification, model estimation, model diagnostic checking, and model's use" [56].

6.2.2.2 Artificial Neural Network Procedure

The simplest neural network model consists of an input layer and an output node. The many forms of neural network designs were described in depth by Bishop [58]. Forward-feed networks, for example, have a one-way connection from the input to the output layer, whereas recurrent neural networks have a forward-backward link that creates a loop around the network. The feed-forward neural network has three layers represented as input I, hidden layer H, and output layer O such that w_{ij} is the weight from x_i input node to H_j hidden node, then the effective input to hidden node H_j is the sum-product of the input nodes and respective weights to the H_j hidden node. This can be represented as $h_j = \sum_i w_i x_i$. An activation function, usually a sigmoid function such as logistic sigmoid or hyperbolic tangent, is applied to each hidden input to produce a hidden output $u_j = f(h_j) = f\left(\sum_j w_{ij} x_i\right)$. In a similar way, the input to the output node O_k is a sum-product of the hidden output and their respective weights w_{jk} such that $v_k = \sum_j w_{jk} u_i$, which are finally transformed again by an activation function to an output $y_k = f(v_k) = \sum_i f(w_{jk} u_j)$ [58].

6.2.2.3 Support Vector Regression Model

Consider the dataset, $D = \{(x_i, y_i) \mid x_i \in \mathbb{R}^d, y_i \in \mathbb{R},\ i = 1,2,\ldots,n\}$, where x_i is a matrix of input forming by of all the independent variables, y_i is the corresponding scalar output, and n is the number of samples. The support vector regression is then defined as

$$y_i = f(x, \omega) = \sum_{i=1}^{T} \omega \phi_i(x) + b) \tag{6.5}$$

where ω is the weight vector corresponding to $\phi_i(x)$, $\phi_i(x)$ is the mapping function, and b is a constant. The task is to estimate the parameters ω and b. The variables y_i-a vector of values for the target variable over the study period and x- is a matrix of variables consisting of the lagged values of y_i are considered as independent variables [43].

The values of ω and b can be obtained based on the structural risk minimization principle:

$$R(C) = \frac{1}{2} \| w \|^2 + C \frac{1}{n} \sum_{i=1}^{n} L_\varepsilon(f(x, \omega)) \tag{6.6}$$

where ε is the predetermined value and $L_{\varepsilon}(x,\omega)$ is the empirical error measured by ε-insensitive loss function and can be defined as

$$L_{\varepsilon}f((x,\omega)) = \begin{Bmatrix} 0 & \text{if } \left|y_i - f(x_i,\omega)\right| < \varepsilon \\ \left|y_i - f\left(x_i,\omega\right)\right|, & \text{otherwise} \end{Bmatrix} \tag{6.6}$$

Thus, minimizing the norm $\|\omega\|^2 = <\omega,\omega>$ involves solving a convex optimization problem:

$$\frac{1}{2}\|\omega\|^2 + C\sum_{i=1}^{n}(\xi_i + \xi_i^*) \tag{6.7}$$

subject,

$$\begin{cases} y_i - f(x_i,\omega) - b \le \varepsilon + \xi_i \\ f(x_i,\omega) + b - y_i \le \varepsilon + \xi^*_i \\ \xi_i \xi^*_i \ge 0 \end{cases}$$

where ξ_i and ξ^*_i are slack variables introduced due to the error in fitting in the optimization problem [43].

The primal problem is then converted into a dual problem and solved as

$$f(x) = \sum_{i=1}^{n_{sv}}(\alpha_i - \alpha_i^*)K(x_i,x) \text{ s.t. } 0 \le \alpha_i^* \le C, 0 \le \alpha_i \le C \tag{6.8}$$

where n_{sv} is the number of support vectors (SVs) and $K(x_i,x)$ is the kernel function. The constant $C > 0$ sets the trade-off between SVR function fitness and the maximum deviations larger than ε that can be allowed. The optimal value of the regularization parameter C can be derived from standard parameterization of SVM using the following expression [43]:

$$\begin{aligned} |f(\mathbf{x})| &\le \left|\sum_{i=1}^{n_{SV}}(\alpha_i - \alpha_i^*)K(\mathbf{x}_i,\mathbf{x})\right| \\ &\le \sum_{i=1}^{n_{SV}}|(\alpha_i - \alpha_i^*)|\cdot|K(\mathbf{x}_i,\mathbf{x})| \le \sum_{i=1}^{n_{SV}} C\cdot|K(\mathbf{x}_i,\mathbf{x})| \end{aligned} \tag{6.9}$$

Moreover, estimation of b can be obtained by exploiting the following conditions:

$$\begin{gathered} \alpha_i(\varepsilon + \xi_i - y_i + \langle\omega, x_i\rangle + b = 0, \alpha_i^*(\varepsilon + \xi_i^* - y_i + \langle\omega, x_i\rangle + b = 0, \\ (C - \alpha_i)\xi_i = 0 \text{ and } (C - \alpha_i^*)\xi_i^* = 0. \end{gathered}$$

This allows only samples (x_i, y_i) with corresponding $\alpha_i^* = C$ to lie outside the ε-insensitive tube around f and $\alpha_i \alpha_i^* = 0$. That is, there can never be a set of dual variables $\alpha_i \alpha_i^*$ that are both simultaneously non-zero as this would require non-zero slacks in both directions [43,59]. Hence, b can be computed as

$$b = y_i - \langle \omega, x_i \rangle - \varepsilon \text{ for } \alpha_i \in (0,1) \text{ and } b = y_i - \langle \omega, x_i \rangle + \varepsilon \text{ for } \alpha_i^* \in (0,1).$$

6.2.3 Evaluating Model Performance

In the literature, there are various summary statistics for assessing the forecast errors of any econometric or time series model [60, 61]. To determine the best models, the following contemporary model selection criteria are used: RMSPE, MPFE, and TIC.

Root Mean Square Percentage Error (RMSPE): The Root Mean Square Percentage Error (RMSPE) can be defined as [56]

$$RMSPE = \sqrt{\frac{1}{T}\sum_{i=1}^{T}\left(\frac{Y_t^{nc} - Y_t^a}{Y_t^a}\right)^2} \times 100. \tag{6.10}$$

Mean Absolute Percent Error (MAPE): The Mean Absolute Percent Error (MAPE) can be calculated by [56]

$$MAPE = \frac{1}{T}\sum_{i=1}^{T}\left|\frac{Y_t^{nc} - Y_t^a}{Y_t^a}\right| \times 100. \tag{6.11}$$

Theil Inequality Coefficient (TIC): The Theil Inequality Coefficient (TIC) is defined as [62]

$$TIC = \frac{\sqrt{\frac{1}{T}\sum_{i=1}^{T}\left(Y_t^{nc} - Y_t^a\right)^2}}{\sqrt{\frac{1}{T}\sum_{i=1}^{T}\left(Y_t^{nc}\right)^2} + \sqrt{\frac{1}{T}\sum_{i=1}^{T}\left(Y_t^a\right)^2}} \tag{6.12}$$

where Y_t^{nc} is the nowcasted value at time t, Y_t^a is the actual value at time t, and T denotes the sample size.

6.3 RESULTS

The descriptive statistics of the selected variables are summarized in Table 6.1. In the case of import variables, the transport services were imported more (average: 64.82% of total service imports, standard deviation (SD): 7.87). The highest and lowest

TABLE 6.1
Summary Statistics of the Selected Variables

	Imports				Exports			
	(% of service imports)			(% of GDP)	(% of service exports)			(% of GDP)
Statistics	Communications, computer, etc.	Insurance and financial services	Transport services	Goods and services	Communications, computer, etc.	Insurance and financial services	Transport services	Goods and services
Mean	16.95	6.46	64.82	17.83	80.80	1.23	11.83	11.03
Median	21.88	6.01	64.62	16.56	81.37	1.60	11.20	11.15
Standard Deviation	6.42	2.77	7.87	4.68	5.64	1.13	4.49	5.09
Kurtosis	-1.61	3.35	-0.53	-0.68	-0.31	-1.44	1.51	-1.29
Skewness	0.00	1.10	0.20	0.62	-0.52	0.15	1.08	0.23
Minimum	11.27	1.17	50.64	11.70	65.77	0.04	4.27	3.40
Maximum	31.79	16.54	80.74	27.95	90.22	3.72	26.10	20.16

percentages of services imported for Transport are 80.74% and 50.64%, respectively. Insurance and financial services account for around 6.5% of service imports, while Communications, computers, etc., account for 19.95%. On an average 17.83% of GDP was imported for Goods and services. On the other hand, Bangladesh exports largely (80.80% of exported services) Communications, computers, etc., to other countries, and a very small portion of exports comes from Insurance and financial services (average: 1.23% of services exported). About 12% of exported services are accounted for transportation-related services. Bangladesh exports just above of 11% of its GDP for Goods and services to other countries (Table 6.1).

The line graphs presented in Figure 6.1 illustrate the changes in the selected variables associated with the percentage of imported services in Bangladesh over the study period. Among the services imported in Bangladesh, Transport services accounted for more than 60% of services imports; though it fluctuated over the study period, it varied between about 50% to around 80%. The highest percentage was observed around the year 2010 and since then a declined trend was observed. A small

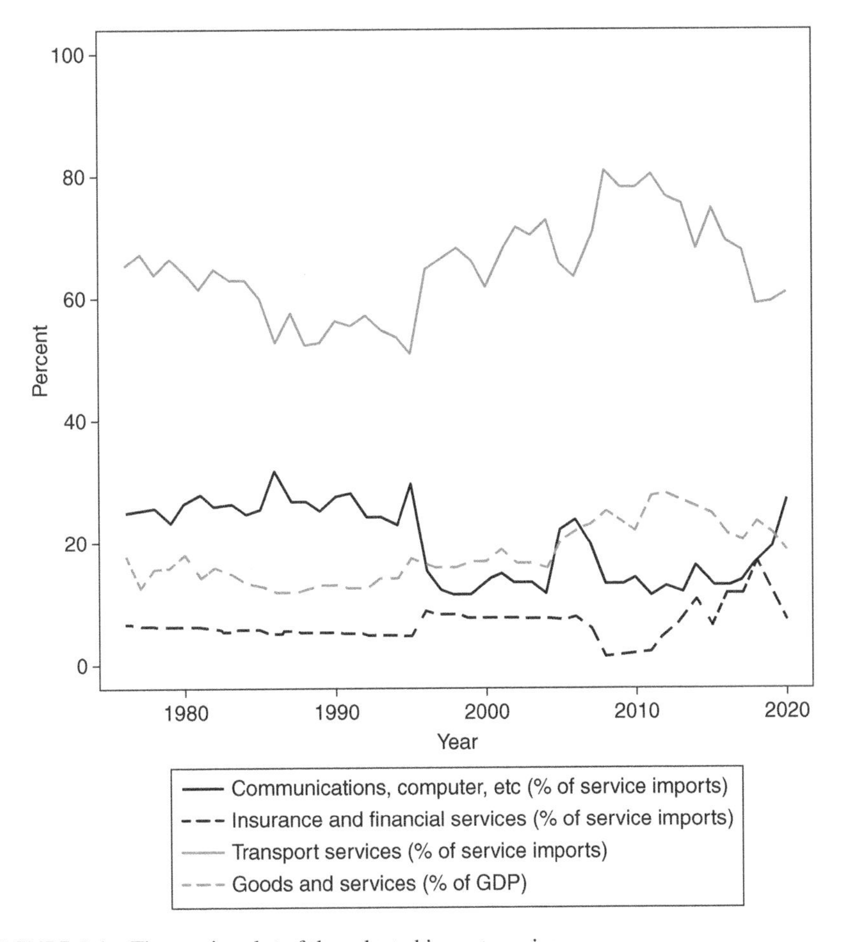

FIGURE 6.1 Time series plot of the selected import services.

portion of Insurance and financial services was imported throughout the period of the study and there was an increasing trend after the year 2010. The percentage of service imports in Bangladesh for Communications, computers, etc., fluctuated throughout the study period. Before 1995 the percentages of services imported for Communications, computers, etc., were more than 20%, however, after then it was almost 15% but in 2020 it again crossed 20% of services imported. The imported goods and services (% of GDP) have a slightly increasing trend up to 2010, but a declining trend was observed then. The pattern of the percentage of services imported for Transport had some similarities with the pattern of the imported Goods and services (% of GDP) in Bangladesh (Figure 6.1).

The time series plots of selected exports considered in this study are shown in Figure 6.2. Bangladesh exports a satisfactory level of Communications, computers, etc. From 1976 to 2020, it accounted for around 70% of total services exported from Bangladesh to foreign countries. Bangladesh began exporting Insurance and financial services in 1983, but just a tiny portion of the total services exported. Up to 2010, there was a downward tendency in transportation-related services, but after that, there

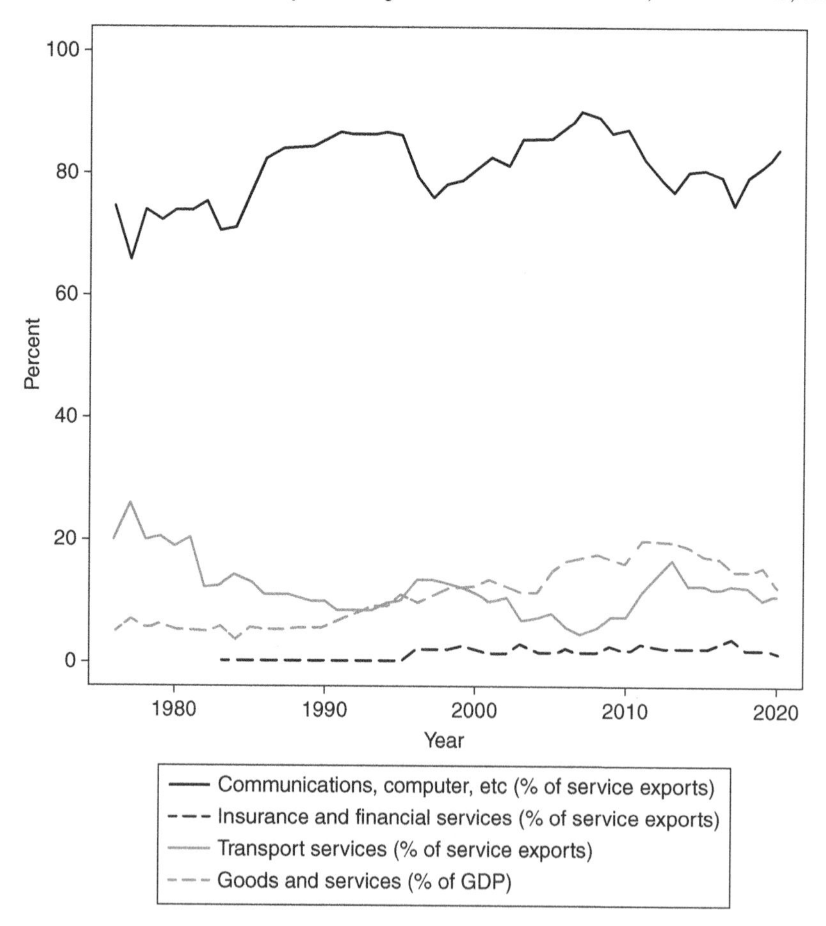

FIGURE 6.2 Time series plot of the selected export services.

was a sharp upward trend. Exports of Goods and services (% of GDP) have been increasing in study period (Figure 6.2).

The best model is determined by the "good forecast" or how well the model delivered the fitted values. The RMSPE, MAPE, and TIC equations (6.10)–(6.12) are used to assess the performance of the chosen models. The lower the value of these model selection criteria, the better the model's performance. The findings demonstrate that while the ARIMA in equation (6.2) and ANN models performed similarly in some circumstances, the SVM model given in equations (6.5)-(6.9) consistently outperformed the other two models for all variables studied (Table 6.2). As a result, the authors suggest that for nowcasting, machine learning models such as SVM be used.

Figure 6.3 shows a graphical comparison of nowcasting by the selected models with observed data points for the four import variables studied in this study. After 2010, the amount of imported Insurance and financial services increased dramatically, from roughly 2% in 2010 to 15% in 2019. Serviced related to Transport and Communications, computers, etc., showed more fluctuations. The SVM model performed statistically well during the period of the study, as evidenced by the graphical comparison of observed and nowcasted values of the target variables (Figure 6.3).

Figure 6.4 illustrates the visualizations of nowcast values using ARIMA, ANN, and SVM models, as well as the observed data for four variables from Bangladesh's list of exported services to foreign countries. From the beginning of the study period, the proportion of exports related to Transport decreased considerably, but after 2010, an upward trend was noted. From 1985 to 2012, exports of Goods and services (as a percentage of GDP) progressively climbed, but then began to decline. For all variables considered in this study, however, there was less discrepancy between the nowcast values by SVM and the observed values compared to the nowcasted values by the ARIMA and ANN models (Figure 6.4).

6.4 CONCLUSION

The comparative performance of the three models based on their root mean square percentage errors, mean absolute percentage error, and Theil inequality coefficient as well as graphical visualizations indicated that the ARIMA and ANN models performed fairly similar. For all variables, however, the support vector regression model performed much better than the other two models investigated in this study. As a result, the authors recommend that machine learning approaches be used for economic growth modelling and nowcasting. The disadvantage of using a secondary annual dataset in this study is the lack of data for a longer period of time. Adding relevant monthly (or higher) frequency variables to nowcast models will likely lead to more accurate estimates. In addition, comparing the models used in this study to other machine learning and econometrics models, such as Bayesian vector auto-regressive model, Elastic Net, XGBoost, and Random Forest, could be a future research topic.

TABLE 6.2
Accuracy Measures of the Models Considered in this Study

Variables	RMSPE			MAPE			TIC		
	ARIMA	ANN	SVM	ARIMA	ANN	SVM	ARIMA	ANN	SVM
Imports									
Communications, computer, etc. (% of service imports)	20.753	18.450	**2.860**	2.390	2.957	**1.293**	0.089	0.079	**0.010**
Insurance and financial services (% of service imports)	65.540	27.802	**5.511**	9.208	6.548	**2.791**	0.135	0.088	**0.009**
Transport services (% of service imports)	6.337	6.707	**0.666**	0.467	0.451	**0.148**	0.032	0.033	**0.003**
Goods and services (% of GDP)	10.553	11.772	**1.972**	0.518	1.306	**1.062**	0.047	0.055	**0.008**
Exports									
Communications, computer, etc. (% of service exports)	3.949	3.884	**0.352**	0.188	0.144	**0.138**	0.019	0.018	**0.002**
Insurance and financial services (% of service exports)	4.974	1.401	**0.875**	3.496	0.838	**0.610**	0.156	0.013	**0.002**
Transport services (% of service exports)	19.284	20.371	**3.076**	1.678	3.474	**1.940**	0.081	0.090	**0.009**
Goods and services (% of GDP)	12.310	16.511	**4.780**	1.536	2.101	**1.200**	0.042	0.054	**0.013**

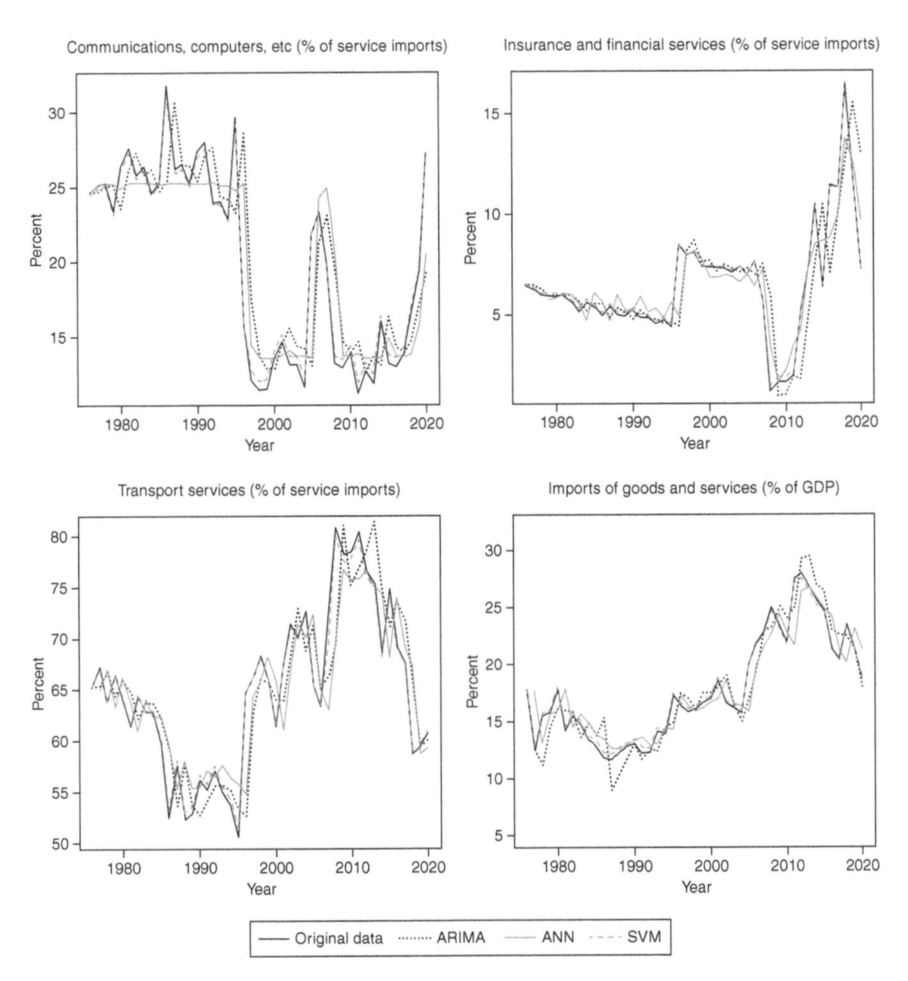

FIGURE 6.3 Comparison among fitted values by ARIMA, ANN, SVM, and observed data for the imports.

AVAILABILITY OF DATA

The data set is available via the following access link https://data.humdata.org/

CONFLICTING INTERESTS

The authors declared no potential conflicts of interest.

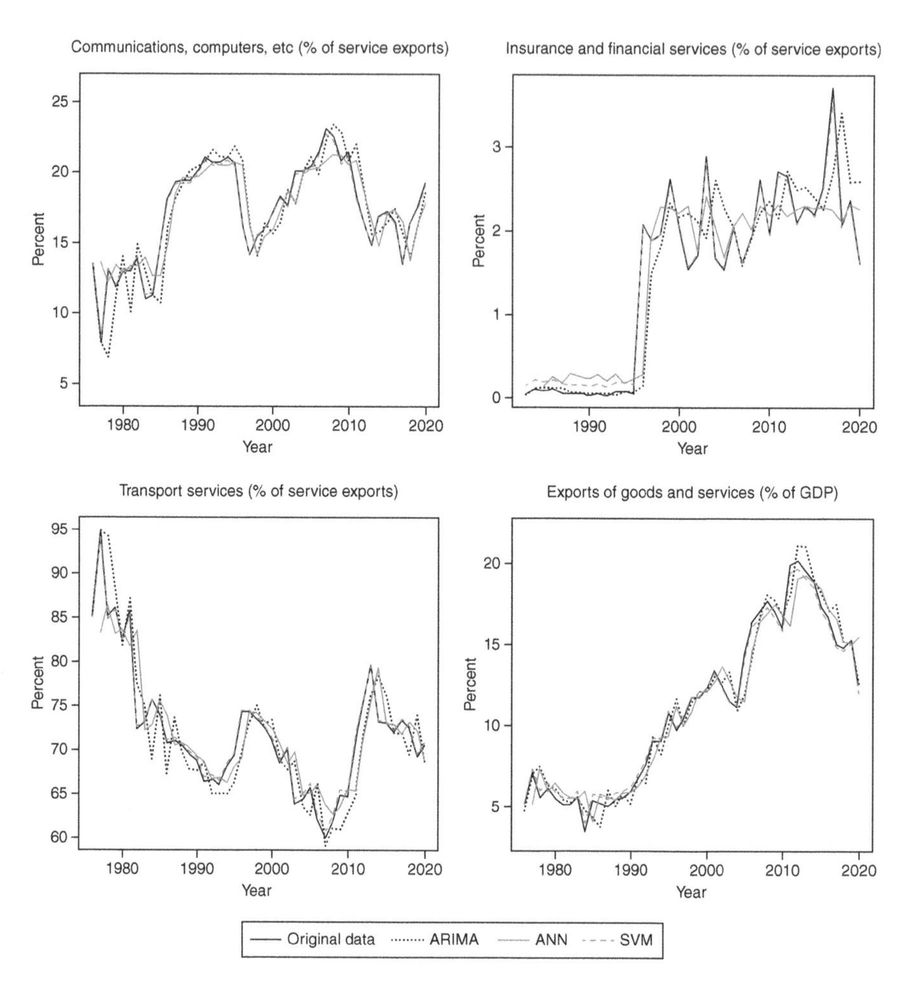

FIGURE 6.4 Comparison among fitted values by ARIMA, ANN, SVM, and observed data for the exports.

FUNDING

The authors received no financial support for the research, authorship, and/or publication of this article from any sources.

REFERENCES

1. Browning K. Nowcasting: Mesoscale Observations and Short-Range Prediction. In: Battrick B, Mort J, editors. Proceedings of an International Symposium [Internet]. Hamburg, Germany; 1981 [cited 2022 Apr 13]. p. ESA SP-165. Available from: https://ui.adsabs.harvard.edu/abs/1981ESASP.165.....B/abstract
2. Hopp D. Economic Nowcasting with Long Short-term Memory Artificial Neural Networks (LSTM) [Internet]. ited Nations Conference on Trade and Development.

Elsevier BV; 2021. Report No.: UNCTAD Research Paper No. 62. Available from: https://papers.ssrn.com/abstract=3855402

3. Wapler K, de Coning E, Buzzi M. Nowcasting. Ref Modul Earth Syst Environ Sci [Internet]. 2019 Jan 1 [cited 2022 Apr 13]; Available from: https://linkinghub.elsevier.com/retrieve/pii/B9780124095489117774
4. Mariano RS, Murasawa Y. A new coincident index of business cycles based on monthly and quarterly series. J Appl Econom [Internet]. 2003;18(4):427–43. Available from: https://onlinelibrary.wiley.com/doi/full/10.1002/jae.695
5. Giannone D, Reichlin L, Small D. Nowcasting: The real-time informational content of macroeconomic data. J Monet Econ. 2008;55(4):665–76.
6. Bańbura M, Giannone D, Reichlin L. Nowcasting [Internet]. 2010. Report No.: ECB Working Paper No. 1275. Available from: www.econstor.eu/handle/10419/153709
7. Guichard S, Rusticelli E. A Dynamic Factor Model for World Trade Growth [Internet]. 2011 [cited 2022 Apr 14]. Report No.: OECD Economics Department Working Papers No. 874. Available from: https://dx.doi.org/10.1787/5kg9zbvvwqq2-en
8. Antolín-Díaz J, Drechsel T, Petrella I. Advances in Nowcasting Economic Activity: Secular Trends, Large Shocks and New Data [Internet]. 2021. Available from: http://econweb.umd.edu/~drechsel/papers/advances.pdf
9. Kuzin V, Marcellino M, Schumacher C. MIDAS vs. mixed-frequency VAR: Nowcasting GDP in the euro area. Int J Forecast. 2011;27(2):529–42.
10. Marcellino M, Schumacher C. Factor MIDAS for Nowcasting and Forecasting with Ragged-Edge Data: A Model Comparison for German GDP*. Oxf Bull Econ Stat [Internet]. 2010;72(4):518–50. Available from: https://onlinelibrary.wiley.com/doi/full/10.1111/j.1468-0084.2010.00591.x
11. Rossiter J. Nowcasting the Global Economy [Internet]. Ottawa, Ontario, Canada: Bank of Canada; 2010. Report No.: Bank of Canada Discussion Paper 2010–12. Available from: www.bankofcanada.ca/2010/09/discussion-paper-2010-12/
12. Bok B, Caratelli D, Giannone D, Sbordone AM, Tambalotti A. Macroeconomic Nowcasting and Forecasting with Big Data. Annu Rev Econom [Internet]. 2018;10:615–43. Available from: www.annualreviews.org/doi/abs/10.1146/annurev-economics-080217-053214
13. Cantú F. Estimation of a coincident indicator for international trade and global economic activity [Internet]. 2018. Report No.: UNCTAD Research Paper No. 27. Available from: https://unctad.org/system/files/official-document/ser-rp-2018d9_en.pdf
14. UN. Transforming our world: the 2030 Agenda for Sustainable Development | Department of Economic and Social Affairs [Internet]. 2015. Available from: https://sdgs.un.org/2030agenda
15. Muktadir-Al-Mukit D, Shaiullah AZM. Export, Import And Inlation: A Study On Bangladesh. Amity Glob Bus Rev [Internet]. 2014;9:38–45. Available from: http://search.ebscohost.com/login.aspx?direct=true&db=bth&AN=94991548&site=ehost-live
16. Khan GU, Chowdhury S, Azdaan S. Does the exchange rate influence the exports? Evidence from Bangladesh. Turkish Econ Rev. 2019;6(4):313–9.
17. Ramos FFR. Exports, imports, and economic growth in Portugal: evidence from causality and cointegration analysis. Econ Model. 2001;18(4):613–23.
18. Akhter M. The Impact of Export Trading on Economic Growth in Bangladesh. World Vis Res J [Internet]. 2015;9(10):67–81. Available from: www.ijebf.com/IJEBF_Vol. 1, No. 10, November 2013/THE IMPACT OF.pdf
19. Ahmed HA, Uddin MGS. Export, Imports, Remittance and Growth in Bangladesh: An Empirical Analysis. South Asia Res [Internet]. 2009;2(2):79–92. Available from: www.researchgate.net/publication/281997371%0AExport,

20. Love J, Chandra R. Testing export-led growth in Bangladesh in a multivarate VAR framework. J Asian Econ. 2005 Jan 1;15(6):1155–68.
21. Clarke J, Ralhan M. Direct and Indirect Causality Between Exports and Economic Output for Bangladesh and Sri Lanka : Horizon Matters [Internet]. Victoria, B.C., Canada; 2005. Report No.: Econometrics Working Paper EWP05012. Available from: https://core.ac.uk/display/24886395
22. Shirazi NS, Manap TAA. Export-Led Growth Hypothesis: Further Econometric Evidence From South Asia. Dev Econ [Internet]. 2005;43(4):472–88. Available from: https://onlinelibrary.wiley.com/doi/full/10.1111/j.1746-1049.2005.tb00955.x
23. Chaudhary MA, Shirazi NS, Choudhary MAS. Trade Policy and Economic Growth in Bangladesh: A Revisit. Pak Econ Soc Rev [Internet]. 2007 Apr 15;45(1):1–26. Available from: www.jstor.org/stable/25825302
24. Hossain MM, Abdulla F. Forecasting the Sugarcane Production in Bangladesh by ARIMA Model. J Stat Appl Probab . 2015;4(2):297–303.
25. Hossain MM, Abdulla F. Forecasting the Tea Production of Bangladesh: Application of ARIMA Model. Jordan J Math Stat. 2015;8(3):257–70.
26. Rahman MA, Hossain MM. Forecasting Rice Production in Jessore, Dinajpur and Kushtia Districts of Bangladesh by Time Series Model. Int J Math Comput. 2019;30(2):105–14.
27. Hossain MM, Abdulla F. On the production behaviors and forecasting the tomatoes production in Bangladesh. J Agric Econ Dev. 2015;4(5):66–074.
28. Hossain MM, Abdulla F. Forecasting the garlic production in Bangladesh by ARIMA Model. Asian J Crop Sci. 2015;7(2):147–53.
29. Hossain MM, Abdulla F. Jute production in Bangladesh: A time series analysis. J Math Stat. 2015 Dec 6;11(3):93–8.
30. Hossian MM, Abdulla F. A Time Series Analysis for the Pineapple Production in Bangladesh. Jahangirnagar Univ J Sci. 2015;38(2):49–59.
31. Hossain MM, Abdulla F. Forecasting Potato Production in Bangladesh by ARIMA Model. J Adv Stat. 2016;1(4):191–8.
32. Faruk M, Hossain M. Forecasting Gold Price: An Application of Auto Regressive Integrated Moving Average Model. Int J Appl Math Stat. 2019;58(4):115–21.
33. Abdulla F, Hossain MM. Forecasting of Wheat Production in Kushtia District & Bangladesh by ARIMA Model: An Application of Box-Jenkin's Method. J Stat Appl Probab. 2015;4(3):465–74.
34. Hossain MM, Abdulla F, Majumder AK. Forecasting of banana production in Bangladesh. Am J Agric Biol Sci. 2016 Jun 16;11(2):93–9.
35. Ahmed S, Karimuzzaman M, Hossain MM. Modeling of Mean Sea Level of Bay of Bengal : A Comparison between ARIMA and Artificial Neural Network. Int J Tomogr Simulation™. 2021;34(1):31–40.
36. Hossain MM, Abdulla F, Majumder AK. Comparing the Forecasting Performance of ARIMA and Neural Network Model by using the Remittances of Bangladesh. Jahangirnagar Univ J Stat Stud. 2017;34:1–12.
37. Hossain MM, Abdulla F, Hossain Z. Comparison of ARIMA and Neural Network Model to Forecast the Jute Production in Bangladesh. Jahangirnagar Univ J Sci. 2017;40(1):11–8.
38. Hamilton JD. A New Approach to the Economic Analysis of Nonstationary Time Series and the Business Cycle. Econometrica. 1989;57(2):384.
39. Granger CWJ. Essays in Econometrics: Collected Papers of Clive W. J. Granger [Internet]. Essays in Econometrics. Cambridge University Press; 2001. Available

from: www.cambridge.org/core/books/essays-in-econometrics/1C53EABABB573 2F7EA2E56E5A9512D5C

40. Rahman A. Statistics for Data Science and Policy Analysis. Statistics for Data Science and Policy Analysis. Springer Singapore; 2020.
41. Sharif O, Hasan MZ, Rahman A. Determining an effective short term COVID-19 prediction model in ASEAN countries. Sci Rep. 2022;12:5083.
42. Rahman A. Statistics-Based Data Preprocessing Methods and Machine Learning Algorithms for Big Data Analysis. Int J Artif Intell. 2019;17(2):44–65.
43. Lagat AK, Waititu AG, Wanjoya AK. Support Vector Regression and Artificial Neural Network Approaches: Case of Economic Growth in East Africa Community. Am J Theor Appl Stat. 2018;7(2):67–79.
44. Cortes C, Vapnik V. Support-vector networks. Mach Learn [Internet]. 1995;20(3):273–97. Available from: https://link.springer.com/article/10.1007/BF00994018
45. Vapnik VN. Statistical Learning Theory [Internet]. New York: John Wiley and Sons; 1998. 736 p. Available from: www.wiley.com/en-gb/Statistical+Learning+Theory-p-9780471030034
46. Peng P, Ma QL, Hong LM. The research of the parallel SMO algorithm for solving SVM. Proc 2009 Int Conf Mach Learn Cybern. 2009;3:1271–4.
47. Dauphin BJ, Dybczak K, Maneely M, Sanjani MT, Suphaphiphat N, Wang Y, et al. Nowcasting GDP: A Scalable Approach Using DFM, Machine Learning and Novel Data, Applied to European Economies [Internet]. 2022. Report No.: IMF Working Paper No. 2022/052. Available from: www.imf.org/en/Publications/WP/Issues/2022/03/11/Nowcasting-GDP-A-Scalable-Approach-Using-DFM-Machine-Learning-and-Novel-Data-Applied-to-513703
48. Richardson A, van Florenstein Mulder T, Vehbi T. Nowcasting GDP using machine-learning algorithms: A real-time assessment. Int J Forecast. 2021;37(2):941–8.
49. Muchisha ND, Tamara N, Andriansyah A, Soleh AM. Nowcasting Indonesia's GDP Growth Using Machine Learning Algorithms. Indones J Stat Its Appl [Internet]. 2021;5(2):355–68. Available from: https://ijsa.stats.id/index.php/ijsa/article/view/824
50. Tiffin AJ. Seeing in the Dark : A Machine-Learning Approach to Nowcasting in Lebanon [Internet]. IMF; 2016. Report No.: IMF Working Paper WP/16/56. Available from: www.imf.org/en/Publications/WP/Issues/2016/12/31/Seeing-in-the-Dark-A-Machine-Learning-Approach-to-Nowcasting-in-Lebanon-43779
51. Mayorova K, Nikita N. Nowcasting Growth Rates of Russia's Export and Import by Commodity Group. Russ J Money Financ. 2021;80(3):34–48.
52. Stratford K. Topical articles Nowcasting world GDP and trade using global indicators 233 Nowcasting world GDP and trade using global indicators. Q Bull. 2013;Q3:233–43.
53. Buell B, Chen C, Cherif R, Tang J, Wendt N. Impact of COVID-19: Nowcasting and Big Data to Track Economic Activity in Sub-Saharan Africa [Internet]. IMF Working Papers. International Monetary Fund; 2021. Report No.: IMF Working Papers 2021, 124. Available from: www.elibrary.imf.org/view/journals/001/2021/124/article-A001-en.xml
54. Jaax A, Gonzales F, Mourougane A. Nowcasting aggregate services trade [Internet]. OECD Publishing, Paris; 2021. Report No.: OECD Trade Policy Papers, No. 253. Available from: www.oecd-ilibrary.org/trade/nowcasting-aggregate-services-trade_0ad7d27c-en
55. HDX. Humanitarian Data Exchange [Internet]. HUMANITARIAN DATA EXCHANGE. 2020 [cited 2020 Mar 12]. Available from: https://data.humdata.org/
56. Box GEP, Jenkins GM, Reinsel GC, Ljung GM. Time series analysis : forecasting and control. 5th editio. Wiley; 2015.

57. Brockwell PJ, Davis RA. Introduction to Time Series and Forecasting [Internet]. 3rd Editio. Switzerland: Springer, Cham, Switzerland; 2016. 1–439 p. Available from: http://link.springer.com/10.1007/978-3-319-29854-2
58. Bishop CM. Neural Networks for Pattern Recognition [Internet]. New York, NY, USA: Oxford University Press; 1995. 482 p. Available from: www.research.ed.ac.uk/en/publications/neural-networks-for-pattern-recognition
59. Bolhuis MA, Rayner B. Deus ex Machina? A Framework for Macro Forecasting with Machine Learning [Internet]. Washington, DC; 2020. Report No.: IMF Working Paper WP/20/45. Available from: www.imf.org/en/Publications/WP/Issues/2020/02/28/Deus-ex-Machina-A-Framework-for-Macro-Forecasting-with-Machine-Learning-49094
60. Rahman, A., & Harding, A. (2016). Small area estimation and microsimulation modeling. Chapman and Hall/CRC.
61. Bhuiyan, M. S. I., Rahman, A., Kim, G. W., Das, S., & Kim, P. J. (2021). Eco-friendly yield-scaled global warming potential assists to determine the right rate of nitrogen in rice system: A systematic literature review. Environmental Pollution, 271(116386), 1–10. [116386]. https://doi.org/10.1016/j.envpol.2020.116386
62. Theil H. Applied Economic Forecasting [Internet]. North-Holland Publishing Company; 1966. 1–474 p. Available from: https://books.google.co.uk/books/about/Applied_economic_forecasting.html?id=D08EAQAAIAAJ&redir_esc=y

Theme 3

Development of the Forecasting Component to the Decision Support Tools

7 An Intelligent Interview Bot for Candidate Assessment by Using Facial Expression Recognition and Speech Recognition System

Sonal Yadav[*1], *Ayu Yaduvanshi*[2], *Shashank Shekhar*[2], *Lokesh Bansal*[2], *Prateek Meena*[2] *and Amit Kumar*[2]
[1]National Institute of Technology Raipur, India
[2]Indian Institute of Information Technology Kota, India
*Corresponding Author: syadav.cse@nitrr.ac.in

CONTENTS

7.1 INTRODUCTION

The Internet is a popular source of information, where obtaining accurate information quickly is very difficult [1, 2, 26]. The present study developed an AI-enabled chatbot for hiring a suitable candidate. Consider a company that wants to conduct an interview. They need to select members of the interview panel, and determine the location of the interview. This is a time consuming and costly endeavor. Using a recruiter chatbot may overcome a number of issues in the recruitment process. A chatbot is a program designed for intelligent communication into a text or spoken word. The chatbot first tokenizes the text into words to interpret the query, and then

DOI: 10.1201/9781003253051-10

selects an answer from a variety of presaved responses. It can be used by employers to perform many time-consuming tasks such as collecting information, evaluating candidates based on relevant metrics, answering FAQs (Frequently Asked Questions), and assisting in organizing people's conversations.

However, chatbots are not intended to replace people, but rather improve their efficiency and reduce time consuming and tedious tasks [3]. Thus, Human Resources (HR[1]) is developing artificial intelligence (AI) [4], because companies want to expand their business operations with new technology and increase profits. Organizations that embrace flexible processes in work processes will reduce staff time and effort and make better use of resources

The rest of the chapter is ordered as follows. Section 7.2 discusses related work. In Section 7.3, the methodology is discussed. The accuracy of chatbots is evaluated in Section 7.4 on experimentation. In Section 7.5, the chapter is concluded with future research directions.

7.2 RELATED WORK

Alan Turing speculated in 1950 what would occur if a computer program communicated with a group of individuals who were unaware that the entity they were attempting to interact with was artificial. Many researchers used this question called the Turing test (Turing, 1950) for developing ideas for the chatbot. The first chatbot named Eliza was created in 1966. Eliza [5] imitated the psychiatrist's operation, returning the user's sentences in the form of an inquiry. It became a source for development of many chatbots (Klopfenstein, Delpriori, Malatini, Bogliolo, 2017) though it has very limited ability to communicate. Instead, it employs pattern matching and a template-based answer selection technique (Brandtzaeg & Flstad, 2017). One disadvantage of ELIZA is that its knowledge is restricted, and therefore it can only discuss a restricted range of topics. Furthermore, it is unable to maintain extended dialogue and is unable to learn or acquire context from the dialogue.

PARRY [6] first emerged in 1972, and it pretended to be a schizophrenia sufferer (Colby, Weber, & Hilf, 1971). PARRY is said to be more advanced than Eliza as it claimed to have a superior personality and better at regulating the structure. It establishes his answers based on a set of rules (i.e., assumptions and emotional reactions triggered by a change in the weight of the user's utterances) (Colby, Hilf, Weber, & Kraemer, 1972). In a 1979 [7] experiment, PARRY was employed in a teletype interview with a patient to determine if he was a computer program or a real schizophrenia patient. As a result, 10 diagnoses were given by psychiatrists. Two right diagnoses were given by the first psychiatrist, whereas two inaccurate diagnoses were given by the second. The third person thought both subjects were actual patients, whereas the other two thought they were chatbots. Because patients with schizophrenia have a degree of incoherence in their speech and the sample size of five doctors is tiny the interpretation of the findings is unclear. PARRY is generally regarded as a chatbot with limited language comprehension and emotional expression abilities. It also has a slow response time and is unable to learn from dialogue.

TABLE 7.1
Existing Models

SN	Author (Year)	Title	Tools/Technology
1.	Joseph Weizenbaum (1966) [5]	ELIZA – a computer program for the study of natural language communication between man and machine	Mad-slip Mac-Time sharing system (MIT)
2.	Kenneth Colby (1972) [6]	PARRY- A Chatbot	Turing Machine, Botsify
3.	D.A. Ferrucci (2011) [8]	Introduction to "This is Watson"	IBM's DeepQA software, Apache UIMA (Unstructured Information Management Architecture)
4.	Winzu Organisation (2018) [10]	A Comparison of AI Conversational Surveys and Traditional Form-Based Surveys (Winzu)	Machine learning basics, Deep learning, Reinforcement learning
5.	Petter Bae Brandtzaeg (2016) [11]	Users' experiences with chatbots: findings from a questionnaire study (Dale)	NLP, POS tagging
6.	Xinyu Le (2019) [12]	Amazon Alexa as a Case Study	ANN language processing techniques
7.	Hao-Hsuan Hsu (2022) [A]	Chinese-based massive open online courses (MOOCs)	NLP, ML and Neural language model

In 2011, IBM introduced Watson [8] (Watson Assistant & IBM Cloud) as a chatbot. Watson was able to read authentic human language well enough to overcome two previous champions on the game show "Jeopardy," where participants were given information in the form of answers and had to anticipate the questions that went with them. Years later, Watson assisted businesses in producing better virtual assistants. Watson Health was also intended to help physicians diagnose illnesses. Watson, however, has the problem of only being available in English.

Early in 2016, artificial intelligence technology underwent a significant transformation that altered the way individuals connect with manufacturers [9]. Developers were able to construct chatbots for the business or service of social media platforms, allowing users to complete certain everyday tasks within their messaging apps [10]. At the end of 2016, 34,000 chatbots covered a wide range of tasks in fields like marketing, support systems, health care, entertainment, education, and cultural heritage. For popular messaging platforms, industrial solutions, and research [11], thousands of text-based chatbots with specialized functionality were built. Furthermore, the Internet of Things (IoT) ushered in a new era of linked smart items, with chatbots facilitating conversation [12–19].

Recently, Hsu and Huang introduced Xiao-Shih, an intelligent chatbot based on questions and answers regarding Chinese massive open online courses [20]. This chatbot was introduced to address the problem of long time waiting by interviewer for answers individually during interview. Xiao-Shih integrates novel natural language processing and machine learning models for expanding the knowledge base through open community-based questions answering. A novel neural network-based model structure using transformer layer to improve the semantic feature extraction was introduced (Xiao, 2022) [21]. An intelligent chatbot for depression detection after having conversation with the human has also been reported recently (Huang, 2022) [23]. This chatbot detects the depression stage by analyzing the articles and posts on their social media. It analyzes amount of words, images, and video streaming and task-based functions of the person. Intelligent chatbots are developed for various tasks as discussed above to understand feelings and emotions. A multimodal conversation bot (Rafiandi, 2021) [25] has been generated to listen and respond to user stories to alleviate stress caused by the pandemic. This bot offers three modes of conversation: chat-based, voice-based, and holographic avatar.

However, chatbots have not yet been implemented for the candidate recruitment process. Our chatbot will ask the candidate questions and the answer will be recorded and evaluated. While evaluating the answers our chatbot will also analyze the sentiments of the candidate while he/she is giving answers and evaluate him/her on the basis of the sentiments also. Many models are working in this area, but none do the same.

7.3 PROPOSED ARTIFICIAL INTELLIGENCE CHATBOT

The proposed approach has two modules: 1) facial recognition module and 2) automatic speech recognition module. Both modules execute consecutively and then combine their individual output. Both of these modules later on generate results that will be passed on through the main file. Here we discuss these modules in detail and how they work and output data.

The proposed chatbot module takes input from both internal modules: automatic speech recognition module and facial detection module. In the automatic speech recognition module, the input is fed through the microphone and received the expected total answer depending on that input. In our facial detection module, the input is given as the expressions of a face detected while answering the question now. This module in turn gives an output of how many times different expressions (nervousness, happy, etc.) were present throughout the time period. Then the system generates the facial score (i.e., if the candidate has positive expression such as happy, confident). Those marks will be added to total score. If the candidate shows negative emotions (nervousness, sadness), certain marks are reduced from the facial score at the end. Thus, the total facial score is dependent on expressions.

Now facial score and score generated by the automatic speech recognition module are available for analysis. In the last step, the average of both of these scores are computed to show it to the candidate or to HR (interviewer) to store it for further evaluation. After that, the chatbot will ask if the candidate wants to proceed further

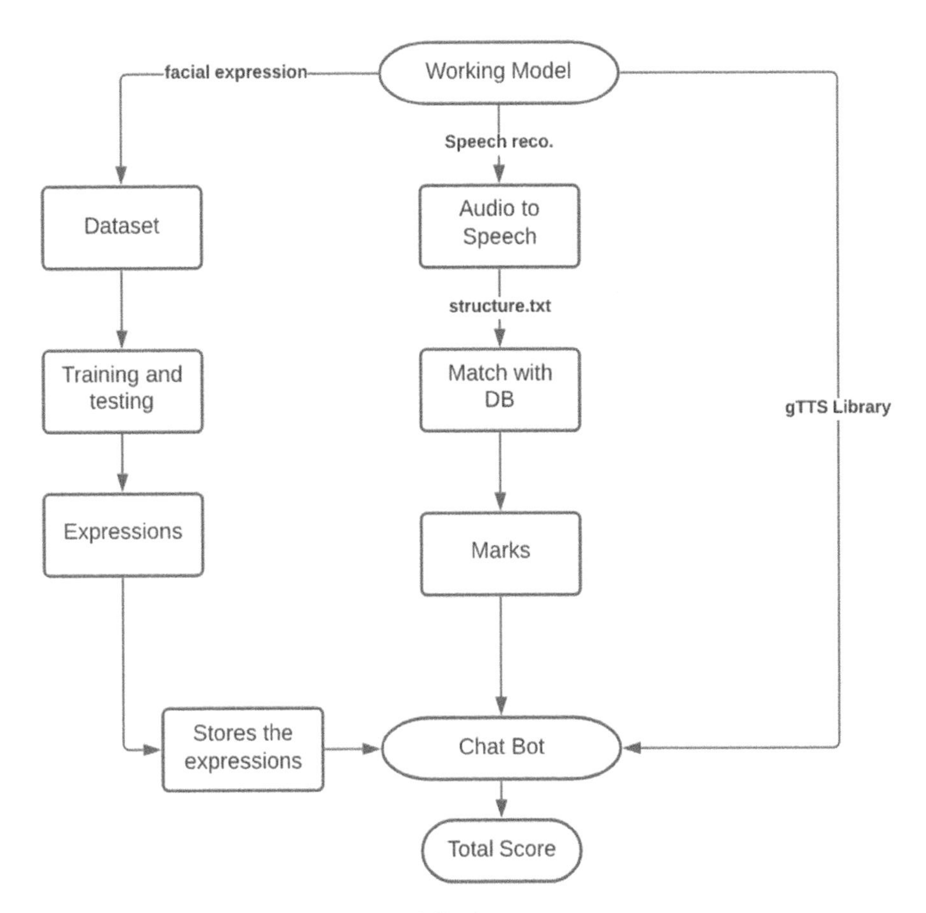

FIGURE 7.1 Flow depicting working of Chatbot.

and the loop will continue. Figure 7.1 depicts the flow of the proposed working of the chatbot on an elementary scale and depicts overall how the chatbot module receives the input from both of these modules and generates a total score later on.

Both modules involved in the proposed methodology are discussed in detail in the following subsections.

7.3.1 Facial Recognition Module

The convolutional neural networks (CNNs) [13] are widely used for image recognition and real-time expression recognition. The benefits of the CNN (i.e., less network parameters, faster training speed, and regularization effect) [14] make it efficient to implement our proposed application. The neural network of the CNN is a model that takes inspiration from the brain. The layered structure is composed of simple connected units or neurons in which at least one layer is a convolutional layer. A typical CNN consists of some combination of the following layers:

1. **Convolutional layers** comprise a layer of a deep neural network. Deep neural networks have convolution layers.
2. **Activation function (ReLU or Sigmoid)** is applied to a weighted sum of input values for evaluating output values of the neuron.
3. **Pooling layers** reduces a matrix, which is created by an earlier convolutional layer into a smaller matrix. It usually involves taking either the maximum or average value across the pooled area. Max pooling takes blocks of every 4 pixels/coefficients and pick the maximum one. Remove other 3. It reduces the size of image. So pooling area is by default the complete matrix.
4. **Dense layers** or a hidden layer where every node in the succeeding hidden layer is linked to each node. Hidden layers usually incorporate a training activation feature (such as ReLU). There is more than one hidden layer in a deep neural network.

The Visual Geometry Group (VGG) architecture [15] is used for neural network model. It is an advanced design of the neural network. The research on neural networks was centered on how the depth of these networks may be increased. The network uses 3x3 filters (kernels); padding always remains the same and the stride size is 1 or 2 (max pooling). Otherwise it is simple; the only additional elements are pooling layers and a completely linked layer. In the final outcome, the softmax function is used to get the probability of classification of expressions.

Figure 7.2 shows the block representation of the VGG architecture of neural networks. Four convolution blocks and two fully connected neural networks are used in this architecture. In each stage the following operations are performed:

- Updating the weights by using the kernel matrix.
- Batch normalization (BN) is applied as given in eq. (7.1)

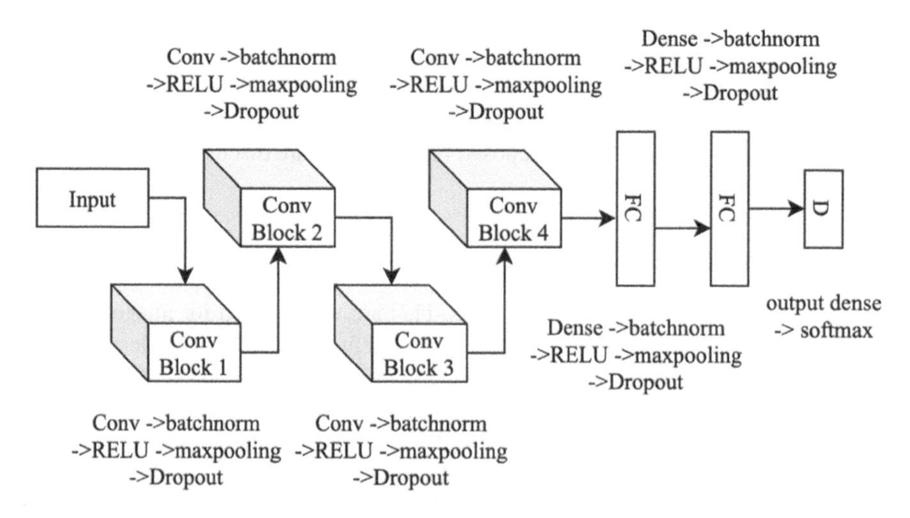

FIGURE 7.2 VVG-16 Architecture.

$$y^{(k)} = BN_{\gamma^{(k)},\beta^{(k)}}\left(x^{(k)}\right) = \frac{\gamma^{(k)}}{\sqrt{Var\left[x^{(k)}\right] + \in}}x^{(k)} + \left(\beta^{(k)} - \frac{\gamma^{(k)}E\left[x^{(k)}\right]}{\sqrt{Var\left[x^{(k)}\right] + \in}}\right) \tag{7.1}$$

Parameters used: Gamma, Beta, Mean, and Variance

- ReLU function (nonlinear activation function) is deployed for the updating of weights. y = max(0, x) if the value of x is negative or 0 then the output will be y=0. If the value of x is positive, then the output will be y=x.
- Max pooling is applied to reduce the size of the output matrix and pull the maximum weights from the output of the activation function layer as shown in Figure 7.3.
- Dropout of 25% will be done (to prevent overfitting problems or to regularize the neural network). Random weights are dropped (25%) from the output of the above layer. This process will repeat four times as the number of convolution blocks are four. In each successive step, the size of the matrix will be half the size of the previous matrix and the number of filters will increase as the input goes from left to right.
- After iterating through four convolution blocks, the output will be flattened and will pass through two fully connected neural networks in which the hidden nodes are (batch normalization of the output, ReLU function is used as an activation function, dropout of 25%).
- In the end, the softmax function is used to get the probability of our expression classes as given in eq. (7.2).

$$\sigma(\vec{z})_i = \frac{e^{z_j}}{\sum_{j=1}^{K} e^{z_j}} \tag{7.2}$$

Thus, VGG architecture is implemented for facial expression recognition. In the next subsection, the speech recognition module is discussed in detail.

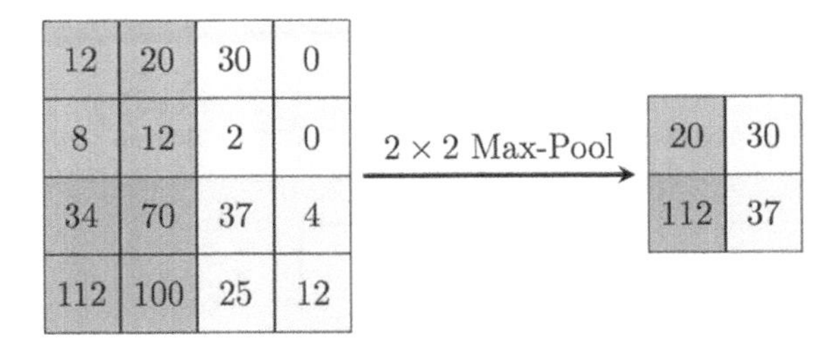

FIGURE 7.3 Max pooling.

7.3.2 Automatic Speech Recognition

The automatic speech recognition (ASR) module [16] is used to take the audio as input and convert it into text for our program. It takes input through a voice channel (i.e., microphone) and then converts this into text that will be used later on. To extract the keywords and match with the database the first part of how the voice is getting converted to text is done through GoogleAPI [17], which is a type of automatic speech recognition system deployed and managed by Google. The architecture of a typical automatic speech recognition system can broadly be divided into four modules [18]: pre-processing module, feature extraction module, classification module and language module as shown in Figure 7.4.

The input given to an ASR is usually captured using a microphone. Since this captured voice can contain noise output pre-processing of signals is done. This processed signal passes through the ASR feature extraction module to extract the essential features required from signals. These features can be sharpness, loudness, frequency, etc. These features are passed through a trained classifiers and acoustic model. Acoustic models are used to estimate the likelihood of the observed features for a given context in speech. These extracted features are related to the spectral characteristics of a speech.

During the training of acoustic models, frames associated with each state are added up and the probability distributions for each state are re-estimated. Based on these, features are extracted. This first system is made up of acoustic and language models, as well as trained classifiers. Trained classifiers are algorithms that have been trained over a certain dataset to recognize a feature algorithm from previous data. The classifier, which in the case of Google API, is a hybrid ANN model that will later convert the voice into text. Hybrid models of ANNs are those in which different activation/transfer functions are used for the nodes in the hidden layer. Several techniques proposed the hybridization of different basis functions, using either one single hybrid hidden layer or several connected pure layers [17].

Using the NLTK library, the text is further reduced to a set of keywords. The NLTK library has a predefined set of filler words (is, are, the, i, am, etc.) that are used on received strings. This input string is received answer in our case. If the candidate answer contains any such filler words, then these words will be removed by

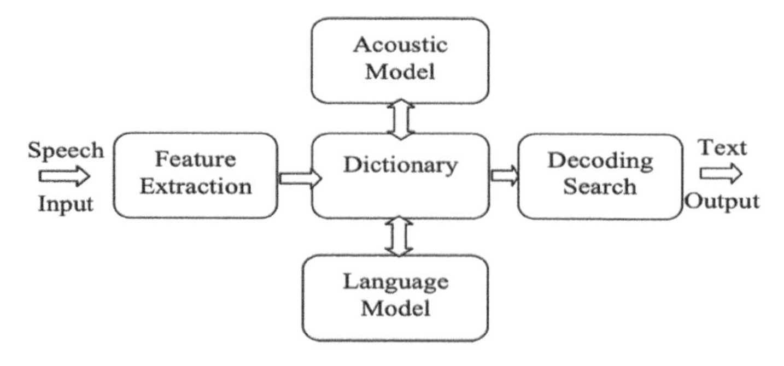

FIGURE 7.4 Automatic speech recognition system.

the library and in the end an answer is received consisting of all the keywords and no filler words. Thus, the desired response is extracted, which has already been saved in the database by the program's administrator, and contains particular keywords that are utilized for matching. Now an answer is received from the candidate and the expected answer, which was extracted from the database. Now both responses are started matching, each term from the anticipated answer is allocated an equal weight for allocating weight [19] using Equation (7.3).

$$Weightperword = \frac{100}{T(k)} \tag{7.3}$$

where T(k) is the total number of keywords in our expected answer. Step-by-step each keyword and its synonyms is checked in the received answer, matching the answer with the database to check how much answer is matched. If it is correct marks of the candidate increased. If the same problem does not encounter, the system proceeds to the next keyword without increasing or decreasing the answer. After checking all the keywords in the received answer, the total score is received from the automatic speech recognition module. This score is passed on for later evaluation by the chatbot by combining the data from both modules and the final score is generated.

As per current scope of the product, certain assumptions are made that includes that the candidate that is giving the interview is a real person and only speaks in English as his/her primary language.

7.4 RESULTS AND EXPERIMENTATION

The chatbot is created by going through the two primary phases. These phases are (1) extracting and storing information (2) conversation flow. A brief discussion on these phases is as follows:

- Extracting and storing information. It is necessary to initially specify the sets of questions and answers. Extraction and representation of knowledge may be accomplished in a variety of ways. The proposed model used tuples of the form <input, response>. The model is trained with respect to certain questions to be asked in the interview and the machine learning is applied for the training process.
- Conversation flow. The conversation between candidate and chatbot can be broken down into smaller ones to further process the information. For example, as soon as the candidate authenticates himself as the correct person, a question will trigger from the dataset of the chatbot and a certain time will be given to the candidate to answer it, following which the candidate will speak his answer, which will be recorded and then interpreted to get final results about the correctness of the answer. In addition, the candidate goes through the following phases:

```
expression_score_total=db['express']['Angry']+db['express']['Disgust']+db['express']['Fear']+db['express']['Happy']+db[
#print(expression_score_total)
expression_score_confident=db['express']['Happy']+db['express']['Neutral']
expression_score_nervous=db['express']['Angry']+db['express']['Disgust']+db['express']['Fear']+db['express']['Sad']+db[
confident_score=100*expression_score_confident/expression_score_total
nervous_score=100*expression_score_nervous/expression_score_total

if (confident_score>=nervous_score):
    last_msg1 = "You were very confident during the interview. Keep it up."
    language = 'en'
    last_msg1=tts("en-US-Wavenet-F",last_msg1,'last_msg1')
    playsound(last_msg1)
    Total=Total+5
else:
    last_msg2 = " You were a little nervous during the interview. Don't be nervous. Relax it's just an interview."
    language = 'en'
    welcome2=tts("en-US-Wavenet-F",last_msg2,'last_msg2')
    playsound(last_msg2)
    Total=Total-5

print("Total",Total)
wel_msg3 = "This is all I wanted to ask. I am done with all the questions. Nice talking to you. Have a nice day."
```

FIGURE 7.5 Code snippet of chatbot module.

(1) Authentication of the given candidate ID. To begin the interview, the user must enter his unique ID, which is assigned to him by the organization and unique to each applicant. Candidates are authenticated by verifying their IDs against the database. The interview may be resumed when authentication has been completed.
(2) Phase of the interview. This phase entails an interview procedure that is equivalent to a standard interview. The only difference is that the interviewer will be a chatbot driven by artificial intelligence. The user's responses are saved, processed, and finally the results are produced.
(3) Phase of assessment. The responses provided by users are compared to the datasets created specifically for each position. Due to the fact that the chatbot is trained on the dataset, the keywords are matched. Each applicant is assigned a score based on his or her performance after the evaluation.
(4) Termination of the session. Once the scores have been computed the session needs to terminate. The backend sends a message terminating the session, and the user is notified. Candidates may be notified of their results by email.

Figure 7.5 shows the main code snippet of the chatbot module in which the score is calculated from the expression received and later used in further modification of the score received from the voice recognition module (i.e., sum of both scores).
Figures 7.6 and 7.7 show the code snippets of the speech recognition module. Figure 7.6 demonstrates how the Google API is used to record voice and then used to convert it into text. Figure 7.7 demonstrates how the keywords are extracted using the NLTK library and later used to find all possible synonyms.

The result of a candidate after giving the interview through the recruiter bot are displayed as demonstrated in Figures 7.8 to 7.11.

First, the output of the individual modules is shown and then later a whole compilation is shown. These results are summarized as follows:

```
    with sr.Microphone() as source:
        print('Speak :')
        audio = r2.listen(source)
        try:
            get = r2.recognize_google(audio)
            print(get)
        #       wb.get().open_new(url+get)
        except sr.UnknownValueError:
            print('error')
        except sr.RequestError as e:
            print('failed'.format(e))
        val=int(input("Answer does't processed correctly?To change input press 1 or to enter input manully press 2
        if val==1 :
            continue
        else:
            if val==2:
                get=string(input("enter answer"))
            cont=False
stop_threads=False
checke = False
```

FIGURE 7.6 Calling of google API.

```
stop_words = set(stopwords.words('english'))
word_tokens = word_tokenize(sent)
text_file= open("synonyms.txt", "w")
text_file.close()
final_keywords=filtered_sentence
length = len(filtered_sentence)
for w in range(length):
    a = 'https://www.lexico.com/synonyms/' + filtered_sentence[w]
    URL = a
    page = requests.get(URL)
    soup = BeautifulSoup(page.content,'html.parser')
    try:
        result = soup.find('div',{'class':'row'})
        #print(result.prettify())
        try:
            job_elems = result.find_all('span', {'class':'syn'})
        except:
            job_elems=[]
    except:
        job_elems=[]
```

FIGURE 7.7 Keywords extraction and synonyms finder.

- The output of the first module (i.e., facial detection module) is shown in Figure 7.8. It shows the array of stored expressions that were present throughout the interview recorded while the candidate was answering. As shown in Figure 7.8, different expressions in the array and number of times the particular

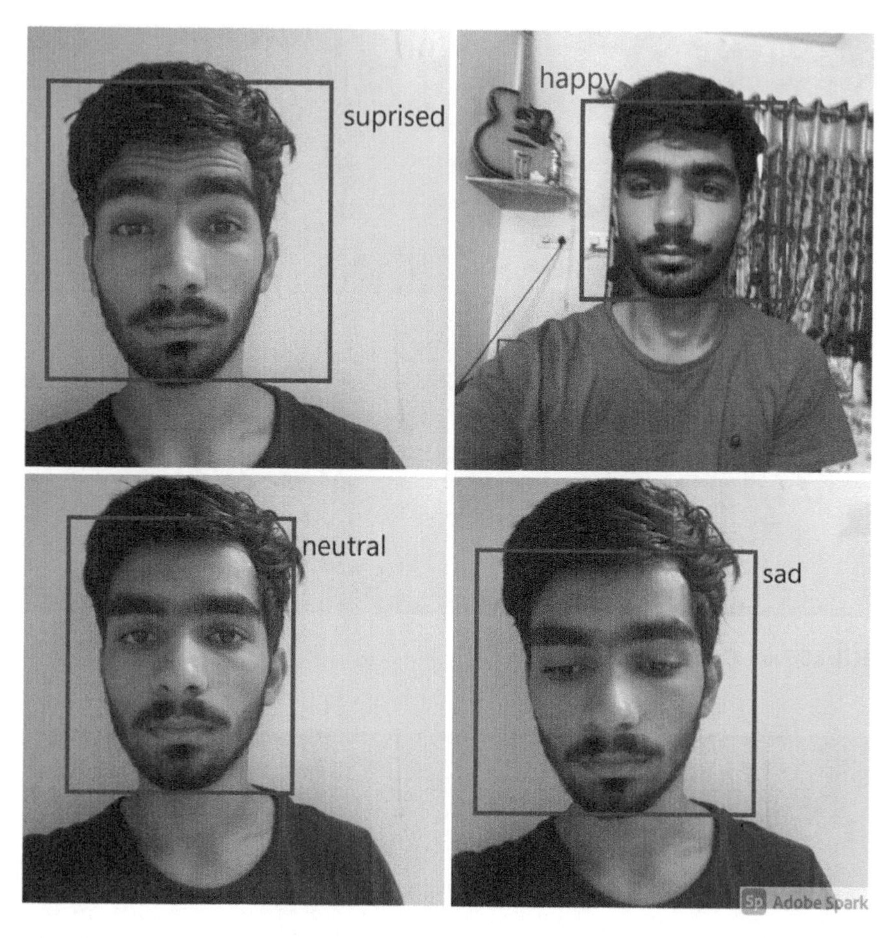

FIGURE 7.8 Facial recognition module sample output.

expression are recorded while answering are given. As seen in Table 7.3, candidate ID 192 was looking angry once, neutral 14 times, and sad 5 times.

- The output of the second module (i.e., automatic speech recognition) is shown in Figure 7.9. The candidate speaking while answering is shown over the terminal along with the initial speech module score. As shown in Figure 7.9, the candidate answer contains all the required keywords and therefore the output score is full (i.e., 5/5).
- Lastly after output from both modules is received, the chatbot integrates them to show the final output as shown in Figure 7.10.

On average, our model works with 81% overall accuracy. The facial emotion recognition accuracy is 71% and speech recognition accuracy is 91%. The reason for the lower accuracy is investigated with the help of a confusion matrix [22, 23].

```
[0:02] Decoding of welcome.mp3 finished.
Speak Anything :
You said : static variable is a variable that has been allocated st
atically meaning that its lifetime is the entire run of the program
Score: 4
Yeah! that's pretty right, The given answer is GOOD
High Performance MPEG 1.0/2.0/2.5 Audio Player for Layer 1, 2, and
3.
Version 0.3.2-1 (2012/03/25). Written and copyrights by Joe Drew,
now maintained by Nanakos Chrysostomos and others.
Uses code from various people. See 'README' for more!
THIS SOFTWARE COMES WITH ABSOLUTELY NO WARRANTY! USE AT YOUR OWN RI
SK!

Playing MPEG stream from welcome.mp3 ...
MPEG 2.0 layer III, 32 kbit/s, 24000 Hz mono
```

FIGURE 7.9 Automatic speech recognition module sample output.

```
[0:02] Decoding of welcome.mp3 finished.
Speak Anything :
Sorry,can u repeat
You said : object oriented programming is a programming paradigm based on the concept of objects which can con
ain data and code
Score: 5
You really have a good knowledge of this topic... i'm impressed
High Performance MPEG 1.0/2.0/2.5 Audio Player for Layer 1, 2, and 3.
Version 0.3.2-1 (2012/03/25). Written and copyrights by Joe Drew,
now maintained by Nanakos Chrysostomos and others.
Uses code from various people. See 'README' for more!
THIS SOFTWARE COMES WITH ABSOLUTELY NO WARRANTY! USE AT YOUR OWN RISK!

Playing MPEG stream from welcome.mp3 ...
MPEG 2.0 layer III, 32 kbit/s, 24000 Hz mono

[0:04] Decoding of welcome.mp3 finished.
[{'studentID': '192', 'express': {'Angry': 1, 'Disgust': 0, 'Fear': 0, 'Happy': 0, 'Neutral': 14, 'Sad': 5, 'Su
rprise': 0}}]
{'studentID': '192', 'express': {'Angry': 1, 'Disgust': 0, 'Fear': 0, 'Happy': 0, 'Neutral': 14, 'Sad': 5, 'Su
prise': 0}}
High Performance MPEG 1.0/2.0/2.5 Audio Player for Layer 1, 2, and 3.
Version 0.3.2-1 (2012/03/25). Written and copyrights by Joe Drew,
now maintained by Nanakos Chrysostomos and others.
Uses code from various people. See 'README' for more!
THIS SOFTWARE COMES WITH ABSOLUTELY NO WARRANTY! USE AT YOUR OWN RISK!

Playing MPEG stream from welcome.mp3 ...
MPEG 2.0 layer III, 32 kbit/s, 24000 Hz mono

[0:04] Decoding of welcome.mp3 finished.
Total 10
```

FIGURE 7.10 Chatbot sample output.

Since the proposed model evaluates seven emotions, our confusion matrix is 7×7, as shown in Figure 7.10. In this, there are seven parameters: Happy, Angry, Disgust, Sad, Neutral, Fear, and Surprised. On the x-axis, there are predicted labels, and on the y-axis there are actual labels. These results are based on the FER 2013 [24–26] dataset. This dataset comprises seven emotions with different numbers of images as follows: Happy (7215), Angry (3995), Sad (4830), Neutral (4965), Disgust (436), Surprise (3171), and Fear (4097). In the dataset, the number of images with happy

TABLE 7.2
Confusion Matrix for N-class Classification

		Predicted Values			
Actual Values		Class 1	Class 2	...	Class N
	Class 1	Cell 1	Cell 2	...	Cell N
	Class 2	Cell N+1	Cell N+2	...	Cell 2N
	...	...	...	...	...
	...	...	...	...	...
	Class N	Cell (N(N-1)+1)	Cell (N(N-1)+2)	...	Cell N^2

TABLE 7.3
Candidate Expressions during Interview

Student			Expression	During		Interview	
ID	Angry	Disgust	Fear	Happy	Neutral	Sad	Surprise
192	01	0	0	0	14	5	0
194	0	2	5	0	1	0	0

expressions is quite high. So the happy expression gives maximum true positive result (i.e., 6101). Whereas, the number of images with disgust expression is much less, and thus the algorithm is not well trained with disgust expression, which results in least true positive result for this expression.

The performance of the proposed methodology in terms of statistical parameters was also evaluated using the confusion matrix for multiclass classification. A confusion matrix, also known as an error matrix, is used to visualize the performance of the proposed algorithm. In our experiments, the face expressions are classified into seven categories so the multiclass confusion matrix is used to depict the required values. In the multi-class classification problem, one cannot get True Positive (TP), False Positive (FP), True Negative (TN), and False Negative (FN) values directly as in the binary classification problem. These classification metrics are calculated for each class. Refer to Table 7.2 for multiclass classification.

The formula to depict the values of TP, FP, TN, and FN for multiclass to calculate precision, recall, F1-measure, and other important parameters is:

$$TPforClass_i = Cell(i-1)*N+i \tag{7.4}$$

$$FNforClass_i = \left(\sum_{j=1}^{N} Cell(i-1)*N+j\right) - \left(Cell(i-1)*N+i\right) \tag{7.5}$$

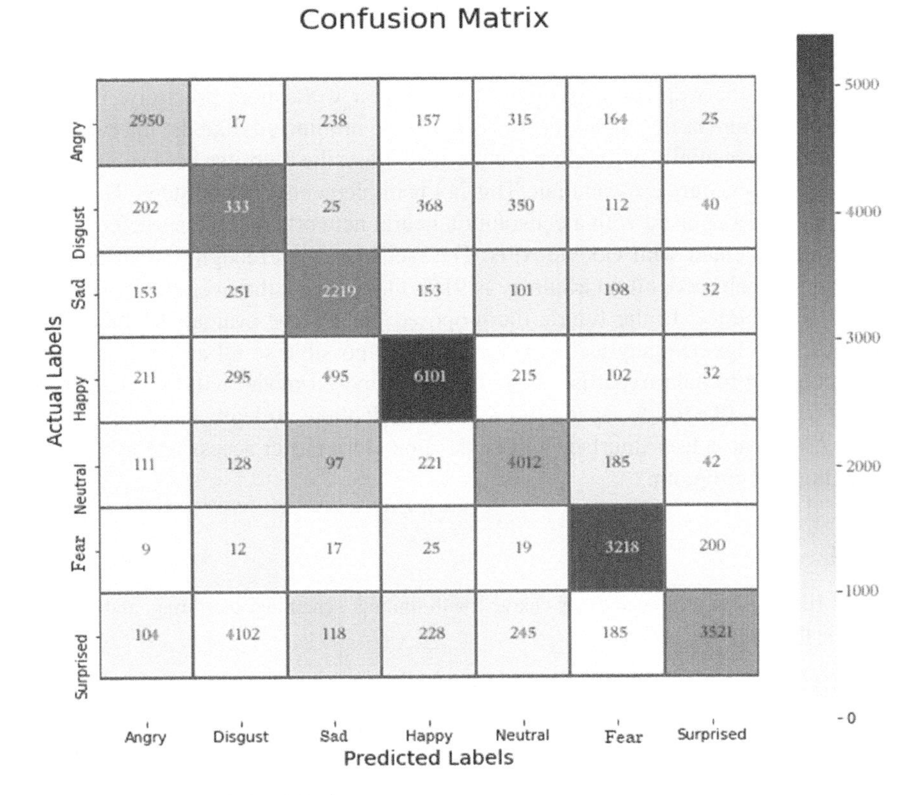

FIGURE 7.11 Confusion Matrix.

$$FPforClass_i = \left(\sum_{j=1}^{N} Cell(j-1) * N + i \right) - \left(Cell(i-1) * N + i \right) \tag{7.6}$$

$$TNforClass_i = \left(\sum_{j=1}^{N} \sum_{k=1}^{N} cell(j,k) \right) - \left(\sum_{j=1}^{N} Cell(j-1) * N + i \right) - \left(\sum_{k=1}^{N} Cell(i-1) * N + k \right) + \left(Cell(i-1) * N + i \right) \tag{7.7}$$

In our study, there are seven classes as shown in Figure 7.11. The values of precision, recall, F1-measure, and other required parameters can be computed by applying the well-known standard formula to visualize the performance. The confusion matrix for the proposed method is shown in Figure 7.11 to observe important parameters such as related correctness and others. The expressions are classified into seven categories and their corresponding outcomes can be seen in the confusion matrix.

7.5 CONCLUSION

Our proposed recruiter chatbot can help companies hire new employees. It improves the quality of the work by helping humans do their work more precisely. The side effects of the human-dependent factors can also be minimized, like the interviewer's mood and personal biases during the interview. Thus, the recruiter bot can make the recruitment procedure easy and fair. This bot is implemented in two steps. The facial emotions are recognised with a convolution neural network, and the speech recognition is implemented with Google APIs. The facial emotion recognition accuracy is 71%, and speech recognition accuracy is 91%. On average, the overall model works with 81% accuracy. In the future, the proposed bot will be evaluated with a large dataset to indulge as many technological fields as possible so all aspirants can prepare according to their expertise. Later on, the proposed model will be extended by adding features like follow-up questions, which will allow in-depth analysis of knowledge. After that, a user interface will be developed for easier access and to navigate throughout the program.

NOTE

1 The HR division of a business is charged with finding, screening, recruiting, and training job applicants.

REFERENCES

[1] Rayna ayu Puspita, "The Impact of Internet Use for Students," in proc. of *IOP Conference Series Materials Science and Engineering*, February 2018.

[2] M. Dahiya, "A Tool of Conversation: Chatbot," *International Journal of Computer Sciences and Engineering*, 23 April 2017.

[3] Ken Braun, "Does The Evolution Of Chatbots Include Replacing Humans?," Forbes Article, October 16, 2018.

[4] Anjali Mary Gomes, Nishad Nawaz, "Artificial Intelligence Chatbots are New Recruiters," in proc. of *International Journal of Advanced Computer Science and Applications (IJACSA)*, Vol. 10, No. 9, 2019.

[5] Joseph Weizenbaum, "ELIZA – a computer program for the study of natural language communication between man and machine," *Communications of the ACM*, 9(101), 36–45, January 1966.

[6] Kenneth Colby, "PARRY: the AI chatbot from 1972," MIT Press, 1972.

[7] Eleni Adamopoulou, Lefteris Moussiades, "Chatbots: History, technology, and applications," *Machine Learning with Applications*, 2, pp. 100006, 2020.

[8] Neha Atul Godse, Shaunak Deodhar, Shubhangi Raut, Pranjali Jagdale, "Implementation of Chatbot for ITSM Application Using IBM Watson," in proc. of *Fourth International Conference on Computing Communication Control and Automation (ICCUBEA)*, IEEE, 25 April 2019.

[9] Blagoj Delipetrev, Chrisa Tsinaraki, Uroš Kostić, "AI Watch Historical Evolution of Artificial Intelligence," Joint Research Centre (European Commission), December 2018.

[10] Wizu Research Paper, "A Comparison of AI Conversational Surveys and Traditional Form-Based Surveys," December 2018.

[11] Asbjørn Følstad and Petter Bae Brandtzaeg, "Users' experiences with chatbots: findings from a questionnaire study," *Quality and User Experience*, 5(1), pp.1–14, 2020.
[12] Xinyu Lei, Guan-Hua Tu, Alex. Liu, Chi-Yu Li, Tian Xie, "The Insecurity of Home Digital Voice Assistants - Amazon Alexa as a Case Study," in proc. of *IEEE Conference on Communications and Network Security (CNS)*, 13 August 2018.
[13] Sakshi Indolia, Anil Kumar Goswami, S.P. Mishra, Pooja Asopa, "Conceptual Understanding of Convolutional Neural Network- A Deep Learning Approach," in proc. of *International Conference on Computational Intelligence and Data Science (ICCIDS)*, 2018.
[14] Manem H, Rajendran J, Rose G.S., "Stochastic gradient descent inspired training technique for a CMOS/nano memristive trainable threshold gate array," *IEEE Transactions on Circuits and Systems*, 2012.
[15] Shaik Asif Hussain, Ahlam Salim Abdallah Al Balushi, "A real time face emotion classification and recognition using deep learning model," *Journal of Physics: Conference Series ICE4CT,* 2019.
[16] Yi-Chen Chen, Jui-Yang Hsu, Cheng-Kuang Lee, Hung-yi Lee, "DARTS-ASR: Differentiable Architecture Search for Multilingual Speech Recognition and Adaptation," Graduate Institute of Communication Engineering, National Taiwan University NVIDIA AI Technology Center, NVIDIA, 2005.
[17] Haoran Miao, Gaofeng Cheng, Changfeng Gao, Pengyuan Zhang, Yonghong Yan, "Transformer-Based Online CTC/Attention End-to-End Speech Recognition Architecture," arXiv:2001.08290v2 [eess.AS], 11 Feb 2020.
[18] Johan Schalkwyk, Doug Beeferman, Fran¸coise Beaufays, Bill Byrne, Ciprian Chelba, Mike Cohen, Maryam Garret, Brian Strope, "Google Search by Voice: A case study," Google, Inc. 1600 Amphiteatre Pkwy Mountain View, CA 94043, USA, 2010.
[19] C. Van Heerden, J. Schalkwyk, and B. Strope, "Language modeling for what-with-where on GOOG-411," 2009.
[20] Hao-Hsuan Hsu and Nen-Fu Huang, "Xiao-Shih: A self-enriched question answering bot with machine learning on chinese-based MOOCs," *IEEE Transactions on Learning Technologies*, 2022. Early Access. DOI:10.1109/TLT.2022.3162572.
[21] Yunze Xiao, "A transformer-based attention flow model for intelligent question and answering chatbot," *14th International Conference on Computer Research and Development (ICCRD)*, Shenzhen, China, 2022. DOI:10.1109/ICCRD54409.2022.9730454
[22] Rahman, A., Nimmy, S. F., & Sarowar, G. (2019). Developing an automated machine learning approach to test discontinuity in DNA for detecting tuberculosis. In J. P. Davim (Ed.), Proceedings of the Twelfth International Conference on Management Science and Engineering Management (pp. 277–286). Springer. https://doi.org/10.1007/978-3-319-93351-1_23
[23] Xiao Huang, "Ideal construction of chatbot based on intelligent depression detection techniques," *IEEE International Conference on Electrical Engineering, Big Data and Algorithms (EEBDA)*, Changchun, China, 2022. DOI:10.1109/EEBDA53927.2022.9744938
[24] Amil Khanzada, Charles Bai, Ferhat Turker Celepcikay, "Facial Expression Recognition with Deep Learning," Improving on the State of the Art and Applying to the Real World, Stanford University - CS230 Deep Learning, 2020.
[25] Rafiandi Ammar Putra, Irsyad Musyaffa, Auzi Asfarian, and Dean Apriana Ramadhan. "CURHAT: Telling Your Story to a Multimodal Conversation Bot to Alleviating the

Stress Caused by Pandemic Fatigue," In Asian CHI Symposium 2021 (Asian CHI Symposium 2021). Association for Computing Machinery, New York, NY, USA, pp. 177–179, 2021.

[26] Rahman, A. (2019). Statistics-based data preprocessing methods and machine learning algorithms for big data analysis. International Journal of Artificial Intelligence, 17(2): 44–65.

8 Analysis of Oversampling and Ensemble Learning Methods for Credit Card Fraud Detection

*Ankit Kumar Singh[1], Rohit Bhaskar Rao Gurijala[1], Utkarsh Kumar Rai[1], Anupam Kumar[*2], Basant Agarwal[1] and Ashish Sharma[1]*
[1]IIIT Kota, Jaipur, Rajasthan, India
[2] National Institute of Technology, Patna, India
*Corresponding Author: anupam.ec@nitp.ac.in; anuanu1616@gmail.com

CONTENTS

DOI: 10.1201/9781003253051-11

8.1 INTRODUCTION

The advancement of technology, new methods of purchasing products, and the increased demand of online shopping has exponentially changed the use of cashless payment methods, especially credit cards [1]. People are using credit cards more often especially for shopping and also the new worldwide trend is to go cashless. The quest of the developing and the already developed nations to popularize the digital payment methods over physical currency is growing at a rapid pace. Therefore, it becomes all the more important to provide the utmost security and privacy to the citizens keeping them safe from being attacked by fraudsters [2]. As most credit card transactions are online, the chances of being attacked or becoming the victim of fraud are significant. Credit card frauds are sometimes difficult to trace and fraudsters are becoming very versed with the technology to hide their identities. Therefore, it very important to identify such malicious attacks and frauds on customers and protect their data. Credit card fraud detection aims to identify such frauds and try to minimize them [3].

Machine learning and artificial intelligence are now being used to provide a clear picture between legitimate and non-legitimate transactions. Data scientists work with companies to build models with high in accuracy to alert companies of suspicious activity in credit card accounts [4]. Machine learning models such as Logistic Regression [5], Decision Trees, K-Nearest Neighbors [6], KMeans, Auto Encoder [7], Random Forest [8], J48 [9], and many more are used for fraud detection [1]. These models have parameters upon which their accuracy and precision depends. The measure of success of any model can be calculated by looking at its Recall, F1 score [10], and the values of false positives and false negatives [11]. These values should be as low as possible. As the field of fraud detection is growing day by day, fraud techniques need to be updated continuously too. This chapter discusses machine learning algorithms that are combined to generate ensemble learning outcomes. It proves to be effective in determining the fraudulent transactions through the credit cards of innocent people. The ensemble models are compared to find the best fraud prevention solution. Classifiers such as Logistic Regression, Random Forest, KNN, SVM, and Naive Bayes are used in ensembles of various groups to find the best results in terms of Recall and F1 Score. The chapter also aims to figure out the best oversampling technique among ADASYN, SMOTE, and Random Over-Sampler for the dataset used. The oversampling of the data forms the backbone of the chapter as the data was highly imbalanced.

The remaining portion of the chapter is structured as follows. In Section 8.2, previous research observations are documented. In Section 8.3, the proposed approach of the chapter is described. In Section 8.4, the experimental observations and results are detailed, discussing the performance increase using the proposed approach. In Section 8.5, conclusion and future scope of work are covered.

8.2 RELATED WORK

Recent advances in data science and machine learning-based models have evolved significantly in recent times [12–15], further for several applications in the financial domain [16]. Data scientists and researchers continually work to find appropriate models for the detection of credit card fraud. There have been many papers published on machine learning algorithms, including KNN, Logistic Regression, Simple K Means, Random Forest, and Naive Bayes [17–20]. Some research focuses on deep learning algorithms as they perform better on larger datasets due to their computational expense [21–23].

In Ref. [7] algorithms such as AutoEncoder and Restricted Boltzmann Machine were used, and the AUC score, ROC Curve, Mean Squared Error (MSE), and Root Mean Squared Error (RMSE) were calculated for three different datasets. The results showed that the AutoEncoder algorithm was best in terms of AUC and Accuracy. In Ref. [6] Logistic Regression, Naive Bayes, and KNN were used with an accuracy of 97.69%, 83%, and 54.86% for the three models, respectively. The authors suggested that Logistic Regression can be used for credit card fraud detection as it had higher accuracy as compared to the others. In Ref. [9], Naive Bayes, Logistic Regression, J48, and Adaboost were compared to find out which algorithm performed better in accurately distinguishing between legitimate and non-legitimate transactions. The Adaboost algorithm outperformed the others with accuracy as high as 100%. The Logistic Regression algorithm also had 100% accuracy, but the time taken by it was higher as compared to that of Adaboost.

Similarly, in Ref. [5], machine learning algorithms such as Logistic Regression, Random Forest, and Naïve Bayes were compared. In this case the Random Forest Classifier proved to be the superior one with accuracy of 99.96%. The ensemble model was used in Ref. [11]. In this study, two or more machine learning models were used together to effectively distinguish legitimate transactions from non-legitimate ones. An architecture consisting of deep neural networks with a multi-layered feed-forward network was evaluated in Ref. [10]. The model was implemented in H2O, an open-source platform that proves very efficient for models that are computationally complex. The MSE was 0.01661334 and the RMSE was 0.1288928. Many researchers also suggested methods such as Genetic Algorithm [24], Hidden Markov Model [25], Big Data methodologies [26], and many more [27–29] for credit card fraud detection.

8.3 PROPOSED APPROACH

Figure 8.1 shows the proposed framework where data preprocessing and oversampling methods are applied on the dataset to handle any issues related to the data. Typical oversampling techniques such as Random Over Sampler, ADASYN, and SMOTE are applied on the dataset for treating the sampling errors. After this previously mention algorithm are used in various groups using the ensemble learning algorithm the different types of sampling techniques analyzed and different machine learning algorithm are used for ensemble learning are briefly explained in the following paragraphs.

Sampling Methods: Sampling is a method that helps us to gather information about the population based on the statistics from a subset of a population (sample), without having the need to investigate every individual.

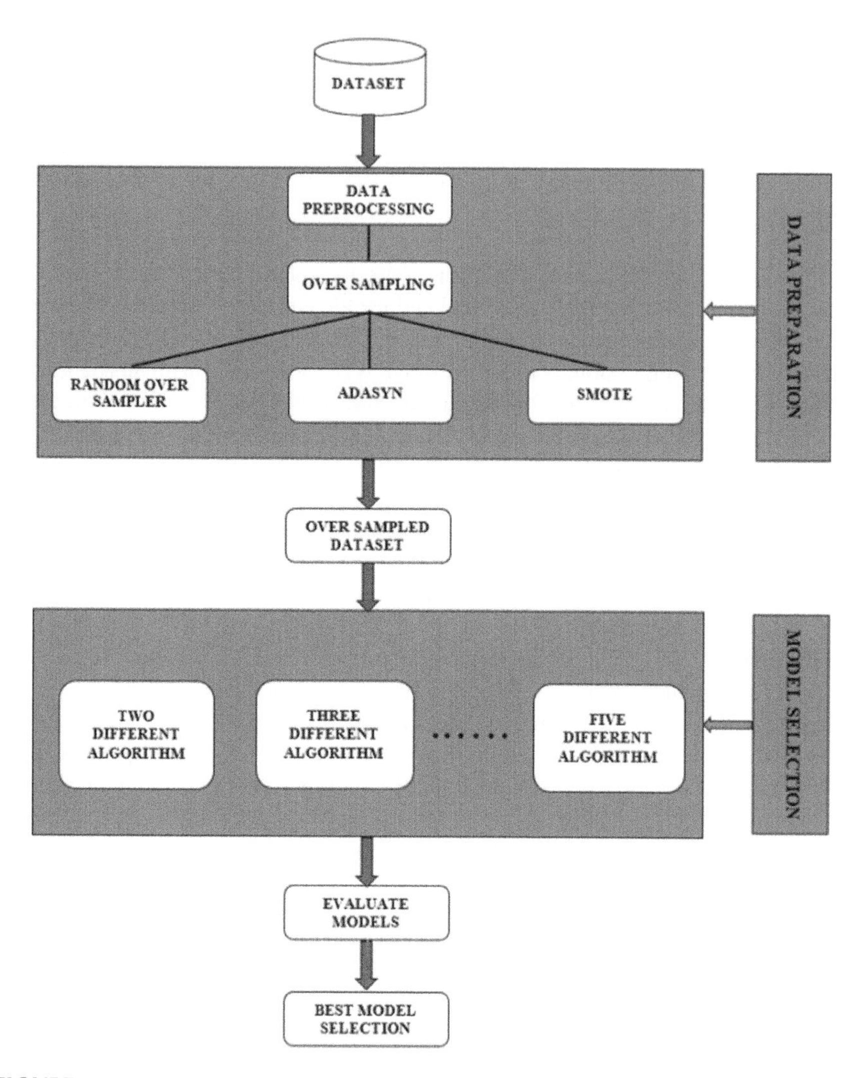

FIGURE 8.1 Proposed Approach Framework.

One of the main problems with classified datasets is that they have imbalanced class distribution, especially for fraud detection. For credit cards, most transactions are legitimate transactions and there are very few non-legitimate transactions. In a European dataset of credit card fraud detection there were around 284,807 transactions out of which only 492 were fraudulent. This shows the need for over-sampling techniques.

There are two main categories of sampling: Under-Sampling and Oversampling. However, this research is mainly confined on oversampling techniques since the tradeoff of the over-fitting factor that comes with oversampling algorithms can be adjusted to a certain level, whereas under-sampling might lead to loss of data points, which in the case of credit card fraud would be a major setback.

We use three types of over-sampling techniques:

1. Random Over-Sampling: Random over-sampling picks samples from a minority class at random with replacement sampling framework. It is a Naïve resampling method. It assumes nothing about the data and is relatively fast and easy to implement. It is highly efficient for algorithms where the data points are clustered towards a particular side of the scale rather being symmetrical.
2. Adaptive Synthetic (ADASYN): ADASYN is an oversampling technique that generates synthetic data, and its greatest advantage is not copying the same minority data, and generating more data for "hard to learn" examples. The major difference between ADASYN and SMOTE is the difference that occurs in the generation of synthetic sample points for minority data points. In SMOTE, there is a uniform weight for all minority points, whereas in ADASYN, a density distribution is considered that decides the total number of synthetic samples that have to be generated for a particular point.
3. Synthetic Minority Over-Sampling Technique (SMOTE): It uses a KNN algorithm for generating synthetic data. SMOTE works by choosing random data from the minority class, then k-nearest neighbors from the data are set. Synthetic data is then made between the random data and arbitrarily selected k-nearest neighbour.

Following are the individual models that are used for the Voting Classifier Ensemble Learning Algorithms:

1. Logistic Regression: Logistic regression model predicts whether something is true or false. For credit card fraud detection there are only two classes: legitimate transactions or non-legitimate transactions. Logistic regression fits an S-shaped curve logistic function, also called a sigmoid function, to the data.

$$\sigma(z) = \frac{1}{1 + e^{(-z)}} \tag{8.1}$$

$$Z = W^T * X(i) + \tag{8.2}$$

The limit of this sigmoid function is between 0 and 1, which is the decision boundary for our logistic regression. This algorithm can be used for the purpose of credit card fraud detection and shows promising results [6].

2. Random Forest: The random forest model is one of the best classifier-based models and is based on the concept of decision trees. Individual decision trees combine to make a random forest and merge together to get accurate prediction. The more decision trees, the more accurate and stable prediction [5].

Firstly, *n* no. (N1.T, N2.T, N3.T) of decision trees are chosen for the creation of a random forest model as shown in Figure 8.2 and for those decision trees they take random *k* no. (K1.D, K2.D, K3.D) of data points from training set points, which means only some data points are selected from the training set for each tree as shown in Figure 8.2. Then, referencing techniques will build a decision tree on the basis

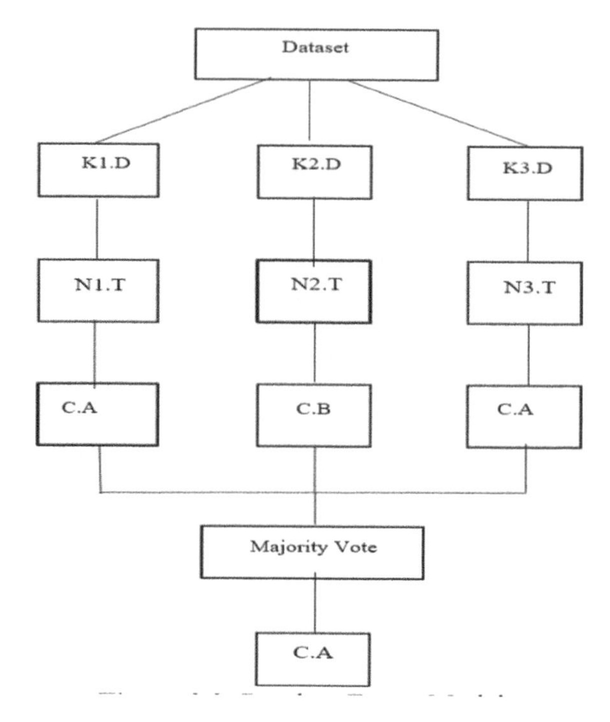

FIGURE 8.2 Random Forest Model.

of k data points for each tree. Then each of the n no. of trees predicts the class of transaction on the basis of its k no. data points. After training the model on the basis of k data points for each n no. of decision trees, it predicts a class for n no. of decision tree. For a new data point, each n no. of trees predicts a class to which the data point belongs and assigns the new data point to the category that wins the majority vote. Suppose there are 11 decision trees in a random forest tree model and each tree predicts two classes (i.e., class A & class B). For a new data point if six trees predict class A and five trees predict class B then the prediction belongs to class A because class A predictions are greater than class B.

3. K-Nearest Neighbor: This algorithm is a supervised learning method and is one of the famous algorithms for classification-based problems [6]. Suppose there are two different categories in the dataset. In KNN a k value must be selected (k is an integer from 1 to n) and this K value indicates how many nearest neighbors are considered in terms of distance from the new data point. Suppose we consider K=5. After our model is trained for a new data point we find five nearest neighbors by calculating the Euclidean distance between the new data point and two classes: legitimate and non-legitimate transaction. If we get three nearest neighbors for legitimate transactions and two nearest neighbor for non-legitimate transactions the new data points are legitimate transactions.

4. Support Vector Machine (SVM): This is a widely used supervised learning algorithm for classification-based problems. In SVM, a hyperplane is selected that

is able to distinguish between classes. Suppose there are two classes, Class A (Fraud Transactions) and Class B (Normal Transactions). SVM considers all the data points and creates a line (hyperplane) that divides these two classes. Anything comes under fraud class belongs to the class A and vice versa. There can be several hyperplanes possible, but the best hyperplane is the one that divides the classes having a large distance from the hyper plane from both the classes' (i.e., maximum margin). Support vectors are the data points that are closest to the hyperplane. In SVM the output is in the form of a linear function. If the output is greater than 1, we classify that class as A and if the output is -1 it is classified as B.

5. Naïve Bayes: This is a type of statistical method that helps in calculating probability that a feature belongs to a class and is based on the application of Bayes theorem. The reason this classifier is called naive is that it assumes the probabilities of the individual features are independent of one another, which is quite hard to apply in the real world. To calculate the likelihood that an event is going to occur, considering the fact that another event has already occurred, the mathematical expression is given as:

$$P\left(\frac{C}{X}\right)=\frac{PrPr\left(\frac{X}{C}\right)*Pr(C)}{Pr(X)} \tag{8.3}$$

where the posterior probability of target class C (i.e., Pr(C/X)) is found using Pr(X/C), Pr(C), and Pr(X). Naive Bayes, Bernoulli Naïve Bayes, and Gaussian Naive Bayes are the three most commonly used Naïve Bayes classifiers.

8.3.1 Ensemble Learning

In ensemble learning, multiple learning algorithms are used together (in a combined form) on the dataset to give the optimum solution and give better prediction than individual models. Through the implementation of individual models like Logistic Regression, Random Forest Classifier, KNN, and SVM, very good accuracy can be achieved, but the problem of over-fitting means models predict some transactions as legitimate but in actuality they are non-legitimate or vice versa. In this chapter, we emphasis on False Positive and False Negative and if FP and FN are higher than there is an uneven class distribution that leads to over fitting of the mode.

F1 score is also used as a parameter. Ensemble learning models give better F1 score and avoid over-fitting. There are two types of ensemble learning models: bagging and boosting. Bootstrap aggregating, also known as the bagging model, is mainly applied for classification problems. There are two approaches in bagging. First is the application of multiple models of the same learning algorithm trained with subsets of a dataset that was randomly picked from the training dataset (e.g., Decision tree and Random Forest). Another method that can be used is a multiple model of different learning algorithm i.e., Voting. Classifier; model trained in parallel in order to get different nature of each different models that are applied on same training set as shown in Figure 8.3, by using this principle we can ensemble different models on

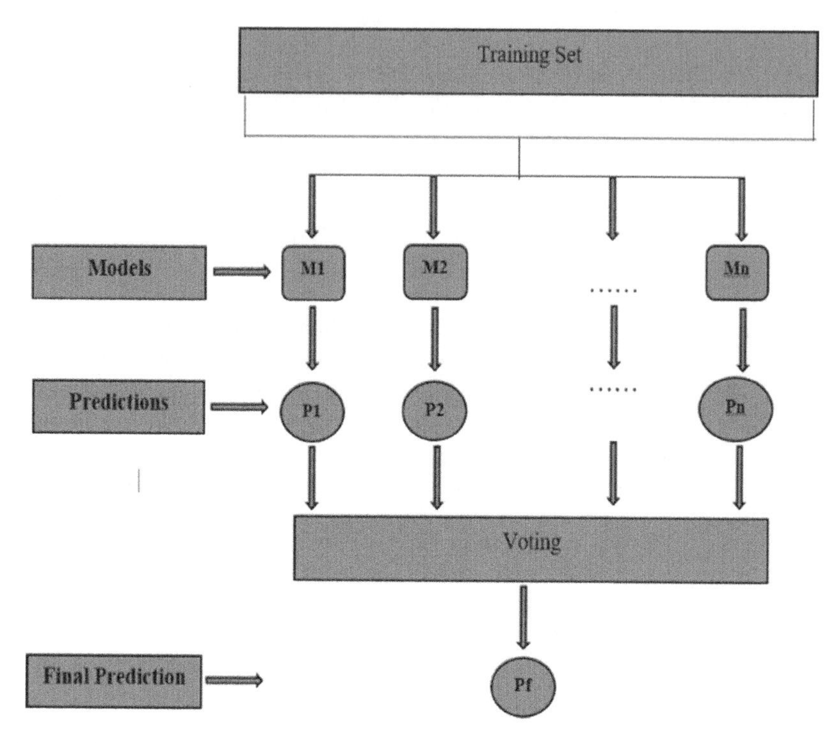

FIGURE 8.3 Proposed Model for Ensemble Learning Algorithm.

same training set to predict the final output. There are two different strategies of voting that are used in voting classifiers for prediction of final output:

1. Hard Voting: This is a simple method of voting. For example, if there are four classifier algorithms ensembled to predict the final class, suppose classifier 1, 2, and 3 predicted legitimate transactions and classifier 4 predicted class non-legitimate transaction and hence we can classify the overall class as Legitimate transactions by Majority Vote.
2. Soft Voting: In soft voting there is one change. Instead of predicting one class soft voting gives probabilities. Suppose there are three ensemble algorithms: M1, M2, and M3.

M1 = 0.9 (class 0) and 0.1 (class 1)
M2 = 0.7 (class 0) and 0.3 (class 1)
M3 = 0.4 (class 0) and 0.6 (class 1)

The average of probabilities of both the classes is then calculated. In this case, class 0 average is 0.666 and class 1 average is 0.333. The class with the highest average is selected so class 0 is selected.

8.4 EXPERIMENT RESULTS

8.4.1 Dataset and Preprocessing

The dataset we used had the data of European cardholders. There were about 492 fraudulent transactions out of 284,407 transactions. Due to the imbalance in the dataset, an oversampling method was needed Also, the model created would be much better trained and tested if there were sufficient data points that could be managed via oversampling as compared to undersampling techniques. The difference in data parity is shown before in Figure 8.4 and after oversampling in Figure 8.5.

As the classification algorithms are sensitive to the input variables, data scaling was required. We used Standard-Scalar to scale the data near a centralized value. The data distribution before and after this implementation is shown in Figure 8.6 and Figure 8.7, respectively.

8.4.2 Evaluation Metrics Ensemble Learning

As discussed above, the dataset was highly imbalanced with only 492 fraud transactions out of a total of 284,407 transactions. Despite having successfully classified all the samples into non-fraudulent categories, the classification accuracy was still extremely high, and therefore a traditional evaluation metric like accuracy was not suitable for this study. Instead, an in-depth analysis of the Confusion Matrix [5, 30] was used to classify the False Positive and False Negative transactions made by each of our ensemble models and then evaluating their overall performance and

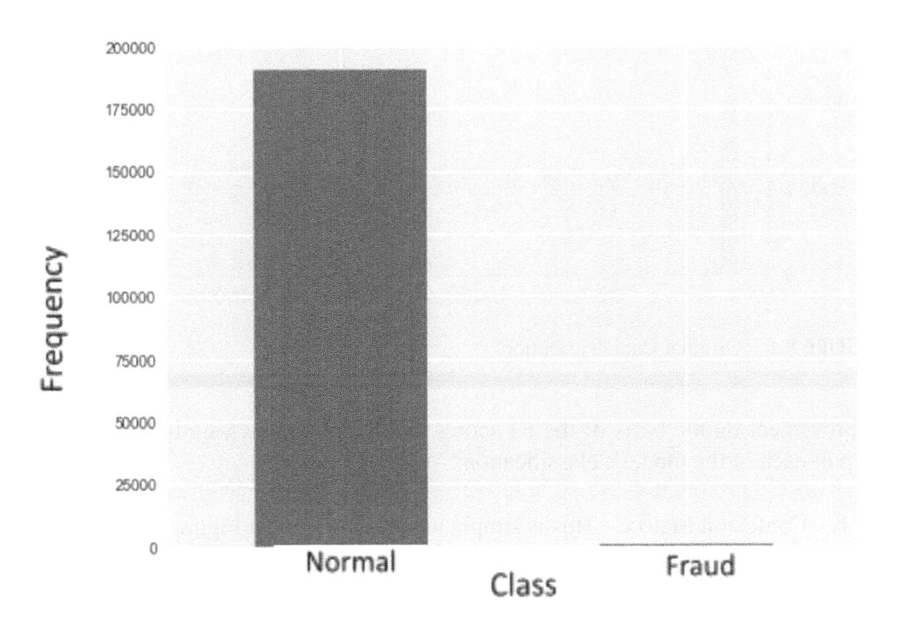

FIGURE 8.4 Class distribution before Oversampling.

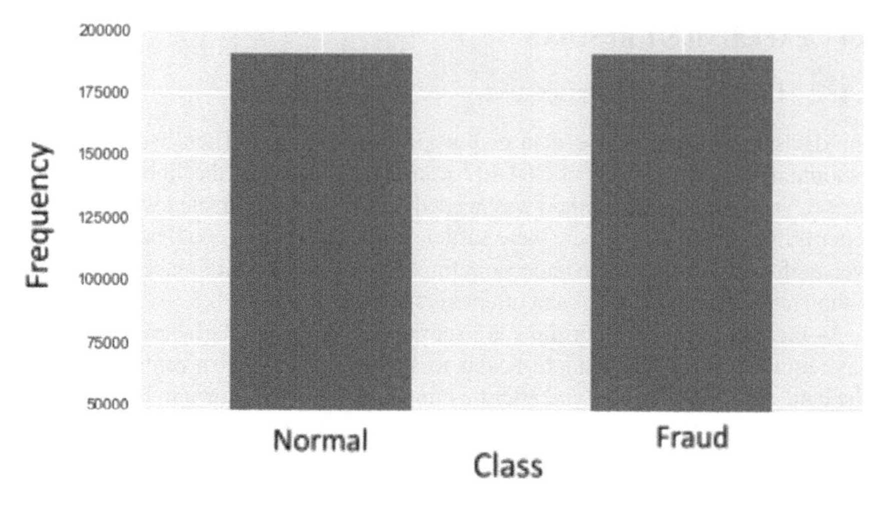

FIGURE 8.5 Class distribution after Oversampling.

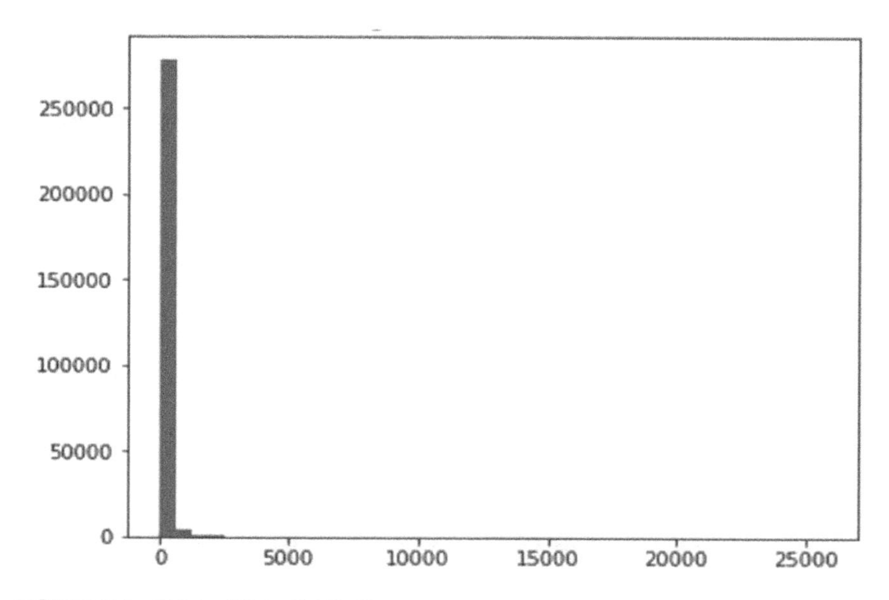

FIGURE 8.6 Original Data distribution.

improvement on the basis of the F1 scores and Recall values we obtained with the help of each of the model's classification.

1. Confusion Matrix – This is simply a table as shown in Figure 8.8 that is used to depict the performance of a classification model on a set of test data for which the true values are known to us in prior. It helps us to visualize the performance of an algorithm. The confusion matrix shows us the ways in which our classification model is confused when it makes predictions.

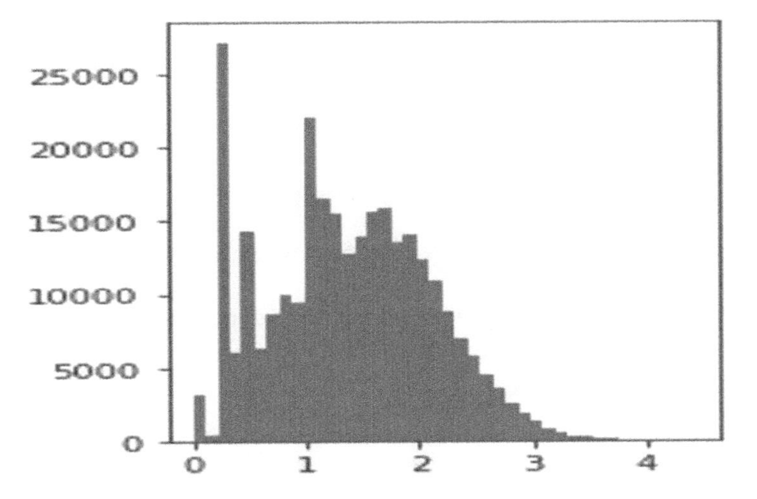

FIGURE 8.7 Log Transformed Amount Distribution.

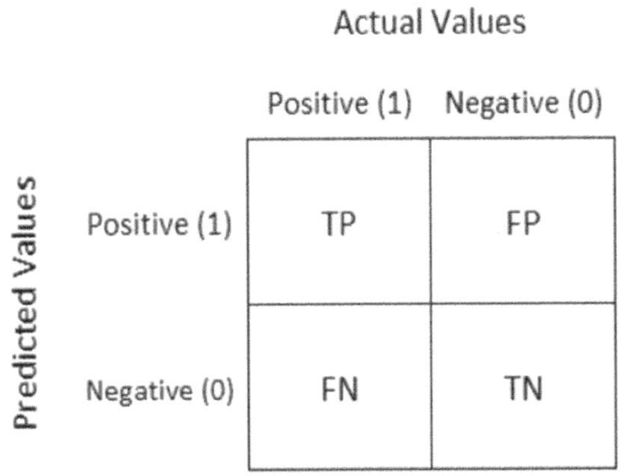

		Actual Values	
		Positive (1)	Negative (0)
Predicted Values	Positive (1)	TP	FP
	Negative (0)	FN	TN

FIGURE 8.8 Simple Confusion Matrix.

2. F1 Score: This represents the balance between the precision and the recall. Mathematically it is represented as,

$$F1_{Score} = 2\frac{Precision * Recall}{Precision + Recall} \tag{8.4}$$

3. Recall – This is the fraction of the total amount of relevant instances that were actually retrieved. In other words, it tells us about the sensitivity. Besides, three different Oversampling techniques namely SMOTE, ADASYN, and Random Oversampling are used in order to evaluate models on the basis of the abovementioned evaluation metrics. Initially during the implementation of these oversampling techniques on individual models such as Logistic Regression and

Random Forest, the F1 scores obtained were in the range of **0.55–0.92** along with quite a high disparity in the false positives and false negatives of the confusion matrix, suggesting the need for an ensemble of prominent supervised learning algorithms in order to improve the overall performance parameters.

The analysis in this chapter includes the ensembles of two, three, four, and five classification algorithms and the discussions regarding improvement in the evaluation parameters by application of three different oversampling techniques.

8.4.2.1 Ensemble of Logistic Regression and Random Forest

A. With ADASYN –
Confusion matrix:

93818	20
24	125

FIGURE 8.9 Confusion Matrix.

This model detected 125 instances of credit card fraud. About 20 instances were false positive (FP) (i.e., predicted as Fraud Transaction) but they were actually Legitimate transactions. On the other hand, 24 (FN) were predicted as Non-Fraud transactions but they were fraudulent transactions as shown in Figure 8.9.

B. With SMOTE –
Confusion matrix:

91817	21
25	124

FIGURE 8.10 Confusion Matrix.

This model detected 124 instances of credit card fraud. About 21 (FP) predicted as Fraud Transaction but they were actually legitimate transactions. Moreover, 25 (FN) is predicted as Non-Fraud transactions but they were fraudulent transactions as shown in Figure 8.10.

C. With Random Oversampling –
Confusion matrix:

91818	20
23	126

FIGURE 8.11 Confusion Matrix.

TABLE 8.1
Classification Report of Ensemble of Two Models

	ADASYN	SMOTE	Random Oversampling
F1_Score	0.93	0.92	0.93
Recall	0.92	0.92	0.91

This model detected 126 instances of credit card fraud, and 20 (FP) predicted as Fraud transactions but they were actually Legitimate transactions. However, about 23 instances were FN i.e., predicted as Non-Fraud transactions but they were actually fraudulent transactions as shown in Figure 8.11.

As seen in Table 8.1, in all three models there was uneven class distribution. On comparing all three models Random Oversampling had the lowest number of false negative instances as compared to ADASYN and SMOTE. Also, the Recall score improved by 0.01 along with F1 score.

8.4.2.2 Ensemble of Logistic Regression and KNN

A. With ADASYN –
Confusion matrix:

93784	54
15	134

FIGURE 8.12 Confusion Matrix.

This model detected 134 instances of credit card fraud. 54 (FP) were predicted as Fraud transactions but were Legitimate transactions. On the other hand, 15 (FN) were predicted as Non-Fraud transactions but were actually fraudulent transactions as shown in Figure 8.12.

B. With SMOTE –
Confusion matrix:

93794	44
15	134

FIGURE 8.13 Confusion Matrix.

This model detected 134 instances of credit card fraud. 44 (FP) were predicted as Fraud transactions but in actuality they were Legitimate transactions. On the other hand, 15 (FN) were predicted as Non-Fraud transactions but were fraudulent transactions as shown in Figure 8.13.

C. With Random Oversampling –
Confusion matrix:

93797	41
15	134

FIGURE 8.14 Confusion Matrix.

This model detected 134 instances of credit card fraud. 41 (FP) were predicted as Fraud transactions were Legitimate transactions. On the other hand, 15 (FN) were predicted as Non-Fraud transactions but were fraudulent as shown in Figure 8.14.

TABLE 8.2
Classification Report of Ensemble of Two Models

	ADASYN	SMOTE	Random Oversampling
F1_Score	0.90	0.91	0.91
Recall	0.95	0.95	0.95

As seen in Table 8.2, all three models have uneven class distribution and have. On comparing all three models there are similar numbers of false negatives in Random Oversampling, ADASYN, and SMOTE. Also, the estimated F1-score has improved by 0.01 along with almost the same Recall in the case of Random Oversampling.

8.4.2.3 Ensemble of Logistic Regression, Random Forest, and KNN

A. With ADASYN –
Confusion matrix:

93784	54
13	134

FIGURE 8.15 Confusion Matrix.

This model detected 134 instances of credit card fraud. 54 (FP) were predicted as Fraud but were Legitimate transactions. On the other hand, 13(FN) were predicted as Non-Fraud transactions but were actually Fraud as shown in Figure 8.15.

B. With SMOTE –
Confusion matrix:

93794	44
15	134

FIGURE 8.16 Confusion Matrix.

This model detected 134 credit card frauds. 44 (FP) predicted as Fraud transactions were Legitimate. On the other hand 15(FN) were predicted as Non-Fraud transactions but were actually fraudulent transactions as shown in Figure 8.16.

C. With Random Oversampling –
Confusion matrix:

93823	15
21	128

FIGURE 8.17 Confusion Matrix.

This model detected 128 credit card fraud instances. 15 (FP) were predicted as Fraud transactions but in actuality were Legitimate transactions. On the other hand 21(FN) were predicted as Non-Fraud transactions but were Fraud transactions as shown in Figure 8.17.

TABLE 8.3
Classification Report of Ensemble of Three Models

	ADASYN	SMOTE	Random Oversampling
F1_Score	0.90	0.91	0.94
Recall	0.95	0.95	0.93

As can be seen in Table 8.3, all three models have uneven class distribution. On comparing all three models on average there are the least number of FN and FP in Random Oversampling as compared to ADASYN and SMOTE. Also, the F1 score improved by a factor of 0.03–0.04, along with an almost comparable Recall score in the case of Random Oversampling.

8.4.2.4 Ensemble of Logistic Regression, Random Forest, KNN, and SVM

A. With ADASYN –
Confusion matrix:

93775	63
18	131

FIGURE 8.18 Confusion Matrix.

This model detected 131 cases of credit card fraud. 63 (FP) were predicted as Fraud transactions but they were Legitimate. On the other hand, 18 (FN) were predicted as Non-Fraud transactions but were fraudulent as shown in Figure 8.18.

B. With SMOTE –
Confusion matrix:

93780	68
19	133

FIGURE 8.19 Confusion Matrix.

This model detected 133 instances of credit card fraud. 68 (FP) were predicted as Fraud transactions but were Legitimate transactions. On the other hand 19 (FN) were predicted as Non-Fraud transactions but were actually Fraud transactions as shown in Figure 8.19.

C. With Random Oversampling –
Confusion matrix:

93768	70
19	130

FIGURE 8.20 Confusion Matrix.

This model detected 130 cases of credit card fraud. 70 (FP) were predicted as Fraud but were actually Legitimate transactions. On the other hand 19 (FN) were predicted as Non-Fraud transactions but were Fraud as shown in Figure 8.20.

TABLE 8.4
Classification Report of Ensemble of Four Models

	ADASYN	SMOTE	Random Oversampling
F1_Score	0.88	0.88	0.87
Recall	0.94	0.91	0.94

As can be seen in Table 8.4, all three models have uneven class distribution. On comparing all three models on average there are a similar number of FN and FP in Random Oversampling, ADASYN, and SMOTE. Also, the F1 score is comparable in all three, with the most efficient Recall score in the case of Random Oversampling along with ADASYN.

8.4.2.5 Ensemble of Logistic Regression, Random Forest, KNN, and Naïve Bayes

A. With ADASYN –
Confusion matrix:

93773	65
20	129

FIGURE 8.21 Confusion Matrix.

This model detected 129 cases of credit card fraud. 65 (FP) were predicted as Fraud transactions but were Legitimate transactions. On the other hand, 20 (FN) were predicted as Non-Fraud transactions but were actually Fraud transactions as shown in Figure 8.21.

B. With SMOTE –
Confusion matrix:

93777	61
19	130

FIGURE 8.22 Confusion Matrix.

This model detected 130 cases of credit card fraud. 61 (FP) were predicted as Fraud transactions but were Legitimate transactions. On the other hand, 19 (FN) were predicted as Non-Fraud transactions but in actuality were Fraud as shown in Figure 8.22.

C. With Random Over Sampling –
Confusion matrix:

93778	70
19	130

FIGURE 8.23 Confusion Matrix.

This model detected 130 credit card fraud instances. 70 (FP) were predicted as Fraud transactions but were Legitimate transactions. On the other hand, 19 (FN) were predicted as Non-Fraud transactions but were actually Fraud transactions as shown in Figure 8.23.

TABLE 8.5
Classification Report of Ensemble of Four Models

	ADASYN	SMOTE	Random Oversampling
F1_Score	0.88	0.89	0.87
Recall	0.93	0.92	0.94

As can be seen in Table 8.5, all three models have uneven class distribution. On comparing all three models on average there are a similar number of FN and FP in Random Oversampling, ADASYN, and SMOTE. Also, the F1 score is comparable in all three, with the most efficient Recall score in the case of Random Oversampling along with ADASYN.

8.4.2.6 Ensemble of Logistic Regression, Random Forest, KNN, Naïve Bayes, and SVM

A. With ADASYN –
Confusion matrix:

93731	107
15	134

FIGURE 8.24 Confusion Matrix.

This model detected instances of 128 credit card fraud. 110 (FP) were predicted as Fraud transactions but were Legitimate transactions. On the other hand 21(FN) were predicted as Non-Fraud transactions but were Fraud as shown in Figure 8.24.

B. With SMOTE –
Confusion matrix:

93786	52
20	129

FIGURE 8.25 Confusion Matrix.

This model detected 134 cases of credit card fraud. 107 (FP) were predicted as Fraud but were actually Legitimate. On the other hand, 15 (FN) were predicted as Non-Fraud transactions but in actuality were Fraud as shown in Figure 8.25.

C. With Random Oversampling –
Confusion matrix:

93728	110
21	128

FIGURE 8.26 Confusion Matrix.

This model detected 129 instances of credit card fraud. 52 (FP) were predicted as Fraud transactions but were Legitimate transactions. On the other hand, 20 (FN) were predicted as Non-Fraud transactions but were actually Fraud as shown in Figure 8.26.

As can be seen in Table 8.6, all three models have uneven class distribution. On comparing all the three models, the least number of FN and FP is obtained in Random Oversampling as compared to ADASYN and SMOTE. Also, the F1 score is best for Random Oversampling, along with the most efficient Recall score in the case of SMOTE.

TABLE 8.6
Classification Report of Ensemble of Five Models

	ADASYN	SMOTE	Random Oversampling
F1_Score	0.83	0.84	0.89
Recall	0.91	0.95	0.93

After successfully analyzing these ensembles, it was observed that the best possible model parameters were obtained from the Ensemble of Logistic Regression, Random Forest, and KNN using the Random Oversampling technique for balancing the dataset.

The complete classification report of the same is given in Figure 8.28:

Confusion Matrix:

93823	15
21	128

FIGURE 8.27 Confusion Matrix.

TABLE 8.7
Classification Report of Ensemble of Final Model

	Precision	Recall	F1_Score	Support
0	1.00	1.00	1.00	93838
1	0.90	0.86	0.88	149
Avg/Total	0.95	0.93	0.94	93987

Finally in Table 8.7, on comparing all the ensemble models implemented on the dataset it was observed that the presented model has the best F1 score (i.e., 0.94) and has lowest values of False Positives and False Negatives.

A comparison of the ensembles of three, four, and five models are depicted with the help of ROC-AUC curve as shown in Figure 8.28.

8.5 CONCLUSION

After the successful implementation of ensembles of all the possible classifiers along with different oversampling methods, it can be concluded that the ensemble of Logistic Regression, Random Forest and KNN gives the best possible F1 score and recall values along with the least values of False Positives and False Negatives in the confusion matrix. Although the analysis was further extended with ensembles of four and five models the scores of the evaluation metrics would improve as such, which is clearly depicted by the experimental values that we obtained after implementation of these ensembles.

For future work, methods like Generative Adversarial Network (GAN) can be worked on along with online learning models. Also, the European dataset that used here has already reached its saturation of performance improvement by most of the existing models that have been implemented to date. So availability of a new balanced

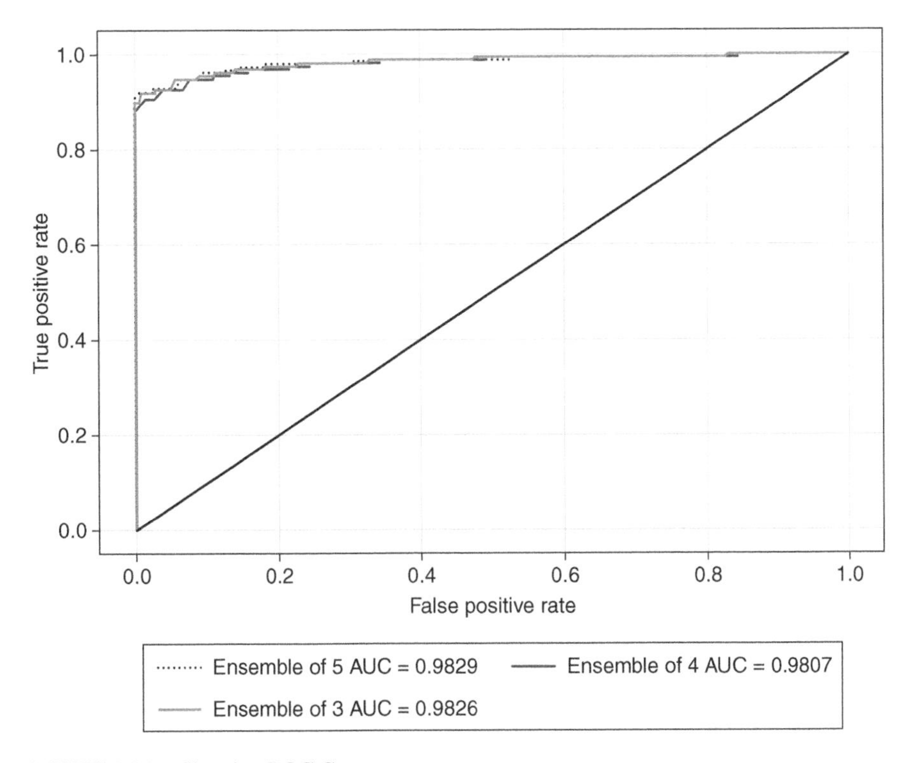

FIGURE 8.28 Showing ROC Curve.

dataset would be really fruitful in order to improve the analysis and efficiency of machine learning models on the problem of credit card fraud, which will in turn help in reducing the number of losses being incurred by financial sectors daily.

ACKNOWLEDGEMENTS

We would like to acknowledge editors for providing the opportunity to present our work.

Funding Statement: The author(s) received no specific funding for this study.

Conflicts of Interest: The authors declare that they have no conflicts of interest to report regarding the present study.

REFERENCES

1. Pankaj Kumar Jadwal, Sonal Jain, Sunil Pathak, Basant Agarwal, "Improved resampling algorithm through a modified oversampling approach based on spectral clustering and SMOTE", In Microsystem Technologies. 2022. doi:https://doi.org/10.1007/s00542-022-05287-8.

2. Pankaj Kumar Jadwal, Sonal Jain, Basant Agarwal, "Financial credit risk evaluation model using machine learning based approach", In World Review of Entrepreneurship, Management and Sustainable Development, Inderscience, 2020.
3. Pankaj Kumar Jadwal, Sonal Jain, Basant Agarwal, "Clustering based hybrid resampling techniques for social lending data", In International Journal of Intelligent Systems Technologies and Applications, Inderscience, 2020.
4. Pankaj Kumar Jadwal, Sonal Jain, Basant Agarwal, "Spectral Clustering and Cost-Sensitive Deep Neural Network-Based Undersampling Approach for P2P Lending Data", International Journal of Information Technology and Web Engineering (IJITWE), 15(4), 37–52, 2020.
5. Varmedja D., Karanovic M., Sladojevic S., Arsenovic M., Anderla A.(2019). "Credit Card Fraud Detection - Machine Learning methods", *International Conference on 18th International Symposium*, Infotech Jahorin, Bosnia, pp . 20–22.
6. M. Safa U., R. Ganga M. (2019). "Credit Card Fraud Detection Using Machine Learning", *International Journal of Research in Engineering, Science and Management*, Volume 2, Issue 11.
7. Bagga S., Goyal A., Gupta N., Goyal A. (2020). "Credit Card Fraud Detection using Pipeling and Ensemble Learning", *International Conference on Smart Sustainable Intelligent Computing and Applications (ICITETM)*, Volume 173.
8. Kumar M. S., Soundarya V., Kavitha S., Keerthika E. S., Aswini E. (2019). "Credit Card Fraud Detection Using Random Forest Algorithm", *3rd International Conference on Computing and Communications Technologies (ICCCT)*, Chennai, India, 149–153.
9. Priscilla C.V., Prabha D.P. (2020). "Credit Card Fraud Detection: A Systematic Review", *Intelligent Computing Paradigm and Cutting-edge Technologies (ICICCT), Learning and Analytics in Intelligent Systems*, vol 8.
10. Pandey Y.(2017). "Credit Card Fraud Detection using Deep Learning", *International Journal of Advanced Research in Computer Science*, 8 (5), 981–984.
11. Sohony I., Pratap R., Nambiar U. (2018). "Ensemble Learning for Credit Card Fraud Detection", *ACM India Joint International Conference on Data Science & Management of Data*, Goa India, 289–294.
12. Agarwal, B., Agarwal, A., Harjule, P., & Rahman, A. (2022). Understanding the intent behind sharing misinformation on social media. Journal of Experimental and Theoretical Artificial Intelligence, 1–15. https://doi.org/10.1080/0952813X.2021.1960637
13. Agarwal, B., Rahman, A., Patnaik, S., & Poonia, R. C. (Eds.) (2022). Proceedings of International Conference on Intelligent Cyber-Physical Systems. (Algorithms for Intelligent Systems). Springer. www.springer.com/gp/book/9789811671357
14. Chowdhury, M.M.H., Rahman, A., & Islam, M. R. (2018). Protecting data from malware threats using machine learning technique. In Proceedings of the 2017 12th IEEE Conference on Industrial Electronics and Applications (ICIEA) (pp. 1691–1694). IEEE, Institute of Electrical and Electronics Engineers. https://doi.org/10.1109/ICIEA.2017.8283111
15. Rahman, A. (2020). Statistics for Data Science and Policy Analysis. Springer.
16. Agarwal, B. Financial sentiment analysis model utilizing knowledge-base and domain-specific representation. Multimed Tools Appl (2022). https://doi.org/10.1007/s11042-022-12181-y.
17. Uddin, M. G., Nash, S., Rahman, A., & Olbert, A. I. (2023). Performance analysis of the water quality index model for predicting water state using machine learning techniques. Process Safety and Environmental Protection, 1–30. https://doi.org/10.1016/j.psep.2022.11.073

18. Rahman, A., & Harding, A. (2017). Small area estimation and microsimulation modeling. CRC Press. https://doi.org/10.1201/9781315372143
19. Chowdhury, M.M.H., Rahman, A., & Islam, M. R. (2018). Protecting data from malware threats using machine learning technique. In Proceedings of the 2017 12th IEEE Conference on Industrial Electronics and Applications (ICIEA) (pp. 1691–1694). IEEE, Institute of Electrical and Electronics Engineers. https://doi.org/10.1109/ICIEA.2017.8283111
20. Rahman, A., & Upadhyay, S. K. (2015). A Bayesian reweighting technique for small area estimation. In U. Singh, A. Loganathan, S. K. Upadhyay, & D. K. Dey (Eds.), Current trends in Bayesian methodology with applications (1st ed., pp. 503–519). CRC Press.
21. Uddin, M. G., Nash, S., Rahman, A., & Olbert, A. I. (2023). A novel approach for estimating and predicting uncertainty in water quality index model using machine learning approaches. Water Research, 1–24. https://doi.org/10.1016/j.watres.2022.119422
22. Chowdhury, M., Rahman, A., & Islam, R. (2018). Malware analysis and detection using data mining and machine learning classification. In J. Abawajy, K-K. R. Choo, & R. Islam (Eds.), International Conference on Applications and Techniques in Cyber Security and Intelligence: Applications and Techniques in Cyber Security and Intelligence (Vol. 580, pp. 266–274). (Advances in Intelligent Systems and Computing; Vol. 580). Springer. https://doi.org/10.1007/978-3-319-67071-3_33
23. Uddin, M. G., Nash, S., Mahammad Diganta, M. T., Rahman, A., & Olbert, A. I. (2022). Robust machine learning algorithms for predicting coastal water quality index. Journal of Environmental Management, 321(11), 1–16. [115923]. https://doi.org/10.1016/j.jenvman.2022.115923
24. I. Trivedi, Monika, M. Mridushi,(2016). "Credit Card Fraud Detection", *International Journal of Advanced Research in Computer and Communication Engineering*, 5, Issue 1.
25. Sadgali I., Sael N. and Benabbou F. (2019). "Fraud detection in credit card transaction using machine learning techniques", *1st International Conference on Smart Systems and Data Science (ICSSD)*, Rabat, Morocco, pp. 1–4.
26. Novakovic J., Markovic S. (2020). "Classifier Ensembles for Credit Card Fraud Detection", *24th International Conference on Information Technology* (IT), Zabljak, Montenegro, 1–4.
27. Muttipati A.S., Viswanadham, S., Senapathi, R. and Rao, K.B. (2021). "Recognizing credit card fraud using machine learning methods", *Turkish Journal of Computer and Mathematics Education*, Vol. 12(12), 3271–3278.
28. Oualid A., Hansali A., Balouki Y., & Moumoun, L. (2021, November). "Application of Machine Learning Techniques for Credit Risk Management: A Survey", In *The International Conference on Information, Communication & Cybersecurity*, Springer, Cham, 180–191.
29. Singh A., Ranjan R. K., & Tiwari A. (2021). "Credit Card Fraud Detection under Extreme Imbalanced Data: A Comparative Study of Data-level Algorithms", *Journal of Experimental & Theoretical Artificial Intelligence*, 1–28.
30. Rahman, A., Nimmy, S. F., & Sarowar, G. (2019). Developing an automated machine learning approach to test discontinuity in DNA for detecting tuberculosis. In J. P. Davim (Ed.), *Proceedings of the Twelfth International Conference on Management Science and Engineering Management* (pp. 277–286). Springer. https://doi.org/10.1007/978-3-319-93351-1_23

9 Combining News with Time Series for Stock Trend Prediction

Ashish Sharma[*1], *Shashwat Singh*[2], *Basant Agarwal*[3], *Vinita Tiwari*[4], and *Niraja Saraswat*[5]
[1]Manipal University Jaipur, India
[2]Columbia University, USA
[3]Central University Rajasthan, India
[4]Indian Institute of Information Technology Kota, India
[5]Dept. of Humanities & Social Science, MNIT, Jaipur, India
*Email: ashishsharma.fitt@gmail.com

CONTENTS

9.1 INTRODUCTION

Stocks represent a kind of ownership or partial ownership of a company. Once you purchase the stocks of a company you become the company's shareholder. You are also eligible to drive profits from it and sell the stocks whenever you see it fit. One can always buy and sell stocks anytime he/she wants. But it is always desired to get maximum profit from the investment. Therefore, it is important to know the trend of the stock you purchased.

DOI: 10.1201/9781003253051-12

Various risks revolve around stocks such as the condition of the economy, inflation, market value, bond, and liquidity risk. The prediction of stock price has been performed by various researchers, with the aim of finding the trend of stocks, since it is more important to get the trend of a stock than its price.

Stock price predictors have minimal bias, along with a way of cross-verifying the results and considering a variety of factors. The models assure consistency, and can even eliminate the need of experts and financial advisors, thus bringing down costs.

Machine learning has been out there since the 1970s but was not commonly used until recently due to low performance and speed of computers. However, advances in technology have paved the way for the use of machine learning techniques in a variety of fields [1–8].

In our experiment we will be converting news samples into numeric vectors using word2vec. We will also use an ensembling technique to combine a model generated from news analysis with a model generated from time series analysis. Ensembling is combining two or more models into one to make weak models grow stronger such that they can give accurate results. Also, models are trained separately with their separate parameters. The final prediction is made by the ensembled model.

We believe that an ensemble model can help us to get better results than before.

9.2 RELATED WORK

In Ref. [9] the author used forward search as a feature selection algorithm along with Logistic Regression and Random Forest Regression to determine important 253 features. Further SVM, Logistic Regression, and Random Forest were used to predict trends. However, this approach was based on classifying methodology and lacked major details in terms of the dataset. Khan et al. [10] used PCA (Principal Component Analysis) to perform Feature Reduction. Then classifiers were applied. The authors performed two-fold cross validation over training data to find out the optimal parameters. In Ref. [11] the authors used four machine learning classifiers and ensembled them together to make one powerful model. They achieved 70% accuracy on testing data. For classifiers they used SVM, Relevance Vector Machine, Random Forest, and K-Nearest Neighbours. Chou et al. [12] did time series analysis and used a hybrid model consisting of a metaheuristic Firefly algorithm (MetaFA) and Least Square Support Vector Regression (LSSVR). In addition there exists another domain-specific limitation where the models failed to achieve good results for long-term investments. One of the major limitations of these algorithms is that they are extremely computationally expensive and are thus extremely slow even with today's computers.

Shen et al. [13] included data of correlated markets along with various commodities such as gold, silver, platinum, etc., to monitor the economy and take into account changes indirectly governed by the financial status of the country. They used Linear Regression, Generalized Linear Model (GLM), and SVM to predict outcomes. In Ref. [14] the author used AdaBoost as an ensemble approach for combining SVM and Na¨ıve Bayes.

Tang et al. [15] used the Moving Average Algorithm to predict the trend from time series and SVM to predict the trend from text mining.

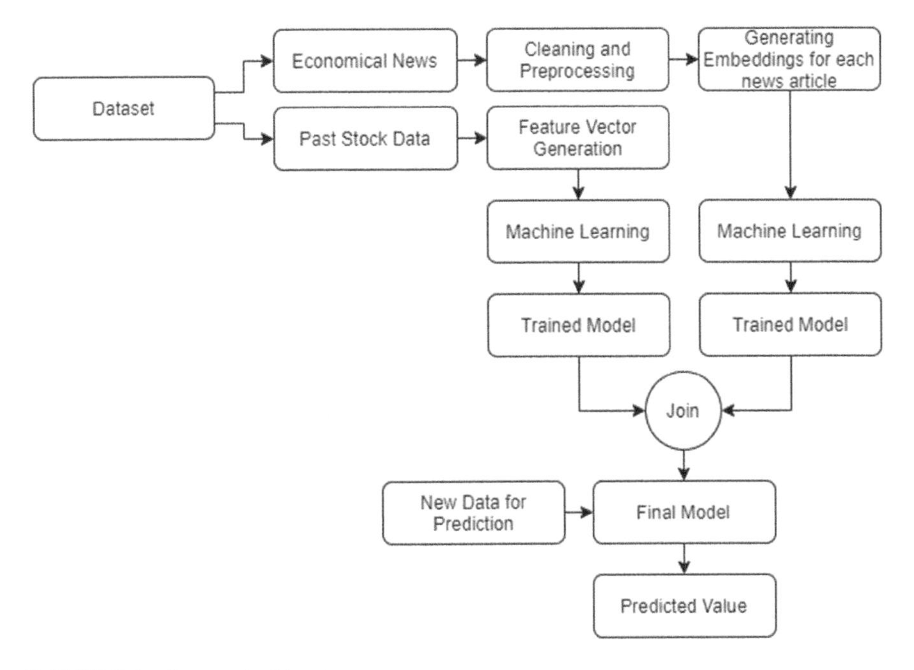

FIGURE 9.1 System Architecture.

While the above studies were helpful, they all lack accuracy, even though they took into account the time series analysis. We also noticed that a wide array of algorithms was not taken into consideration. Therefore, we decided to test the data for a wide range of algorithms.

News from all spheres of life has some role in deciding the fate of stocks. Recently, papers that came in with combining these both had a lot of feature vectors due to which time taken by models were fairly large. Recent papers had a lot of input features creating high dimensional input space, which is both computationally expensive and results in a non-robust model. Thus, we tried to select optimal input features using multiple feature selection algorithms. Figure 9.1 describes the architecture of our system.

9.3 METHODOLOGY

Our system has three parts: Time Series Prediction, Text Mining and Prediction, and Ensembling Prediction models.

9.3.1 Time Series Prediction

In time series prediction we performed some pre-processing of both training and testing data. This includes interpolation for missing values in data, the addition of new features in data, and normalization to bring entire data in the same range. The data consisted of date (Date of entry), open (Opening value), high (Highest value), low

(Lowest value), close (Closing value), adj close (Adjusted close value), and volume (No. of trades in the day). The newly added features were prev day diff (Difference in close values of today and yesterday), 50 day moving avg (The avg. close value for last 50 days), and 10 day volatility (the standard deviation of close value for last 10 days). We took the date column as the index column and not as a feature. Our objective was to predict the actual close values of the stock, thus used regression algorithms. Before the machine learning algorithms were run on this feature vector, we applied a feature selection algorithm to find the six most useful features for prediction. We applied feature selection on the data as it enabled the machine learning algorithm to train faster. If the right subset of data is chosen the accuracy of a model improves. We applied Recursive Feature Elimination (RFE) as the feature selection algorithm and generated the feature vectors for each machine learning algorithm. The machine learning algorithms used were Decision Trees, Gaussian Process, Gradient Boost, K Neighbor, Linear Regression, Random Forest, and SVR (Support Vector Regression). For hyperparameter tuning we used Grid Search with algorithms. Grid search makes combinations of values of parameters and trains and tests the models. This process is time-consuming because it creates different models for a different combinations of parameters and tests them. To speed up this process we distributed the computation to all the cores present on the machine's processor. Instead of doing a random split on data for training and validation we used time series split, which maintains data in order of time. This helps in training ordered sets of data and not from random data. After getting the best parameters from the grid search, we trained the model with those parameters and predicted the close values for test data, after which we calculated the Mean Absolute Error (MAE) of the model from the test data. For each model a graph was made comparing the actual values of the stock and predicted values of the stock. Also, each model was saved to avoid building models again in the future.

9.3.2 Text Mining and Prediction

In the text mining phase we performed the cleaning of textual data. Cleaning is an important phase in text mining. It helps in improving the performance of a model. Our text cleaning includes converting the text to lowercase, changing abbreviations to their full forms removing unnecessary characters from text, and removing words that have little or no meaning (e.g., a, and, the, etc.). This data cleaning process is done on both training and testing data. Next, we found the word embeddings for the textual data in both train and test sets. Embeddings are numerical representations of words. For this we used the Word2vec approach to tokenize the words from the sentences. This is also a time-consuming process. It converts the text to a 300 feature vector upon which we applied machine learning algorithms. The machine learning algorithms used were Decision Trees, Gaussian Process, Gradient Boost, K Neighbor, Linear Regression, Random Forest, and SVR (Support Vector Regression). We again used Grid Search with algorithms for hyperparameter tuning. To speed up the process we distributed the computation to all the cores present on the machine's processor. Also, we again used time series split to train from ordered sets of data and not from random data. After getting the best parameters from the grid search, we

TABLE 9.1
MAE for Time Series Prediction Models and Text Mining and Prediction

Models	Time Series Prediction (MAE)	Text Mining and Prediction (MAE)
Decision Tree	2.508177	33.198084
Gaussian Process	0.705862	29.869217
Gradient Boost	29.70073	29.324229
K Neighbour	8.921798	28.639458
Linear Regression	0.736712	27.558388
Random Forest	19.89880	29.353095
SVR	1.185542	26.298480

trained the model with those parameters and predicted the close values for test data, after which we calculated the Mean Absolute Error (MAE) of the model from the test data (Table 9.1). For each model a graph was made comparing the actual values of stock and predicted values of the stock. Also, each model was saved to avoid building models again in the future.

9.3.3 Ensembling Prediction Models

In the ensembling phase, we combined prediction models from time series prediction and prediction from online news articles. The ensembling was done using a weighted average according to this formula:

$$FinalPrediction = \frac{Prediction_{timeseries} * Error_{news} + Prediction_{news} * Error_{timeseries}}{Error_{timeseries} + Error_{news}} \tag{9.1}$$

We ensembled the corresponding models from time series prediction phase and prediction from text phase (i.e., SVM with SVM, Decision Tree with Decision Tree, etc.). Along with this we ensembled the model giving least mean absolute error in time series prediction with the model giving least mean absolute error in prediction from the text. For each ensembled model we calculated the predictions using the above equation. After this we determined the MAE of the ensembled models (Table 9.2) and also plotted the graphs for each ensemble. Then we performed Mann-Whitney U on the ensembled model to compare the ensembled predictions with the original prices in the test data.

TABLE 9.2
MAE for Ensembled Models

Models	Combined Prediction (MAE)	Mann Whitney U Test (p – value)
Decision Tree	4.220013	0.016168
Gaussian Process	0.642904	0.399295
Gradient Boost	28.24038	3.995903e-49
K Neighbor	12.90226	5.155562e-12
Linear Regression	0.672736	0.433741
Random Forest	23.29260	2.159102e-45
SVR	1.762893	0.181592
Time series model: Gaussian Process Text model: SVR	0.647215	0.408679

9.4 EXPERIMENT AND RESULTS

We collected a data set that includes past stock of Bank of Baroda [16] and news related to the money and banking sector [17]. This data spanned November 2015 to November 2019, which we divided into training and test data sets. The data of three years was taken as a training data set and data of the last year was taken as a test data set.

Note: In all the figures (Figures 9.2–9.21), the Red Line is the Original Price while the Blue Line is the Predicted Price.

9.4.1 Inference from Graphs

Graphs of algorithms like Linear Regression, Gaussian Process in time series, and SVR in text mining and analysis phases give the best results. Gaussian Process uses probability densities and is a good algorithm for time series problems. It was only after feature selection that linear regression could perform this well. Feature selection is therefore an important phase in improving the predictions of an algorithm. Ensembled models of these algorithms have also performed well.

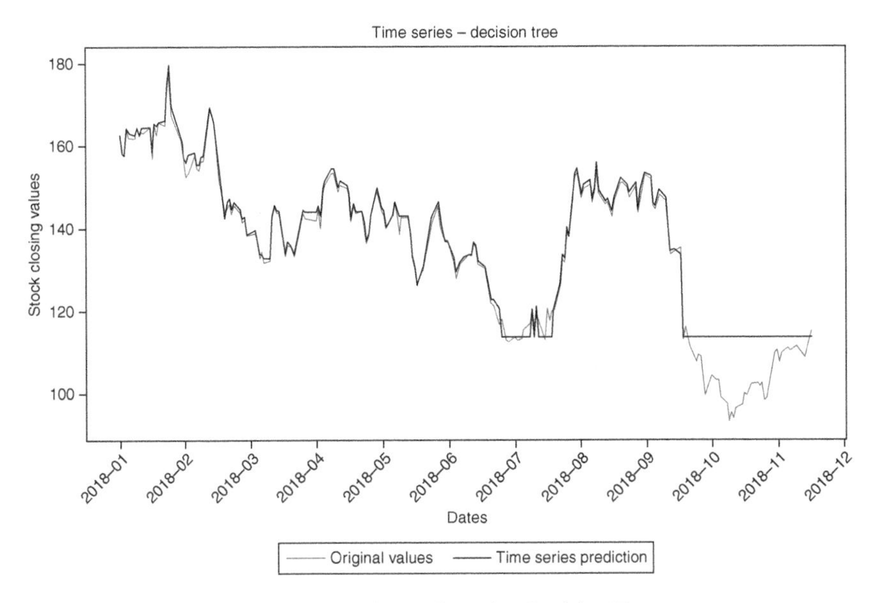

FIGURE 9.2 Prediction of the stock close price using Decision Tree.

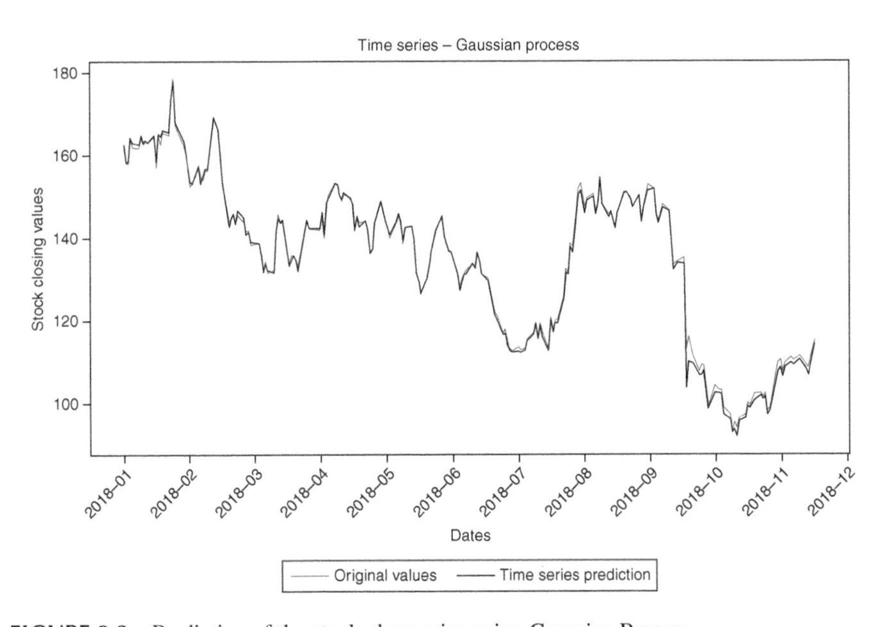

FIGURE 9.3 Prediction of the stock close price using Gaussian Process.

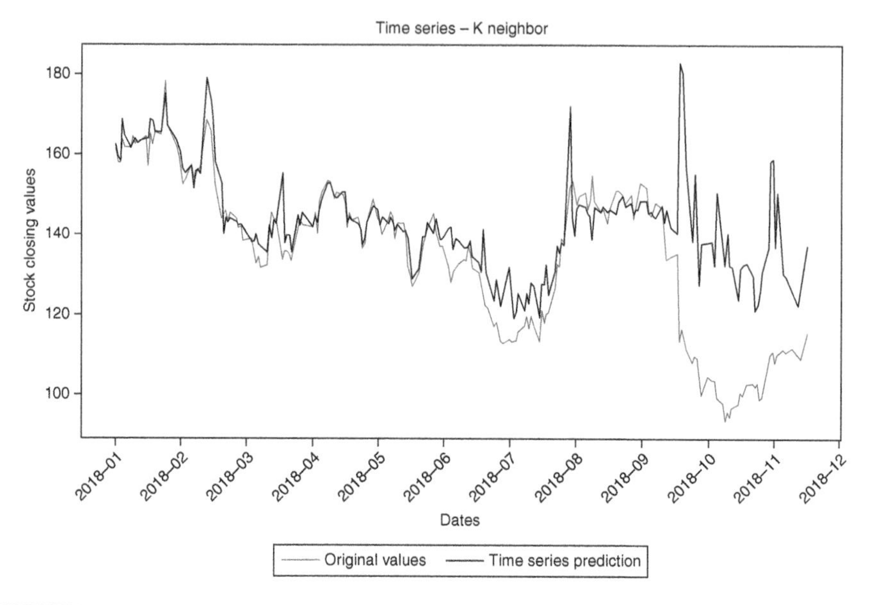

FIGURE 9.4 Prediction of the stock close price using K Neighbour.

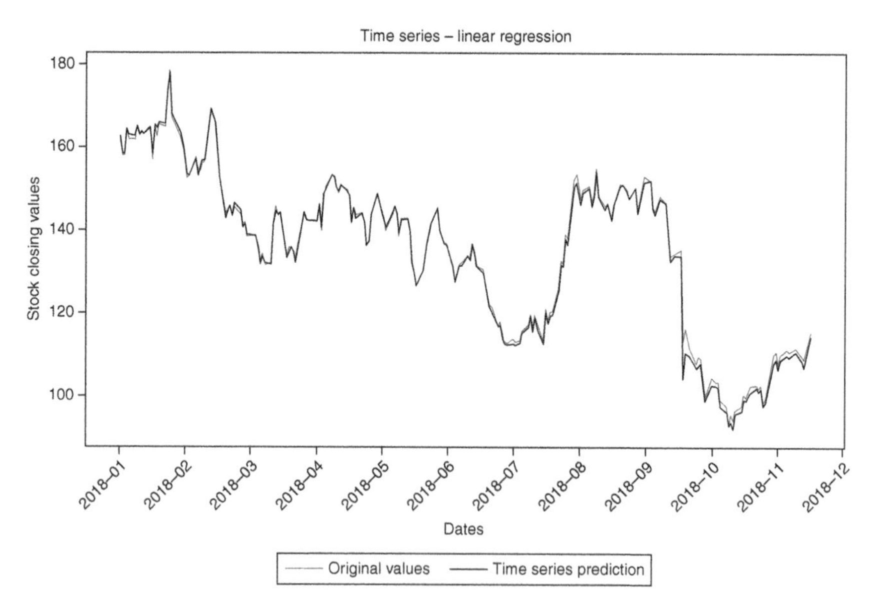

FIGURE 9.5 Prediction of the stock close price using Linear Regression.

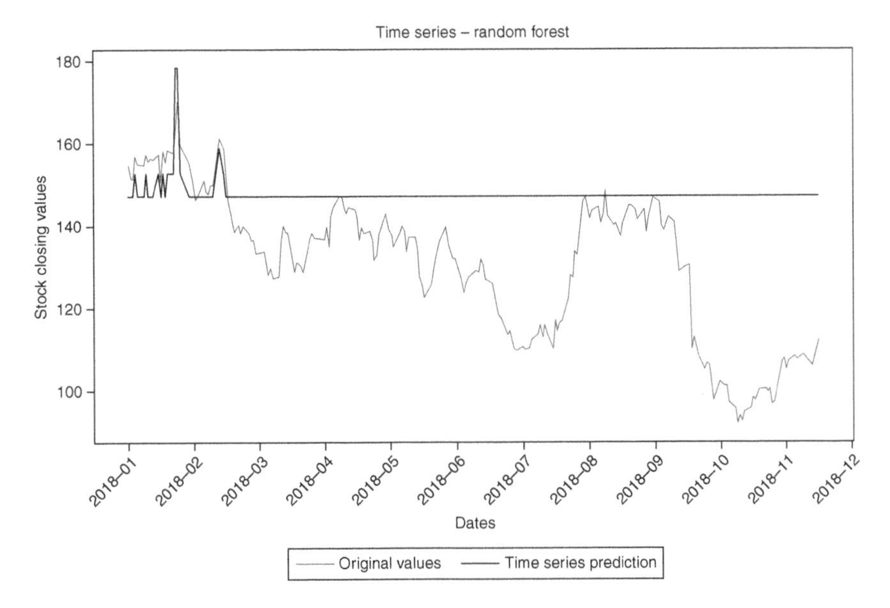

FIGURE 9.6 Prediction of the stock close price using Random Fores.

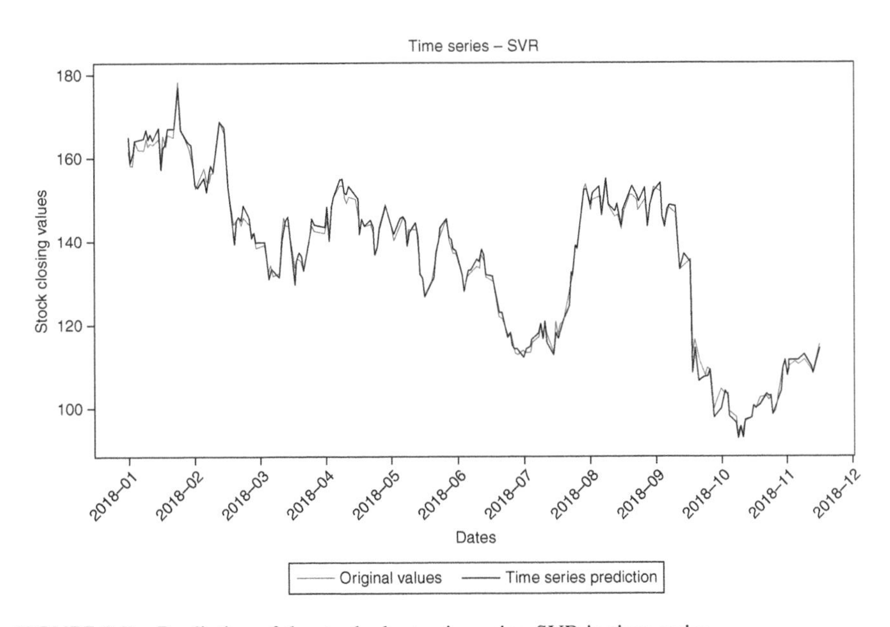

FIGURE 9.7 Prediction of the stock close price using SVR in time series.

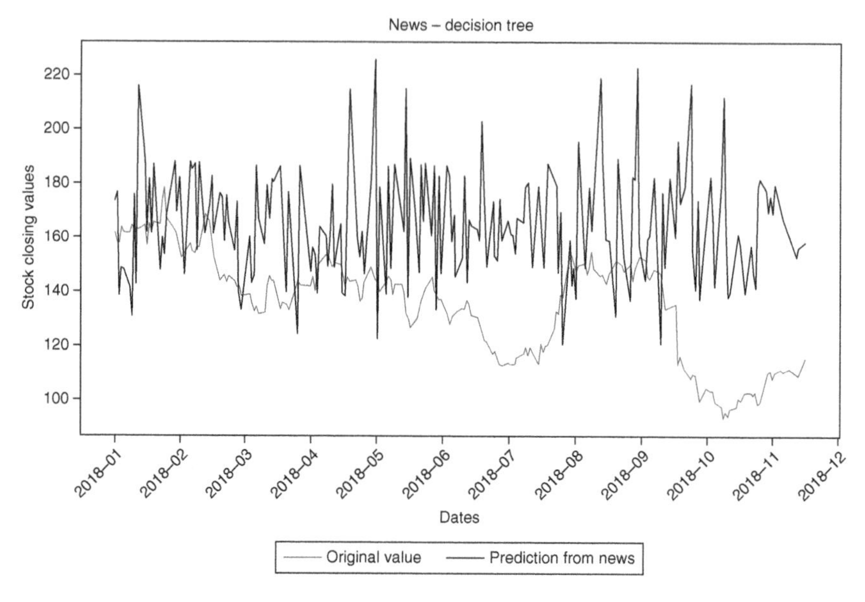

FIGURE 9.8 Prediction of the stock close price using Decision Tree with News.

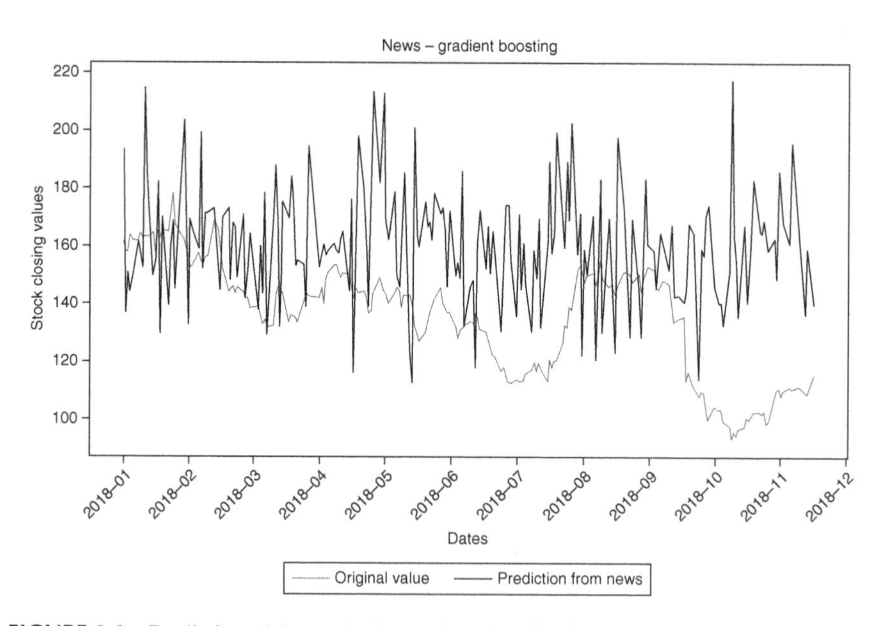

FIGURE 9.9 Prediction of the stock close price using Gradient Boosting with News.

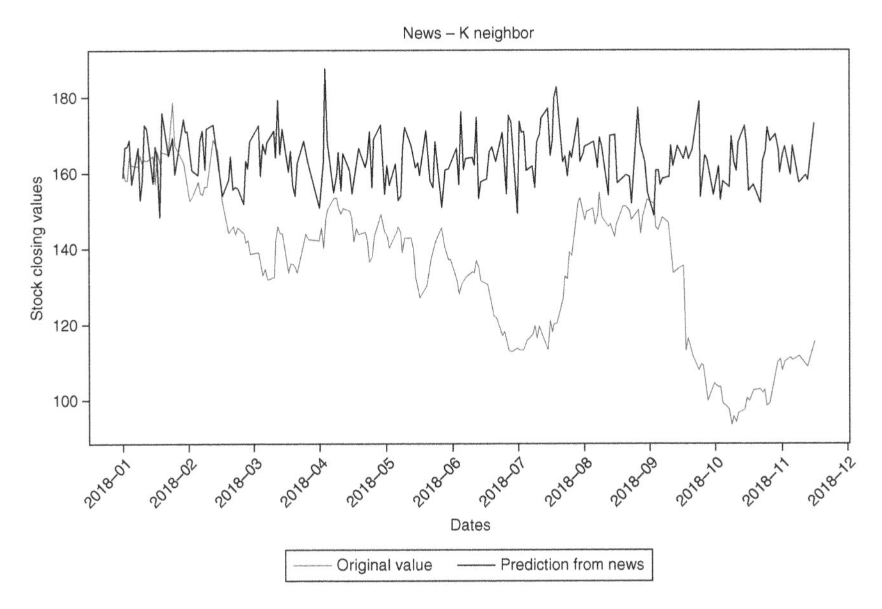

FIGURE 9.10 Prediction of the stock close price using K Neighbour with News.

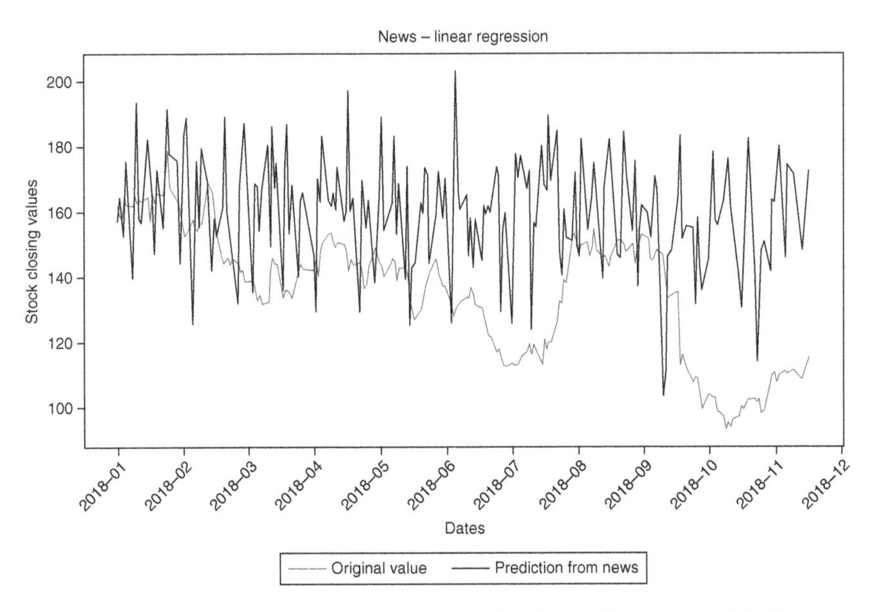

FIGURE 9.11 Prediction of the stock close price using Linear Regression with News.

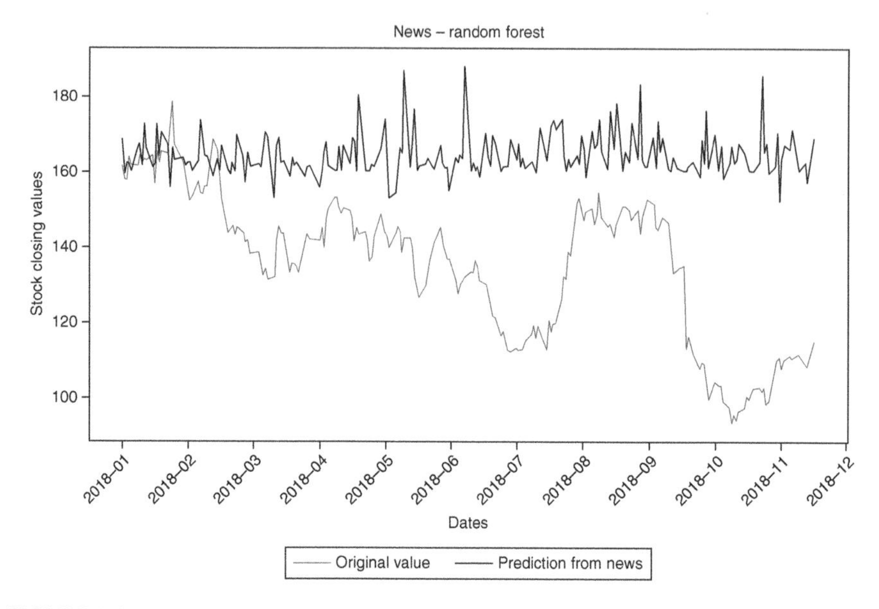

FIGURE 9.12 Prediction of the stock close price using Random Forest with News.

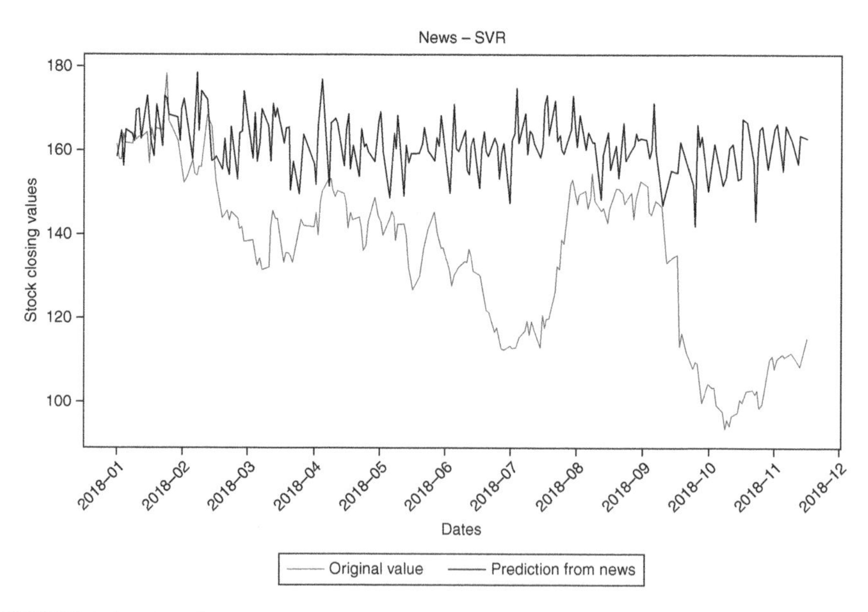

FIGURE 9.13 Prediction of the stock close price using SVR in prediction using news.

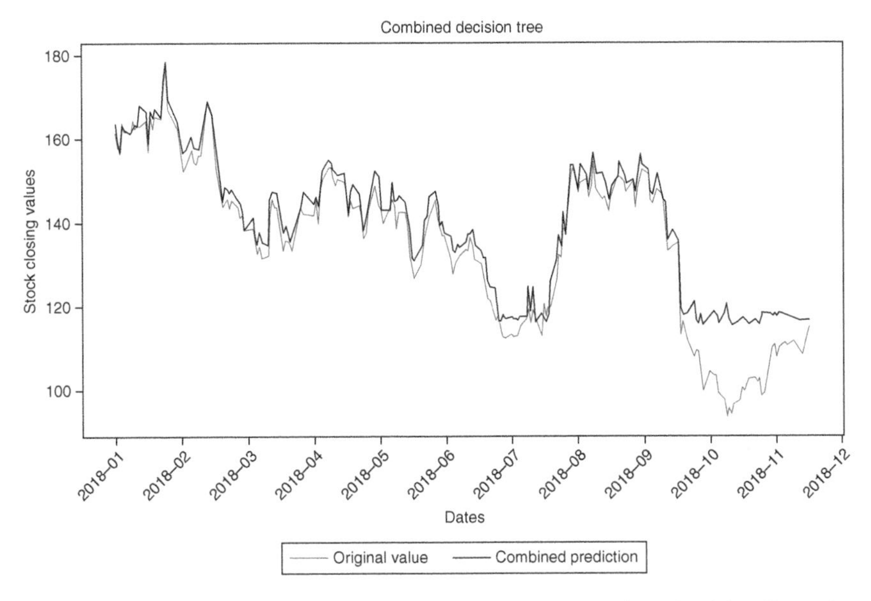

FIGURE 9.14 Prediction of the stock close price using Ensembled Decision Tree of text analysis.

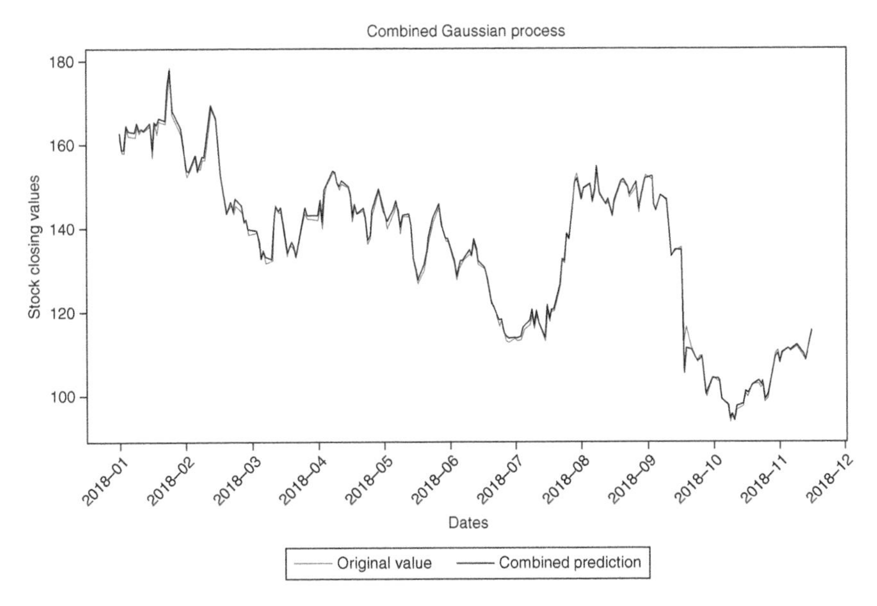

FIGURE 9.15 Prediction of the stock close price using Ensembled Gaussian Process of text analysis.

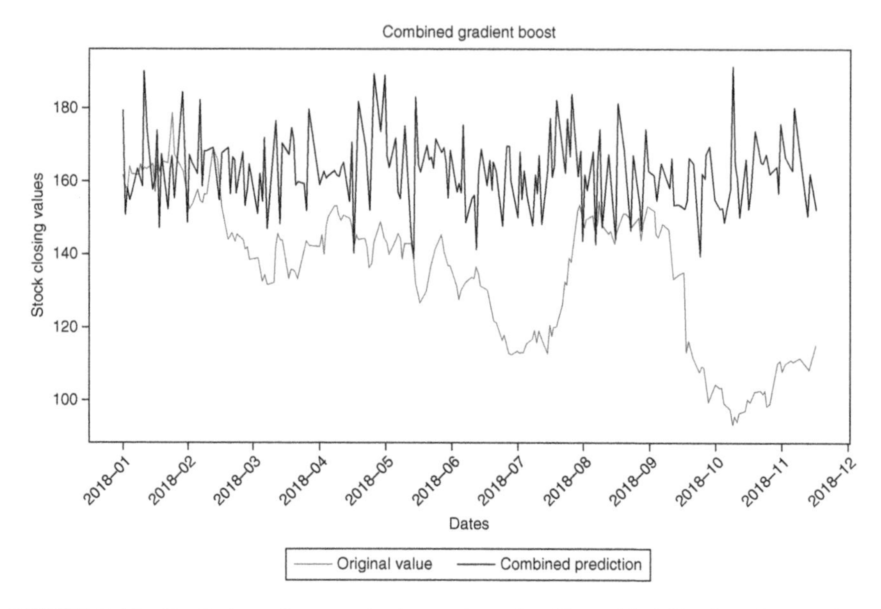

FIGURE 9.16 Prediction of the stock close price using Ensembled Gradient Boost of text analysis.

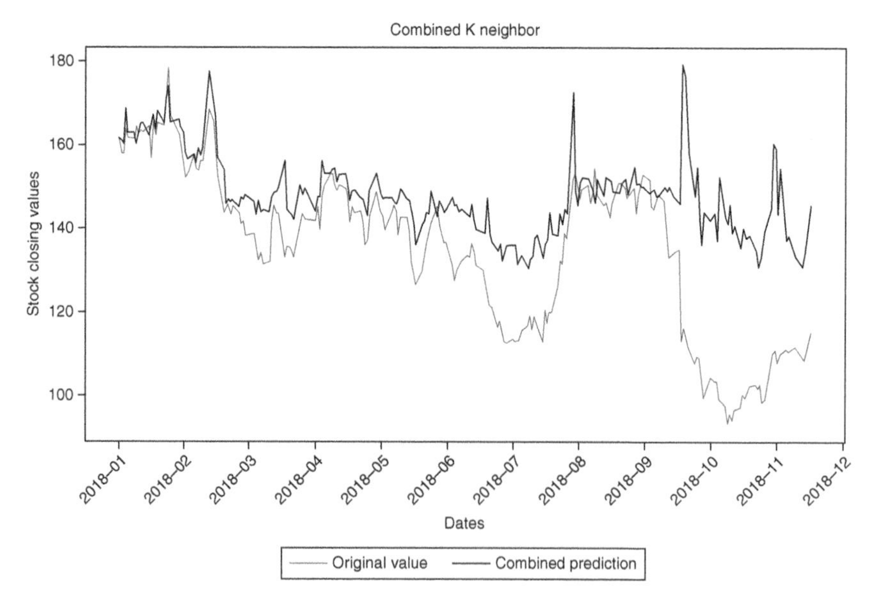

FIGURE 9.17 Prediction of the stock close price using Ensembled K Neighbour of text analysis.

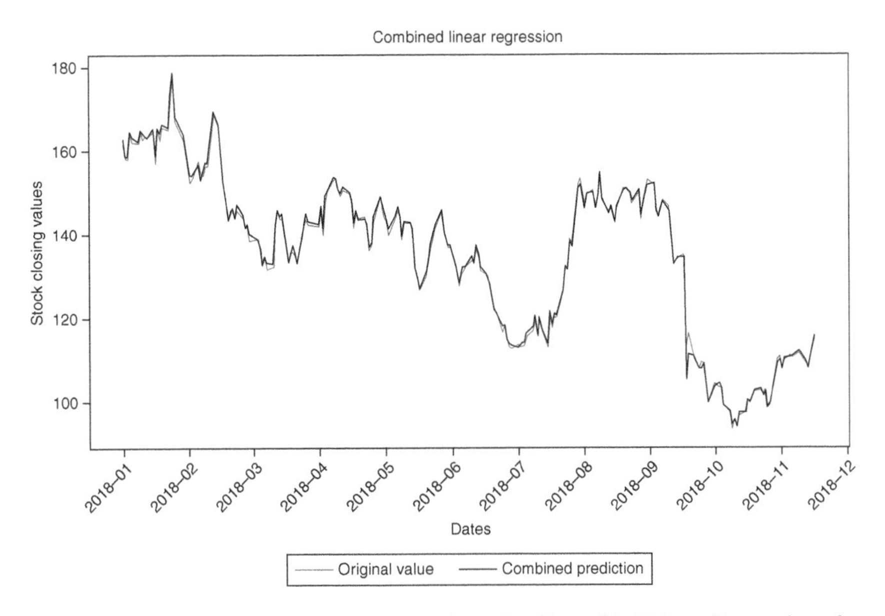

FIGURE 9.18 Prediction of the stock close price using Ensembled Linear Regression of text analysis.

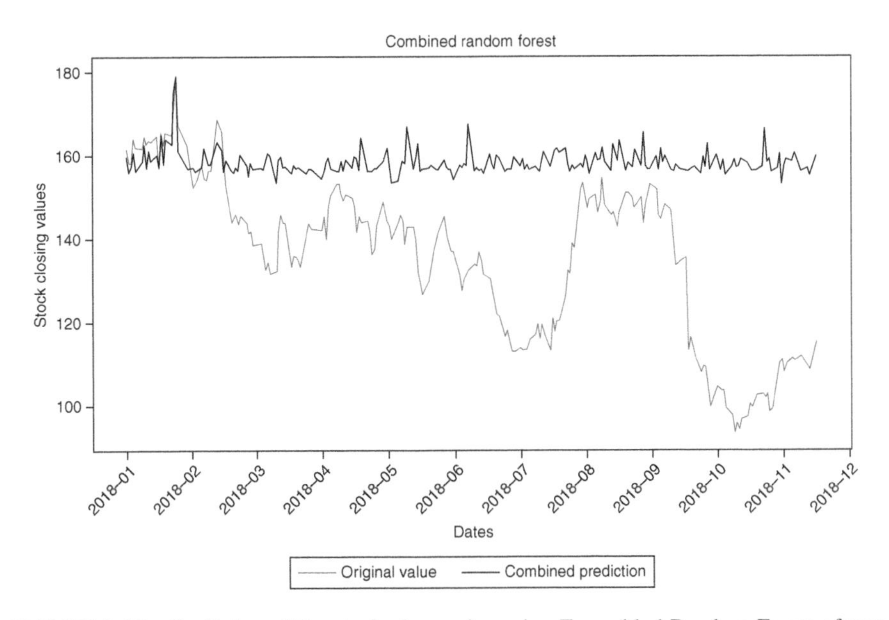

FIGURE 9.19 Prediction of the stock close price using Ensembled Random Forest of text analysis.

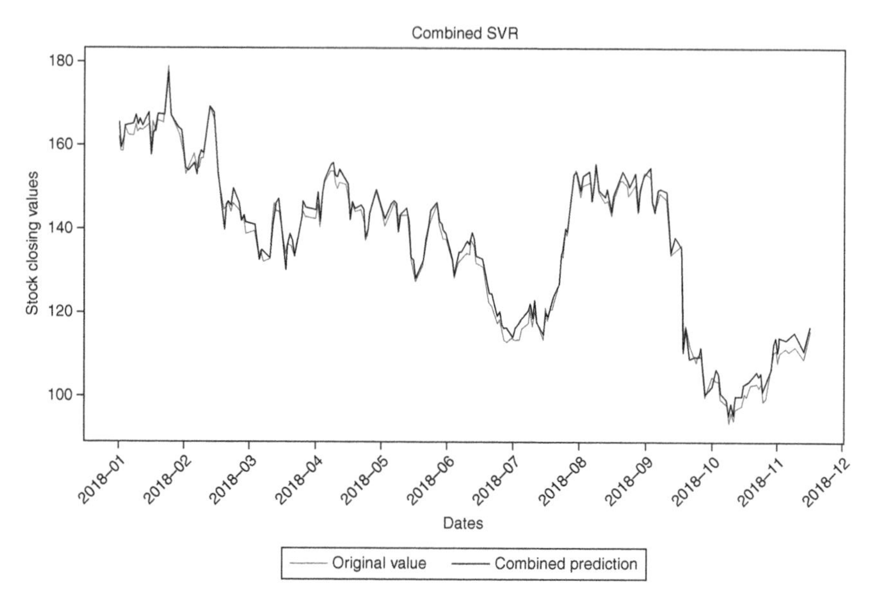

FIGURE 9.20 Prediction of the stock close price using Ensembled SVR of text analysis.

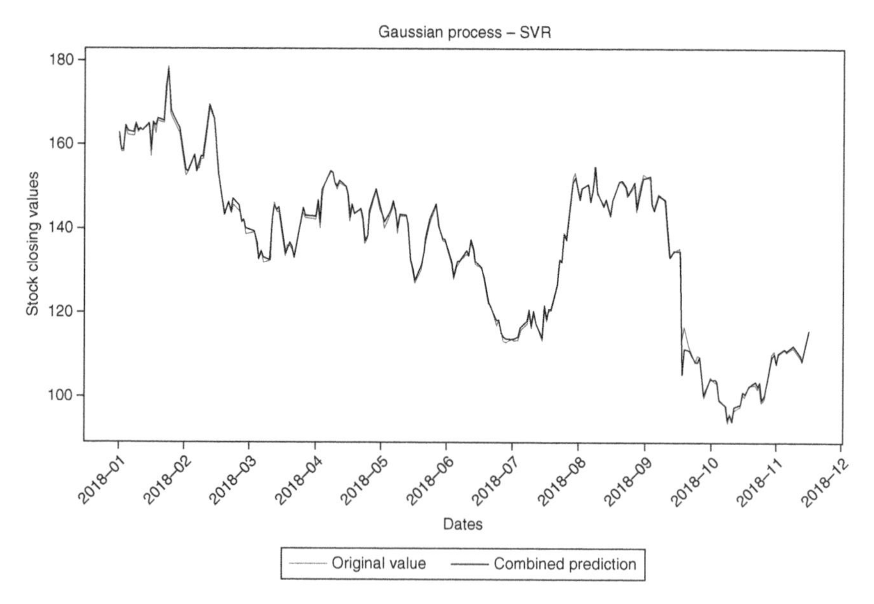

FIGURE 9.21 Prediction of the stock close price using Ensembled Gaussian Process model of time series and SVR model of text analysis.

9.5 CONCLUSION

In this project we used a combined approach to forecast the stock price of Bank of Baroda. We combined text mining and time series forecasting approaches. For the textual data, we used the online economic news articles of most of the stocks present in the exchange and their respective companies. Prediction of the stock price using time series prediction or text mining analysis and prediction independently is not able to predict with complete accuracy. From the results it can be concluded that combined models can decrease error. Also, the combined models can imitate the trends of the stock market better than the independent time series and text mining and analysis models.

Mann Whitney U test shows to what degree two graphs are similar in trends. Values near 0.5 mean that graphs are very similar and values away from 0.5 show that the graphs are quite different. Ensembled graphs of linear regression and that of the Gaussian process in time series and SVR outcomes have high degrees of similarities. This is because the individual predictive models performed better as compared to the other models.

9.6 FUTURE WORK

We have not used approaches like deep learning that may improve the individual correctness of prediction of stock trends from the news. We could also increase the size of the data set in order to be able to train our model with more data than that of 3 years. Also, the addition of some more useful features in the data set like news related to other sectors that have an influence on the banking sector may improve the predictions. So, our future work includes applying a deep learning approach, the addition of more features in the data set, and finding a better formula to be used in ensembling for predicting the final value.

REFERENCES

[1] Uddin, M. G., Nash, S., Rahman, A., & Olbert, A. I. (2023). A novel approach for estimating and predicting uncertainty in water quality index model using machine learning approaches. Water Research, 1–24. https://doi.org/10.1016/j.watres.2022.119422

[2] Chowdhury, M., Rahman, A., & Islam, R. (2018). Malware analysis and detection using data mining and machine learning classification. In J. Abawajy, K-K. R. Choo, & R. Islam (Eds.), International Conference on Applications and Techniques in Cyber Security and Intelligence: Applications and Techniques in Cyber Security and Intelligence (Vol. 580, pp. 266–274). (Advances in Intelligent Systems and Computing; Vol. 580). Springer. https://doi.org/10.1007/978-3-319-67071-3_33

[3] Uddin, M. G., Nash, S., Rahman, A., & Olbert, A. I. (2023). Performance analysis of the water quality index model for predicting water state using machine learning techniques. Process Safety and Environmental Protection, 1-30. https://doi.org/10.1016/j.psep.2022.11.073

[4] Rahman, A., & Harding, A. (2017). Small area estimation and microsimulation modeling. CRC Press. https://doi.org/10.1201/9781315372143

[5] Uddin, M. G., Nash, S., Mahammad Diganta, M. T., Rahman, A., & Olbert, A. I. (2022). Robust machine learning algorithms for predicting coastal water quality index. Journal of Environmental Management, 321(11), 1–16. [115923]. https://doi.org/10.1016/j.jenvman.2022.115923
[6] Chowdhury, M.M.H., Rahman, A., & Islam, M. R. (2018). Protecting data from malware threats using machine learning technique. In Proceedings of the 2017 12th IEEE Conference on Industrial Electronics and Applications (ICIEA) (pp. 1691–1694). IEEE, Institute of Electrical and Electronics Engineers. https://doi.org/10.1109/ICIEA.2017.8283111
[7] Rahman, A., & Upadhyay, S. K. (2015). A Bayesian reweighting technique for small area estimation. In U. Singh, A. Loganathan, S. K. Upadhyay, & D. K. Dey (Eds.), Current trends in Bayesian methodology with applications (1st ed., pp. 503–519). CRC Press.
[8] Rahman, A. (2019). Statistics-based data preprocessing methods and machine learning algorithms for big data analysis. International Journal of Artificial Intelligence, 17(2): 44–65.
[9] X. Guo, "How can machine learning help stock investment?"
[10] W. Khan, M. A. Ghazanfar, M. Asam, A. Iqbal, S. Ahmad and J. A. Khan, "Predicting Trend in Stock Market Exchange using Machine Learning Classifiers," Science Interna- tional, vol. 28, 2016.
[11] M. Asad, "Optimized Stock market prediction using ensemble learning," in 2015 9th International Conference on Application of Information and Communication Technologies (AICT), 2015.
[12] J. Chou and T. Nguyen, "Forward Forecast of Stock Price Using Sliding-Window Metaheuristic-Optimized Machine-Learning Regression," IEEE Transactions on Indus- trial Informatics, vol. 14, pp. 3132–3142, 7 2018.
[13] S. Shen, H. Jiang and T. Zhang, "Stock market forecasting using machine learning algorithms," Department of Electrical Engineering, Stanford University, Stanford, CA, pp. 1–5, 2012.
[14] B. Narayanan and M. Govindarajan, "Prediction of Stock Market using Ensemble Model," International Journal of Computer Applications, vol. 128, pp. 18–21, 2015.
[15] X. Tang, C. Yang and J. Zhou, "Stock price forecasting by combining news mining and time series analysis," in Web Intelligence and Intelligent Agent Technologies, 2009. WI-IAT'09. IEEE/WIC/ACM International Joint Conferences, 2009.
[16] "Yahoo Finance," [Online]. Available: https://in.finance.yahoo.com/.
[17] "Hindu Business," [Online]. Available: www.thehindubusinessline.com/archive/.

10 Influencing Project Success Outcomes by Utilising Advanced Statistical Techniques and AI during the Project Initiating Process

*Jennifer Hayes, Azizur Rahman and Champake Mendis
Charles Sturt University, Australia
*Corresponding Author: jehayes@csu.edu.au

CONTENTS

DOI: 10.1201/9781003253051-13

10.1 INTRODUCTION: BACKGROUND AND DRIVING FORCES

Conceptualising the evolution of businesses against the backdrop of Industry 4.0 highlights an era where the pace of change is rapid across technology, industries, and society [1,2]. Increasing connectivity and automation is resulting in visible impacts to the IT project management domain as businesses evolve; embrace agile ways of working; and expand into ecosystem partnership models, becoming increasingly dynamic in a landscape that is technologically complex and in which speed to market is paramount [3–6]. Projects within technology generally, and the retail domain specifically, are becoming increasingly complex to design, build, and deliver [7–17,53], In this fast-moving consumer goods (FMCG) market there is little room for mistakes, design and implementation processes are complicated, and the pace of development has increased in line with consumer expectations and market impacts [5,6,60].

The evolution of the ecosystem business structure in retail as a value proposition is additionally influencing and impacting on the technology and projects that support it [18]. Requiring technological responses to modularity, customisation, multilateralism, and coordination, technology projects that have previously been aligned to single, or well-established stakeholder groups and domains are now multi-faceted and require streams of design, development, integration, and deployment that previously did not exist [18]. Adapting to dynamic and evolving business structures requires a corresponding adaptation of the methods in which technology projects are delivered. Enabling both "...high variety and a high capacity to evolve" (17,18], these structures require capabilities (including technology) that are scalable, flexible, resilient, and reliable. Project management in this environment is inherently more complex and challenging than delivery within a single business entity, with the IT landscape that supports this structure also being technologically complex and sensitive [19–21].

Project initiation and choosing an appropriate delivery methodology are two of the most important factors to consider when establishing a new project in this domain, however, there exists a small amount of research on formal methods of categorising projects according to theories on complexity, prior to commencing activities under the Initiating Process Group (PMI) [22–24]. Projects are facing situations of deep complexity and uncertainty where existing delivery methods are frequently inadequate and in which the project manager is increasingly required to implement multiple combinations of "...often contradictory methods" to achieve a successful outcome [48]. Such methods will increasingly be required to rely on applications of incremental and potentially disruptive change, utilising AI to make sense of the big data involved in modern projects [49–52]. Complexity and Difficulty Assessment Tools (CATs/DATs) are sparsely utilised in project environments and a modernised approach to these tools is required to accommodate changes to the technology landscape and ecosystem business structures, thus improving on the validity and value of such tools to project managers [22].

The need for simple-to-use and vendor-agnostic AI tools capable of analysing large historical datasets to arrive at a decision is evidenced by the proliferation of self-created spreadsheets in organisations that enable the user to customise scenarios and unique situations [55, 59, 61]. The capabilities of large-scale Decision Support Systems (DSS) may be beyond the reach of many organisations, with feature-rich offerings accompanied by steep licensing and integration costs to access the dataset. Such DSS applications historically manage structured decision making more than capably, with descriptive, prescriptive, satisficing, and optimising models adequately catered for. Their ability to manage unstructured data and decisions is not as advanced and the reliance on specific complexity-based frameworks and models such as Cynefin is not advanced. It is in this area where the proposed CDAT tool will allow organisations to quickly apply a complexity-based RPA tool to existing datasets within an optimising model. A considered and statistically validated tool for understanding the initial complexity structure of the project prior to handover to a project manager is expected to provide non-subjective insights into the project itself, allowing for respective management, governance, and delivery structures to be put in place.

The management of technology projects in times of complexity and for which complexity is inherent in the delivery requires a methodology that has itself been derived from complexity [21]. With a statistically validated understanding of the initial complexity structure of the proposed project, suitable delivery methodologies unique to the project requirements may be implemented [25]. The "default" waterfall methodology is predicated on the belief that requirements are known and understood upfront and that delivery follows a structured, linear approach [26]. As the organisation moves to "Agile ways of working," project managers are increasingly pressured to implement an Agile delivery methodology, in some instances where categorisation of complexity and technology requirements are ill-suited. Without an understanding of the complexity of the project prior to Initiation, the project manager may "default" to their preferred, or business requested, methodology. Resultant issues throughout the project lifecycle may be directly attributable to the suitability, or lack thereof, of the chosen methodology. "One infamous reason for a project failure is the wrong choice of approach in managing the project" [26,27] and it is this intersection of tacit knowledge [54] applied by individuals to new and innovative project situations based on previous expectations of "what worked well" that serves to increase the complexity of project delivery.

Complementary to a model that aligns the project complexity category to suitable delivery methodology, an understanding of the assigned project manager competencies is required to extend the tool to predict the likelihood of a successful delivery based on their reported hard and soft skills prior to commencing management of the project [28–30]. As business and technological complexity increases, and artificial intelligence applications emerge, the industry is pivoting to a focus on those "soft" skills that are not machine-replicable, and that are increasingly being attributed to improve project success outcomes, with leadership ranking of skills' preferences overwhelmingly favouring soft skills as "Most Important" on a Likert scale of 15 competencies [29,30]. A mapping of the individual project managers hard and soft skill effectiveness in complexity domains will ultimately align with the project categorisation and methodology and produce a matrix of project managers matched to the project whose experience and skills are statistically likely to result in a successful project delivery.

Cynefin, and Liminal Cynefin, providesa "...conceptual framework for making sense of the different landscapes faced within and by projects" [31] and will form the complexity framework in this model both to the IT project landscape itself [32] and to the individual project manager competencies [28]. Technology projects, delivery methodologies, and project manager competencies will all be individually aligned to this framework with respective business rules developed and forming input to the solution.

The Complexity and Decision Assessment Tool (CDAT) developed in this research will provide a project categorisation and matching tool, incorporating suitable project management methodologies, adaptable governance structures, and guidance on aligning the assignment of project managers, underpinned by historical assessment of results against relevant theories and application of a combination of processes grounded in complexity theory and Liminal Cynefin. This augmentation at a task level, in this instance data analysis, of a project management function is situated within the analytical intelligence capabilities of existing AI solutions and seeks to move the current state from descriptive assistance to predictive application [53]. With research attributing project failures to the attempt to apply a standardised leadership model regardless of the "...prevailing operative context." [26,33,44] the need for a modernised, automated CDAT is essential to assist project managers and key project stakeholders in identifying and applying contextually appropriate management methods in each scenario. The development and deployment of AI tools to assist the project manager with decision support and problem-solving allows the analysis of large volumes of data to be incorporated into the decision-making process [46]. The project and portfolio management market value are increasingly exponentially with a key driver being the AI automation of project management processes, aiming to reduce time to delivery and subsequent operational costs, coupled with the drive to adopt Agile practices across the organisations [47].

Advanced statistical techniques are required to discover and explore the complex multivariate relationships among the variables and cases. A latent variable model, Factor Analysis, and a partitioning method, Cluster Analysis, in combination with linear regression are considered for this model. While both multivariate statistical tools have different goals, their use as complementary tools in this analysis and model development will work to confirm the existence of a simple structure within the data or provide awareness of complex data structures that may not be revealed by either tool alone [34].

The CDAT will be implemented through Robotic Process Automation (RPA), allowing the automation of the proposed process and for the purpose of this study, the R programming language will be utilised for model validation, followed by a transition to UiPath through a multi-environment development and testing landscape to productionise within the organisational context.

The rest of this chapter proceeds as follows: Section 10.2 describes the data collection, cleansing, and analysis procedures. Section 10.3 discusses the statistical methods used to analyse the datasets. Section 10.4 addresses future experimental analysis setup while Section 10.5 provides concluding remarks and future research directions.

10.2 DATA COLLECTION

Multiple disparate sources of quantitative and qualitative datasets require a mixed approach to collection and determination of population and sampling size as illustrated in Figure 10.1. Data collection approval is subject to a level of abstraction to any conditions that will protect organisational commercial interests, including de-identification of project and survey data, and non-attribution to specific projects or the organisation.

10.2.1 Quantitative Data Collection

The primary location of key project and program metrics, status, risks, issues, milestones, financial tracking, and links to project artefacts is a commercial portfolio and program management software application with key integrations back to core and complementary organisational applications. Full access to this dataset has been provided with conditions related to privacy, confidentiality, and abstraction of data prior to data analysis required to be adhered to. Data privacy and confidentiality are important issues in any contemporary research [55–57]. Individual project identifying information will be coded to allow analysis to progress without identification of specific projects. This code will then be removed once analysis is completed, rendering the data and responses anonymous and unidentifiable.

Data collection procedures for this dataset are known, current, and the data has meaningful application to the population being studied. Analysing data gathered for alternative purposes risks the introduction of systematic, researcher, or random biases [35] for which accuracy of analysis outcomes should be constantly questioned.

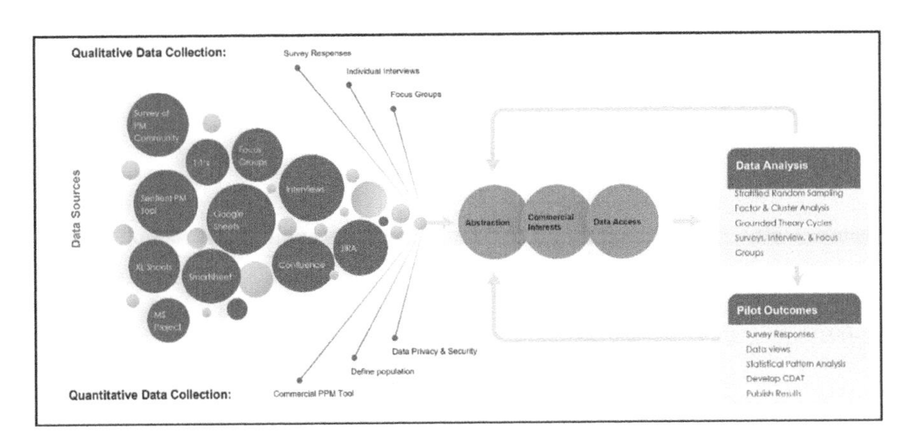

FIGURE 10.1 Data Collection Process.

Continual review of the data will progress throughout the research program to weed out contaminating influences (e.g., unknowns, inaccuracies).

10.2.2 Stratified Random Sampling

Stratified random sampling is justified where the entire population is available for sampling, and where the population may be clearly delineated into stratum [36]. The population size for this dataset is N=1,099 records across financial years 2013 to 2022 inclusive, with calculated sample size of n=633 for the pilot study, based on proportion of 0.5 and confidence interval of 0.05. The pilot study will validate the data to be extracted from the database and assist in identifying where missing data may cause an issue in the primary research study requiring an imputation method such as K-Nearest Neighbour (KNN) to replace the missing values with one most frequently found among the k number of most similar data [37]. Stratification of the population aligns to distinct business units within the Technology domain as illustrated in Table 10.1.

10.2.3 Qualitative Data Collection

A comprehensive structured survey distributed to all Technology project and program managers in the organisation, population size N=450 with sample response required of n=50 forms the primary source of qualitative data for the complexity and decision assessment tool. Survey questions are designed to obtain information on complexity and experience of the project managers, measuring responses to leadership styles and project management experience in both hard and soft skills [28–30], with responses taking a range of forms – Likert scale, closed ended, binary, rating, and multiple-choice questions, and open comments.

Due to the nature of the privacy constraints required by the organisation, neither the name of the organisation, nor the data is given in the study. The stages of the methodology are explained in detail in the following section.

TABLE 10.1
Stratified Random Sampling Population

Business Unit [1]	Total number in population	Random Sample	% of Population Sampled
BU-001	84	69	82%
BU-002	267	158	59%
BU-003	493	216	44%
BU-004	145	105	72%
BU-005	110	85	77%
Total	**1099**	**633**	

[1] Business unit names coded in accordance with data access and abstraction provisions.

10.3 PROPOSED METHOD

An integrated approach to the project complexity categorisation problem requires intelligent data analysis of existing quantitative datasets combined with the in-progress collection of qualitative survey results, focusing on enhancing and transforming large volumes of operational data into meaningful insights for project complexity assessment prior to, or as a component of, the Initiation Process of the project. The following primary stages of analysis were planned to achieve this purpose:

Stage 1 – Factor Analysis of qualitative surveys and quantitative dataset
Stage 2 – Cluster Analysis of qualitative surveys and quantitative dataset
Stage 3 – Assign factors and clusters to Cynefin quadrants
Stage 4 – Complexity and Decision Assessment matrix – project complexity quadrant aligned with methodology and project manager competencies with highest determination of project success. This approach is illustrated in Figure 10.2.

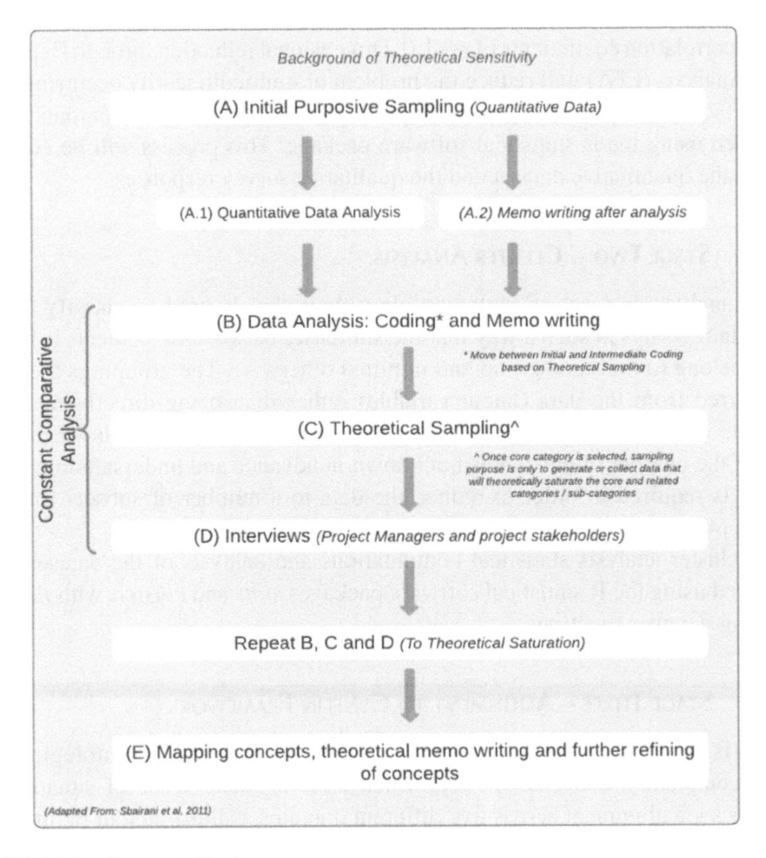

FIGURE 10.2 Research Design.

Adapted from Shairani et al., 2011.

10.3.1 Stage One – Factor Analysis

Factor analysis is a multivariate statistical approach, expected to identify which dimensions or clusters are relevant, grouping similar variables into dimensions in order to reduce the dataset from a large number of variables into fewer numbers of factors [38]. Iterations of factor analysis will serve to condense the variables in the dataset and to identify clusters of responses in the questions. Condensing the variables will identify factor-dependent variables, or latent variables. This will result in representative dimensions of variables with weightings for each respondent. Identification of clustering in the dataset (or survey questions that move together), allows attribution of components of common latent variables to other latent variables.

Rotation will be applied to show the estimated correlation between each of the identified variables from the survey and the factors. Pearson's correlation coefficient is calculated to measure the strength of the relationship between variables and their association with each other. It is expected that the majority of items will have some correlation with each other, inclusive of some very high correlations. Multicollinearity is expected within the dataset, with several independent variables capable of demonstrating correlation coefficients of +/- 1.0. Dimensional reduction through Exploratory Factor Analysis (EFA) will reduce the problem of multicollinearity occurring [39].

The factor analysis statistical computations and graphing of outputs will be completed using the R statistical software package. This process will be completed for both the quantitative dataset and the qualitative survey responses.

10.3.2 Stage Two – Cluster Analysis

Cluster analysis is a set of tools and algorithms that is used to classify different objects into groups in such a way that the similarity between two objects is maximal if they belong to the same group and minimal otherwise. The groupings themselves are inferred from the data (latent variable) rather than being directly measurable [40]. The use of cluster analysis is justified for use in large datasets where knowledge of the number of clusters is not known in advance and understanding of these clusters is required in order to reduce the data to a number of subsets for further analysis [41].

The cluster analysis statistical computations and analysis of the dataset will be completed using the R statistical software packages *stats* and *cluster*, with *factoextra* providing wrapper functions.

10.3.3 Stage Three – Alignment to Cynefin Framework

Cynefin is a phenomenological framework based on three distinct ontological states (order, complexity, and chaos) [31], which aims to make sense of situations and challenges via alignment across five different domains as illustrated in Figure 10.3:

- Simple (Known) domain
- Complicated (knowable) domain
- Complex (Unknowable) domain

FIGURE 10.3 The Cynefin Framework.

Source: https://cynefin.io/wiki/Cynefin_Domains.

- Chaotic (Unknowable) domain
- Disorder (Centre) domain [21,31,33]

The Cynefin sense-making model aims to bring chaos, complexity, and constraints together into a simple to visualise and apply framework, responding to a modern environment that is "...fluid, ambiguous and uncertain." [42] and in which complex organisational structures are the norm [43].

10.4 CYNEFIN AND THE QUALITATIVE DATASET

A significant percentage of the Survey questions have been designed to generate initial mapping of respondent-chosen projects across the respective Cynefin domains throughout the project's lifecycle. The aim of these questions is to determine the project management populations current state understanding of complexity and responses to it based on an existing framework, although the terminology in the survey is kept deliberately agnostic of the proposed tool. The hypothesis is that a majority of respondents will select the fifth domain of "Disorder" in the centre of the framework, described as the "Cause of fleeing to the 'safe' context of the Obvious while failing to realise that uncertainty and chaos are merely learning tools…" while also calling it the "central pivot – an indicator of its potential for destruction" [33].

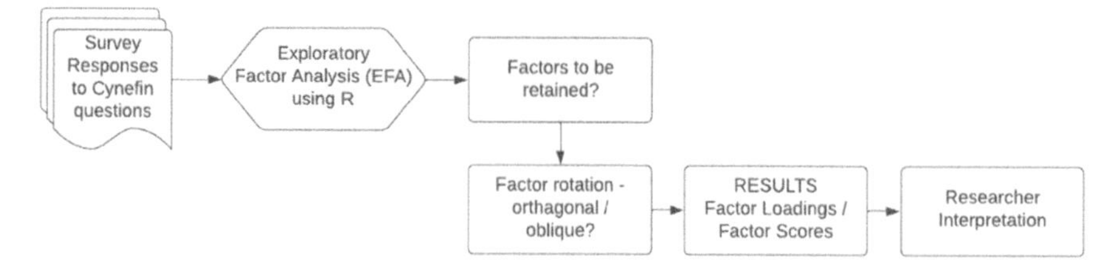

FIGURE 10.4 Respondent understanding of complexity movement and categorisations.

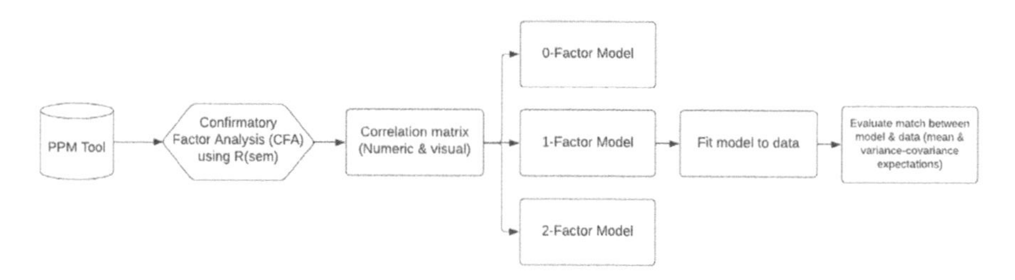

FIGURE 10.5 Confirmatory Factor Analysis of PPM Tool for project movement due to complexity.

Exploratory Factor Analysis (EFA) as discussed in Section 10.3.1 will be extended to test for this scenario as illustrated in Figure 10.4.

10.5 CYNEFIN AND THE QUANTITATIVE DATASET

A parallel Confirmatory and Exploratory Factor Analysis will be performed on the portfolio and program management database, analysing the population of N=1,455 closed projects to confirm notions about the structure of the content in relation to the movement of the project across complexity domains. The hypothesis is that the timing of risks, issues, and variations will correlate to category movement, which should necessitate a revised management and leadership approach to the project. The Confirmatory factor analysis process requires the use of Structural Equation Modelling (SEM) packages in R for quantitative data analysis as illustrated in Figure 10.5.

10.6 COMPLEXITY AND DECISION ASSESSMENT MATRIX

The final stage in development of the Retail IT project Complexity and Decision Assessment model involves combining outcomes from all analysis stages and developing business rules based on these outcomes. This process will result in a matrix that reflects the project complexity quadrant aligned with delivery methodology and project manager competencies with the highest determination of project success.

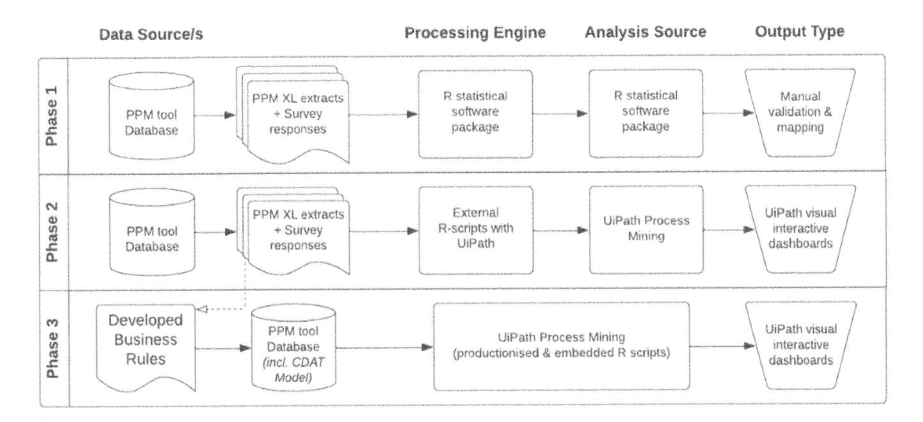

FIGURE 10.6 Phases of future development of CDAT Tool.

10.7 ROBOTIC PROCESS AUTOMATION (RPA)

Phase 1 of the tool will be developed and tested using the R language and environment that contains a suite of advanced statistical techniques to analyse and graph large datasets. R is capable of cleansing the input data, performing complex statistical analysis techniques, and producing a coherent graphical representation for output and further analysis. Phase 2 and future iterations of the tool will be implemented into UiPath, a process automation software licensed by the study organisation. This implementation will follow the software development lifecycle across a four-environment landscape – Development, SIT, UAT, and Production, enhancing the UiPath Process Mining platform with R-scripts to implement external data processing.

Phase 3 will involve the complete solution developed within UiPath, with internal data extraction, analysis, and outputs completed according to a pre-defined schedule and output into Tableau for management reporting purposes. Commencing with lower-order decisions and automating the processes assists in gleaning valuable insights and patterns from data that are not normally visible to the organisation. Future development of the tool to incorporate natural language processing capabilities will have the ability to expand the robot to handle cognitive processes. Refer to Figure 10.6.

10.8 LIMITATIONS AND RESTRICTIONS OF THE PROPOSAL

The use of AI, as with technology more generally, is subject to "interpretive flexibility" [45], with modification of its intended use having the potential for erroneous application and interpretation. With concurrent research in progress to determine the existence of chaos and complexity in the projects in this domain, this application of automation may provide guidance on recommended project management approaches and resources that based on historical data will result in future project success. The interpretation of the results may be misunderstood as a representation of chaos inherent in the organisation when complexity is rather the prevailing operative context [46]. Critical ethical consideration is required to prevent the tool from being used as a performance management indicator for project managers.

10.9 CONCLUSION

The project initiation process in Information Technology is significant to the overall success of the project as decisions made here directly influence and impact on the methodology, management, and outcomes of the project. Not understanding the starting state of the project based on the Cynefin domain and assuming every project starts out straightforward and simple is expected to have directly resulted in persisting with methodologies that were misaligned to the unique requirements of the project or failed to determine when critical changes or movements in the project necessitated a corresponding change in management styles and delivery techniques. The ability of the PM to deliver in each of these quadrants has not been analysed to date, although there has been some recent research to identify the types of skills required to deliver projects in each of the Cynefin quadrants [29–33].

The proposed model will utilise a combination of advanced statistical techniques to identify and align patterns and dimensions across disparate datasets, seeking combinations that are not easily identifiable to the human respondent.

Development of a project categorisation model that maps new projects to the Cynefin framework, determines the most appropriate methodology based on that categorisation, and then assigns a project manager skilled in that project delivery style to the respective project, is expected to result in improved project success outcomes when compared to the traditional assignment methods currently in place in the industry and the organisation. The model may be developed in future iterations to incorporate Lessons Learned, identifying, and applying those deemed relevant, again via the use of advanced statistical techniques within an RPA toolset. Future development will seek to implement this approach in the organisation, developing a robotic process automation tool and applying it to new projects within the Enterprise Technology domain.

Of additional interest is the ability of RPA to detect and identify the stages when an in-flight project moves between Cynefin domains as it is in these transitions that the greatest risk to project delivery is incurred. These areas are proposed for future research.

REFERENCES

1. Tromp, J. G., Le, D.-N., & Le, C. V. (2020). Emerging extended reality technologies for industry 4.0: early experiences with conception, design, implementation, evaluation and deployment. Hoboken, NJ: Wiley-Scrivener.
2. André, J.-C. (2019). Industry 4.0: paradoxes and conflicts. London: ISTE.
3. Echeberria, A. L. (2020). Digital framework for Industry 4.0: managing strategy (1st ed. 2020. ed.). Cham, Switzerland: Palgrave Macmillan.
4. Matt, D. T., Modrák, V., & Zsifkovits, H. (2020). Industry 4.0 for SMEs Challenges, Opportunities and Requirements (1st ed. 2020. ed.). Cham: Springer International Publishing.

5. Hayes, J., & Rahman, A. (2020). How to determine the existence of chaos in retail IT projects?: Charles Sturt University.
6. Bolat, B., Çebi, F., Tekin Temur, G., & Otay, İ. (2014). A fuzzy integrated approach for project selection. Journal of enterprise information management, 27(3), 247–260. doi:10.1108/JEIM-12-2013-0091
7. Efebeli, T. V. (2021). Strategies for Reducing Project Cost Overruns in the Oil and Gas Construction Industry. (D.B.A.). Walden University, Ann Arbor. Retrieved from https://ezproxy.csu.edu.au/login?url=https://www.proquest.com/dissertations-theses/strategies-reducing-project-cost-overruns-oil-gas/docview/2601524585/se-2?accountid=10344. Accessed: 01 April 2022
8. Williamson, D. J. (2011). A correlational study assessing the relationships among information technology project complexity, project complication, and project success. ProQuest Dissertations Publishing,
9. Abel, D. L. (2009). The capabilities of chaos and complexity. International journal of molecular sciences, 10(1), 247–291. doi:10.3390/ijms10010247
10. Malgorzata Ali, I. (2014). Methodological Approaches for Researching Complex Organizational Phenomena. Informing science, 17, 59–073. doi:10.28945/1949
11. Birks, M., & Mills, J. (2011). Grounded theory: a practical guide. London: SAGE.
12. Birks, M., Hoare, K., & Mills, J. (2019). Grounded Theory: The FAQs. International Journal of Qualitative Methods, 18, 160940691988253. doi:10.1177/1609406919882535
13. Burnham, R. (2020). An overview of Complexity Theory for Project Management. Retrieved from www.academia.edu/6908046/An_Overview_of_Complexity_Theory_for_Project_Management?email_work_card=view-paper. Accessed: 01 April 2022
14. Camci, A, and T Kotnour. "Technology Complexity in Projects: Does Classical Project Management Work?" IEEE, Vol. 5, 2006, pp. 2181–2186, https://doi.org/10.1109/PICMET.2006.296806.
15. Courtney, J., Merali, Y., Paradice, D., & Wynn, E. (2008). On the Study of Complexity in Information Systems. International journal of information technologies and systems approach, 1(1), 37–48. doi:10.4018/jitsa.2008010103
16. Dahlberg, R. (2015). Resilience and Complexity Conjoining the Discourses of Two Contested Concepts. Culture Unbound: Journal of Current Cultural Research, 7, 541–557. https://doi-org.ezproxy.csu.edu.au/10.3384/cu.2000.1525.1572541
17. Meyer, A. D., & Loch, C. H. (2002). Managing project uncertainty: from variation to chaos. MIT Sloan management review, 43(2), 60.
18. Ulrich, P., Martin, R., & Balázs, Z. (2022). What Is Your Business Ecosystem Strategy? In. Boston: Boston Consulting Group Boston, MA
19. Geraldi, J. G. (2008). The balance between order and chaos in multi-project firms: A conceptual model. International journal of project management, 26(4), 348–356. doi:10.1016/j.ijproman.2007.08.013
20. Gorod, A., Hallo, L., & Nguyen, T. (2018). A Systemic Approach to Complex Project Management: Integration of Command-and-Control and Network Governance: A Systemic Approach to Complex Project Management. Systems research and behavioral science, 35(6), 811–837. doi:10.1002/sres.2520
21. Hayes, J., Rahman, A., & Islam, M. R. (2020). Shaping the Future of Multidimensional Project Management in Retail Industry Using Statistical and Big-Data Theories. In (pp. 347–360). Singapore: Springer Singapore.
22. Beale, D., Tryfonas, T., & Young,M. (2017). Evaluating Approaches for the next Generation of Difficulty and Complexity Assessment Tools. IEEE, pp. 227–233, https://doi.org/10.1109/TEMSCON.2017.7998381.

23. Kim, C. (2011). Data Mining-Based Predictive Model to Determine Project Financial Success Using Project Definition Parameters. Paper presented at the 28th International Symposium on Automation and Robotics in Construction (ISARC 2011).
24. Balsera, J. V., Montequin, V. R., Fernandez, F. O., & González-Fanjul, C. A. (2012). Data Mining Applied to the Improvement of Project Management. In (Ed.), Data Mining Applications in Engineering and Medicine. IntechOpen. https://doi.org/10.5772/48734
25. Jüngen, F. J., & Kowalczyk, W. (1995). An intelligent interactive project management support system. European journal of operational research, 84(1), 60–81. doi:10.1016/0377-2217(94)00318-7
26. Najdawi, A, and Shaheen, A. (2021). "Which Project Management Methodology Is Better for AI-Transformation and Innovation Projects?" IEEE, pp. 205–210.
27. Camci, A, and T Kotnour. "Technology Complexity in Projects: Does Classical Project Management Work?" IEEE, Vol. 5, 2006, pp. 2181–2186, https://doi.org/10.1109/PICMET.2006.296806.
28. Pollack, J. (2007). The changing paradigms of project management. International journal of project management, 25(3), 266–274. doi:10.1016/j.ijproman.2006.08.002
29. Stevenson, D. H., & Starkweather, J. A. (2010). PM critical competency index: IT execs prefer soft skills. International journal of project management, 28(7), 663–671. doi:10.1016/j.ijproman.2009.11.008
30. Gokhale, D. (2005). PM competency mapping. Paper presented at the PMI Global Congress 2005, Asia Pacific, Singapore.
31. Kurtz, C. F., & Snowden, D. J. (2003). The new dynamics of strategy: Sense-making in a complex and complicated world. *IBM systems journal, 42*(3), 462–483. doi:10.1147/sj.423.0462
32. Shalbafan,S., Leigh,E., Pollack,J., and Sankaran,S. (2017). Decision-making in project portfolio management: using the Cynefin framework to understand the impact of complexity. International Research Network on Organizing by Projects (INROP) 2017, UTS ePRESS, Sydney:NSW, pp.1–20. Retrieved from http://doi.org/10.5130/pmrp.irnop2017.5775. Accessed: 01 April 2022
33. Shalbafan, S. & Leigh, E. (2018). Design Thinking: Project Portfolio Management and Simulation – A Creative Mix for Research. 10.1007/978-3-319-91902-7_1.
34. Gorman, B. S., & Primavera, L. H. (1983). The Complementary Use of Cluster and Factor Analysis Methods. The Journal of experimental education, 51(4), 165–168. doi:10.1080/00220973.1983.11011856
35. Baldwin, J.R., Pingault, JB., Schoeler, T. et al. Protecting against researcher bias in secondary data analysis: challenges and potential solutions. Eur J Epidemiol 37, 1–10 (2022). https://doi.org/10.1007/s10654-021-00839-0
36. Frey, B. (2018). The SAGE encyclopedia of educational research, measurement, and evaluation (Vols. 1–4). Thousand Oaks, CA: SAGE Publications, Inc. doi: 10.4135/9781506326139
37. Cunningham, P., & Delany, S. J. (2021). k-Nearest neighbour classifiers-A Tutorial. ACM Computing Surveys (CSUR), 54(6), 1–25.
38. Williams, B., Onsman, A., & Brown, T. (2010). Exploratory factor analysis: A five-step guide for novices. Australasian Journal of Paramedicine, 8(3). doi:10.33151/ajp.8.3.93
39. Kim, J. H. (2019). Multicollinearity and misleading statistical results. Korean journal of anesthesiology, 72(6), 558–569. doi:10.4097/kja.19087
40. Erbacher, M., Atkinson, P., Delamont, S., Cernat, A., Sakshaug, J. W., & Williams, R. A. (2020). Cluster analysis. London: SAGE Publications Ltd.

41. Cooper, B., & Glaesser, J. (2011). Using case-based approaches to analyse large datasets: a comparison of Ragin's fsQCA and fuzzy cluster analysis. International journal of social research methodology, 14(1), 31–48. doi:10.1080/13645579.2010.483079
42. Vasilescu, C. (2011). Strategic Decision Making Using Sense-Making Models: The Cynefin Framework. Defense Resources Management in the 21st Century.
43. Fierro, D., Putino, S., & Tirone, L. (2018). The Cynefin Framework and Technical Competencies: a New Guideline to Act in the Complexity. INCOSE International Symposium, 28, 532–552. doi:10.1002/j.2334-5837.2018.00498.x
44. Guo, Y., & Lyu, X. (2021). Strategic Management Model of Network Organization Based on Artificial Intelligence and Modular Enterprise Big Data. Mobile information systems, 2021, 1–13. doi:10.1155/2021/1987430
45. Stahl, B. C., Andreou, A., Brey, P., Hatzakis, T., Kirichenko, A., Macnish, K., Laulhe, S., Patel, A., Ryan, M., Wright, D. (2021). Artificial intelligence for human flourishing – Beyond principles for machine learning. Journal of business research, 124, 374–388. doi:10.1016/j.jbusres.2020.11.030
46. Pozo-Puértolas, R. (2020). Creative Chaos Theory Inductive Method for Viewing Information from an Applied Research. *New Trends in Qualitative Research*, *2*, 13–26. https://doi.org/10.36367/ntqr.2.2020.13-26
47. "Project Portfolio Management (PPM) Market to Hit USD 5.74 Billion by 2027 Owing to Increasing Adoption of Artificial Intelligence, States Fortune Business Insights: Companies in the Global Project Portfolio Management (PPM) Market Are Planview, Integrated Project Management Company, Changepoint, Wrike, Hexagon AB, Logic Software, ProductDossier Solution, ServiceNow, UMT 360 LLC and Others." *NASDAQ OMX's News Release Distribution Channel*, 2022.
48. Wachnik, B. (2022). Analysis of the use of artificial intelligence in the management of Industry 4.0 projects. the perspective of Polish industry. Production Engineering Archives, 28(1), 56–63. doi:10.30657/pea.2022.28.07
49. Johnsonbabu, A. (2017). Reinventing the role of Project manager in the Artificial Intelligence era. Paper presented at the Project Management National Conference, India.
50. "New PwC Report Outlines How Artificial Intelligence Will Disrupt Project Management and Change the Role of Project Managers." *Al Bawaba*, 2019.
51. Niederman, F. (2021). Project management: openings for disruption from AI and advanced analytics. Information technology & people (West Linn, Or.), 34(6), 1570–1599. doi:10.1108/ITP-09-2020-0639
52. Rahman, A. (2020). Statistics for Data Science and Policy Analysis. Springer.
53. Huang, M.-H., & Rust, R. T. (2018). Artificial Intelligence in Service. Journal of service research: JSR, 21(2), 155–172. doi:10.1177/1094670517752459
54. Zhanru Li & Tingting Mo (2020): Early warning of engineering project knowledge management risk based on artificial intelligence, Knowledge Management Research & Practice, DOI: 10.1080/14778238.29020.1834885
55. Błaszczyk, T., & Błaszczyk, P. Contracting Decisions in Project Management – An Outline of the Dedicated Decision Support System. In (pp. 347–356). Berlin, Heidelberg: Springer Berlin Heidelberg.
56. Rahman, A. (2019). Statistics-based data preprocessing methods and machine learning algorithms for big data analysis. International Journal of Artificial Intelligence, 17(2): 44–65.
57. Chowdhury, M.M.H., Rahman, A., & Islam, M. R. (2018). Protecting data from malware threats using machine learning technique. In Proceedings of the 2017 12th IEEE Conference on Industrial Electronics and Applications (ICIEA) (pp. 1691–1694).

IEEE, Institute of Electrical and Electronics Engineers. https://doi.org/10.1109/ICIEA.2017.8283111
58. Rahman, A., & Harding, A. (2016). Small area estimation and microsimulation modeling. Chapman and Hall/CRC.
59. Harjule, P., Rahman, A., & Agarwal, B. (2021). A cross-sectional study of anxiety, stress, perception and mental health towards online learning of school children in India during COVID-19. Journal of Interdisciplinary Mathematics, 24(2), 411–424. https://doi.org/10.1080/09720502.2021.1889780
60. Agarwal, B., Agarwal, A., Harjule, P., & Rahman, A. (2022). Understanding the intent behind sharing misinformation on social media. Journal of Experimental and Theoretical Artificial Intelligence, 1–15. https://doi.org/10.1080/0952813X.2021.1960637
61. Agarwal, B., Rahman, A., Patnaik, S., & Poonia, R. C. (Eds.) (2022). Proceedings of International Conference on Intelligent Cyber-Physical Systems. (Algorithms for Intelligent Systems). Springer. www.springer.com/gp/book/9789811671357
62. Sbaraini, Alexandra, et al. (2011). How to Do a Grounded Theory Study: A Worked Example of a Study of Dental Practices. *BMC Med Res Methodol*, 11(1), 128–128. https://doi.org/10.1186/1471-2288-11-128

Theme 4

Socio-economic and Environmental Modelling

11 Computational Statistical Methods for Uncertainty Assessment in Geoscience

Scott McManus[*1], *Azizur Rahman*[1],
Jacqui Coombes[2] *and Ana Horta*[1]
[1] Charles Sturt University, Australia
[2] AMIRA
*Corresponding Author: smcmanus@csu.edu.au

CONTENTS

11.1 INTRODUCTION

In the context of the minerals mining and exploration industry, public reporting is subject to reporting codes that require a full assessment of all possible errors and uncertainties. Mineral Resource Estimation (MRE) is a key activity that is publicly reported and consists of four phases (Coombes, 2008). A key phase is the interpretation of geology and the sub-setting of geology into spatial domains. This requires

DOI: 10.1201/9781003253051-15

a complex iterative process that includes expert opinion. Due to this, assessment of uncertainty of spatial domains is challenging.

Studies of failure in minerals project feasibility and development have found that 17% of failures are attributable to problems with the geological spatial domains (McCarthy, 2014). In completed projects, 5 to 70% of development costs could be saved with more accurate spatial domains (Vallee, 2000). Recent stock exchange releases continue to report downgrades or project failures due to flawed spatial domains (Sterk et al., 2019).

In the minerals industry, the slow adoption of quantitative methods for uncertainty assessment is attributed to a lack of relevant case studies, cost of sampling, lack of education, or the time required to carry out the assessments (Sterk et al., 2019). This is particularly true for early-stage minerals projects. At the time of publication, only one mining software application has included geological uncertainty assessment using machine learning tools (Sullivan, 2021; Sullivan, 2022); however, this methodology is focused on mid-to-late stage projects with high density of data (Agarwal et al. 2022a; Agarwal et al. 2022b).

An initial review of industry practice and peer-reviewed papers looking for a method of assessing early-stage spatial domain uncertainty found a lack of methodology and case studies in minerals projects (McManus et al., 2021b). What is missing is a robust and practical approach that can be adopted across different minerals in the mining industry, particularly for early-stage mining projects, to ensure that uncertainty related to geological interpretation is quantified and effectively communicated from the start of the mining life cycle.

With an improved understanding of project uncertainty, there should be an improvement in the number of failed projects due to geological interpretation. Valid uncertainty assessment methods and approaches have already been developed in other disciplines; recent work (McManus et al., 2021b) has reviewed these methods in the context of the minerals industry and the following is a summary of methods: manual statistical difference of spatial domains (Lark et al., 2013; Scheidt & Caers, 2009; Siler et al., 2018; Witter et al., 2016); spatial uncertainty from geostatistical simulation methods (Adeli & Emery, 2021; Chiles et al., 2007; Deraisme & Field, 2006; Jones et al., 2013; Mery et al., 2017; Mustapha & Dimitrakopoulos, 2010; Yarus et al., 2012); weights of evidence (Hill et al., 2014; Nielsen et al., 2019; Peters et al., 2017); spatial bootstrap (Journel & Bitanov, 2004); and interpretation uncertainty using Bayesian approximation (McManus et al., 2020; Wellmann et al., 2010). The implementation of the machine learning method of Bayesian approximation (Rahman, 2008, 2020; Rahman & Harding, 2016) for interpretation uncertainty in the mining context is a new application. It is particularly useful in early-stage projects where many of the geostatistical methods of uncertainty assessment are problematic due to the need for more data than is usually missing at the start of the project (Rahman, 2019). Additionally, the method was tested using high-quality laboratory measurements (using the Inductively Coupled Plasma analytical method) and the results compared to low-quality, but cost- and time-effective portable X-ray Fluorescence measurements (McManus et al., 2021a) where it was found that the uncertainty models using the machine learning methods provided equivalence. Meaning that the method could be successfully used in early-stage projects with cheap multielement data. The industry importance of the method and its novel approach in assessing early-stage interpretation uncertainty has been acknowledged (McManus et al., 2022).

Sequential indicator simulation (SIS) is one of the computationally intensive geostatistical methods that can be used to assess spatial uncertainty of the geological spatial domains. The objective of this work was to consider if the Bayesian approximation methodology of assessing interpretation uncertainty identified equivalent issues in the borehole interpretation as the spatial uncertainty provided by spatial uncertainty using the SIS methodology. By understanding if interpretation uncertainty is captured by the SIS method a modified workflow for total project uncertainty can be recommended to improve industry best practice for early, mid- and late-stage projects with recommendations if one can be used or if both are needed once there is enough data to run simulations and hence to quantify spatial uncertainty.

11.2 METHODS

11.2.1 Case Study Description

The deidentified project data is from a gold project in North Eastern Australia. The dataset includes multielement data (pXRF and ICP) from gold assays and geological qualitative borehole data (recorded by the mining companies' geologists). The mineralization has been subset into spatial domains defining the rock codes for use in MREs. A subset of this data has been used in this work and the subset includes seven rock codes as well as a rock code for non-mineralized material. The project is mid-stage and has enough data to produce robust variogram models.

11.2.2 Bayesian Approximation of Interpretation Uncertainty

11.2.2.1 Selection of Important Variables

Due to many variables available in late-stage projects, it is important to make sure that the variables presented to the model are statistically the most useful to explain the variance and hence increase prediction accuracy and reduce overfitting by removing predictors that contribute less or that have strong multicollinearity with another predictor. To carry out the selection, an ensemble of statistical and machine learning algorithms to evaluate variable importance was utilized. This included Principal Component Analysis (Abdi & Williams, 2010), Generalized Linear models (Ripley et al., 2018), ANOVA (Fisher, 1992), and Random Forest (Breiman, 2001). The Wrapper method and the Filter method (Kuhn & Johnson, 2013) were used to select the final variables. Akaike information criterion (Agresti, 2013; Akaike, 1974; Kuhn & Johnson, 2013), Bayesian information criterion (Kuhn & Johnson, 2013), and Kappa and model performance (Leave one out Cross-validation) (Gelman, 2013) metrics were used to determine the final models used in the Bayesian model, using the workflow presented in McManus et al. (2019). From 78 variables, which included 35 ICP, 25 pXRF and 18 categorical geological observations, 12 variables were selected. The variables were Vein4, Vein1, and Minerals from the categorical observations,[1]

Gold, Sulphur, and Molybdenite from the pXRF variables, and Sulphur, Sodium Molybdenite, Magnesium, phosphorus, and manganese from the ICP variables.

11.2.2.2 Bayesian Approximation

Bayesian approximation uses a set of techniques and methods of approximating the credibility of the parameters from simulation of an extensive dataset without the need to specify or solve the integrals of complex models using Markov Chain Monte Carlo (MCMC) methods (Kruschke, 2015). Point predictions were drawn from the posterior distributions of the simulations, to provide realizations of each sample interval for each of the boreholes. The variance of the realizations provided an assessment of the interpretation uncertainty for each of the borehole sample intervals.

The Bayesian approximation was carried out using *Stan* (Carpenter et al., 2017), *R statistics* (R Core Team, 2018), and the R package *BRMS* (Bürkner, 2017) using a weakly informative prior normal (0,8) (Gelman et al., 2008). A run in time of 1,000 iterations was used followed by 10,000 iterations for 3 chains (Bürkner, 2017). The MCMC sampling passed convergence tests (Brooks et al., 2011), had Rhat values less than 1.05 (Gelman et al., 2013; Gelman & Rubin, 1992), and final testing of the models using leave-one-out (LOO) and K-folds Pareto K diagnostics confirmed the models were appropriate (Vehtari et al., 2019) for use.

Regression models were determined using the following multinomial Equation (11.1);

$$\lambda_k \; where \; log \; log\left(\frac{\varnothing_k}{\varnothing_r}\right) = log \; log\left(\frac{exp \; exp\left(\beta_{0,k} + \beta_{1,k}x\right)}{exp \; exp\left(\beta_{0,r} + \beta_{1,r}x\right)}\right) = \beta_{0,k} + \beta_{1,k}x. \tag{11.1}$$

where λ_k is the prediction of one class k compared to not being class r and are the coefficients for predictors for each class using the Softmax equation provided in *BRMS*.

11.2.3 Conditional Indicator Simulation

11.2.3.1 Variogram Parameters

To run geostatistical simulations, the covariance or variogram that describes the spatial autocorrelation of the data needs to the be estimated and modelled. It describes a positive definite function, which may be spherical, exponential, Gaussian, or a hole effect model (Deutsch, 2003).

Stationarity is the basis for geostatistics theory and assumes that there is the same amount of variance from one point to another point in space where the mean is 0 and the covariance is C(h), with h being the separation between values in distance and direction. If the mean is not constant, then the covariance is non-existent and an assumption from Matheron (1965) where intrinsic stationarity is invoked with the expected differences are set to zero and covariance is set to have the variance of the differences, h which is also known as semivariance or, γ(h) the variogram (Oliver & Webster, 2014). This need for stationarity is one of the main reasons for the subsetting of the geological models into spatial domains.

The experimental variogram is estimated from the data z(x) and is defined by the following Equation (11.2)

$$\hat{\gamma}(h) = \frac{1}{2m(h)} \sum_{j=1}^{m(h)} \left\{ z(x_j) - z(x_j + h) \right\}^2 \tag{11.2}$$

where $m(h)$ is the number of pairs that have been compared at a lag (distance) h. The lags are incremented in steps by ordered values and plotted as a curve of variance versus lag steps. The variance and distance or range of each portion of the model along with the direction vector are used in the Kriging equation embedded in the simulation algorithm to estimate a point (Oliver & Webster, 2014).

In this study, all variograms were modelled using *Snowden Supervisor 8.13.0.2* through a process of first obtaining the direction of maximum continuity in the horizontal plane, then in the vertical plane and then in the plunge to define the direction using contours of variance (Coombes, 1996). Once the direction is found, the down hole or omnidirectional variogram is used to estimate the nugget structure, which is the variance at two points with no separation. The directional experimental variogram is then modelled, to obtain the range and sill. The model is checked using the cross-validation technique that uses the leave-one-out method and selecting the model that produces the mean squared error closes to the mean Kriging variance (Oliver & Webster, 2014). Figure 11.1 is an example of the semivariogram model for indicator data to be used in with the downhole variogram to determine the nugget effect or model, and then the major, minor, and semi-minor directional variograms. Variography modelling for the SIS was performed on the indicator transformation of each of the eight rock code variables.

11.2.3.2 Simulation

Stochastic simulation is a process where alternative, equally probable joint realizations are drawn from a random function model. All draws from the model honor the data values at their locations and thus are known as conditional to the data values. The method was initially devised to correct for smoothing in Kriging as Kriging aims to minimize variance. Thus, stochastic simulation trades variance minimization for ensuring the reproduction of the spatial variability of the model (Deutsch & Journel, 1997; Goovaerts, 1997). Sequential indicator can be applied to both categorical data or numerical data (after binning the numerical data into integers). It has been used to simulate geological features and rock type models to enable the assessment of spatial uncertainty of the feature investigated within a spatial domain. In this study, the categorical method using an indicator random function model is used, which states the probability of the category being present using two-point statistics using the RF *I*(u) with the indicator variable *I*(u) set to 1 if present or 0 if absent.

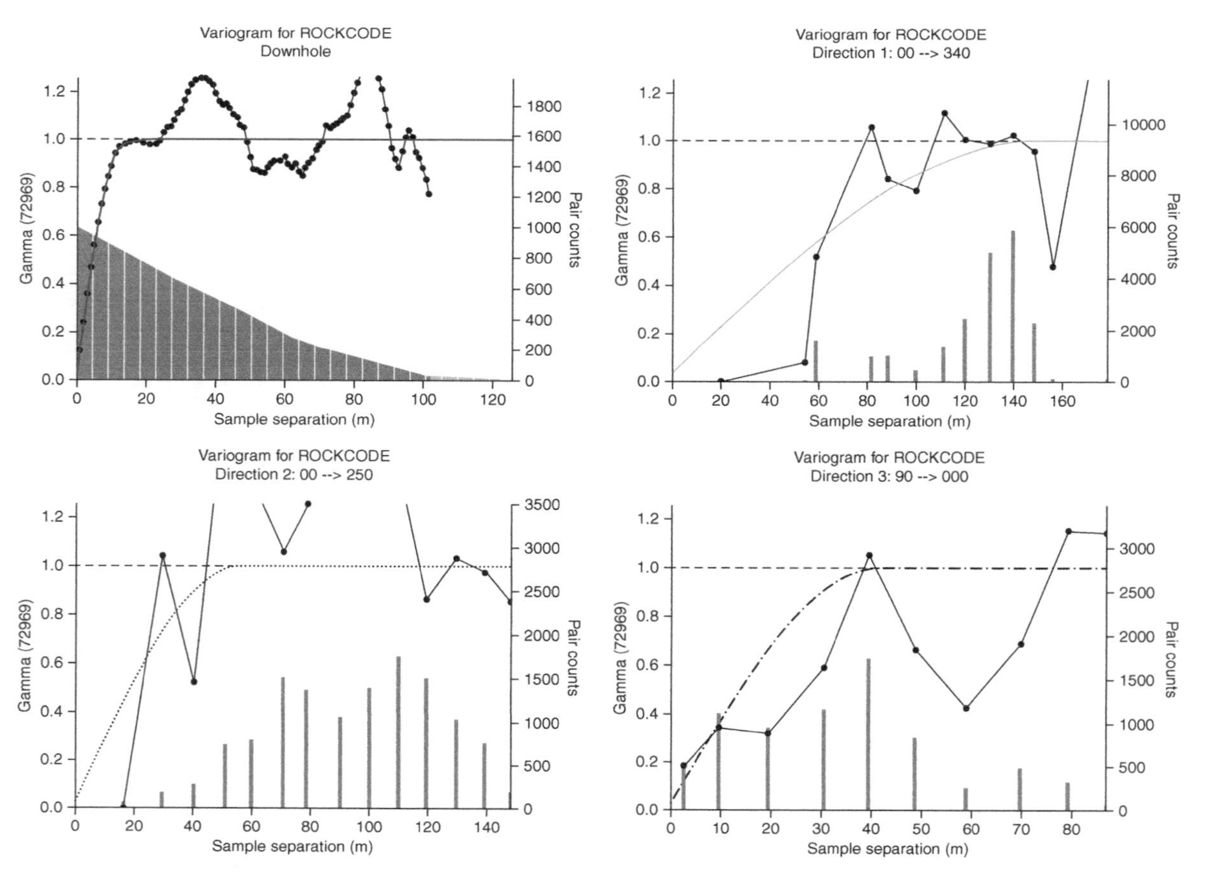

FIGURE 11.1 Downhole variogram and directional variogram model for indicator data for all samples.

For the binary categorical case the following Equation (11.3) holds for $i(\mathrm{u})$:

$$\mathrm{Prob}[\mathrm{I(u)} = 1|(\mathrm{n})] = \mathrm{E}[\mathrm{I(u)}|(\mathrm{n})] \tag{11.3}$$

Or the following for probabilities of a continuous function $z(\mathrm{u})$:

$$\mathrm{Prob}\ [\mathrm{Z(u)} \leq \mathrm{z}|(\mathrm{n})] = \mathrm{E}[\mathrm{I(u; z)}|(\mathrm{n})] \tag{11.4}$$

with

$$\mathrm{I(u; z)} = 1 \text{ if } \mathrm{Z(u)} \leq \mathrm{z}, \text{ or } = 0 \tag{11.5}$$

with the expectation that a point can only belong to one category, the spatial distribution of K mutually exclusive categories is s_k, $k = 1, \ldots, K$.

Simple indicator kriging is used to provide an estimate of the probability that s_k exists at that location u using

$$Prob^*\left\{I\left(u;s_k\right)=1|\left(n\right)\right\} = p_k + \sum_{\alpha=1}^{n} \lambda_\alpha\left[I\left(u_a;s_k\right)-p_k\right] \tag{11.6}$$

where $p_k = E[I(u;s_k]$ ∈ [0, 1] is the marginal frequency of the category s_k, and the weights λ_α are provided by a simple kriging system using the indicator covariance of the category s_k.

At each point along a random path, indicator kriging provides the probabilities from conditioned data that includes all previous estimates as well as the original data, a point is drawn from the set ordered CDF type scaling of the probability interval. The new point is added to the set of conditioned data and the estimator is moved along the path to the next position. The variogram parameters are provided from an indicator variogram model for each category in the spatial domain (Deutsch & Journel, 1997).

The SIS was undertaken using the *Stanford Geostatistical Modelling Software* (*SGeMS*) using the SISIM: sequential indicator simulation module (Remy et al., 2009). One hundred realizations for each model were imported into a Microsoft SQLServer database and managed through *R statistics* using the *RODBC* connection package (Ripley & Lapsley, 2020) and the realizations summarized and analyzed to assess spatial uncertainty using the variance between the simulation realizations. The quality of the simulations was assessed subjectively by viewing 2D sections of plans of each realization and raw point data and assessing the histograms of simulated and raw data. Variography of simulations were compared to variograms of the raw data.

11.2.4 Comparison of Intervals

Output from the models was in the form of Confusion Matrix tables (Rahman et al., 2018) and metrics including Cohens Kapa (Kuhn & Johnson, 2013); overall Accuracy (Liu et al., 2007), Receiver Operating Characteristic (Hand & Till, 2001; Mahmudah et al., 2021; Pham et al., 2021); Specificity (Sokolova & Lapalme, 2009); and

Sensitivity (Sokolova & Lapalme, 2009) as well as the Watanabe-Akaike information criterion (WAIC) (Vehtari et al., 2017), a modification of the AIC for Bayesian simulation data. The uncertainty was displayed in borehole trace sections and uncertainty band plots (Croke, 2009).

To compare Bayesian and expert interpretation uncertainty, which is point data along a borehole trace, with geostatistical spatial uncertainty, which is gridded 3D block data, the following method was used. Block predictions and uncertainty assessments from the SIS have been coded back to the borehole sample. This was carried out using GEMCOM Desktop Edition (version 4.11) mining software by intersecting the borehole sample points with the simulated block three-dimensional space. This allowed the use of model metrics using confusion matrices to compare block predictions and original interpreted classification for the borehole intercept, and to compare block predictions versus Bayesian approximation point predictions on a point-by-point basis along the trace of the boreholes.

Uncertainty values from both methods were visually compared in downhole plot graphs and in correlation and regression analysis (Gelman & Hill, 2007).

11.3 RESULTS

11.3.1 Uncertainty Assessment Using Bayesian Approximation

The rock code models displayed acceptable values for Overall Accuracy and Kappa, and had high specificity and low sensitivity. Two of the rock code models, A103 and A205, had poor accuracy and two rock codes, A101 and A203, had moderate accuracy, which affected the overall metrics.

Figure 11.2 shows the uncertainty band graph for the Bayesian Approximation simulation. Table 11.1 shows the Confusion Matrix for the model and Table 11.2 shows the model and confusion matrix metrics.

The combined model had a reasonable overall accuracy of 0.73 and a moderate Kappa value of 0.49. The specificity was high at 0.93, but the model suffered from poor sensitivity at 0.43. The overall model metrics were lowered due to the lower performing classes A103 and A205. It is possible that they would perform better if modelled individually using a binomial model with a model trained specifically for

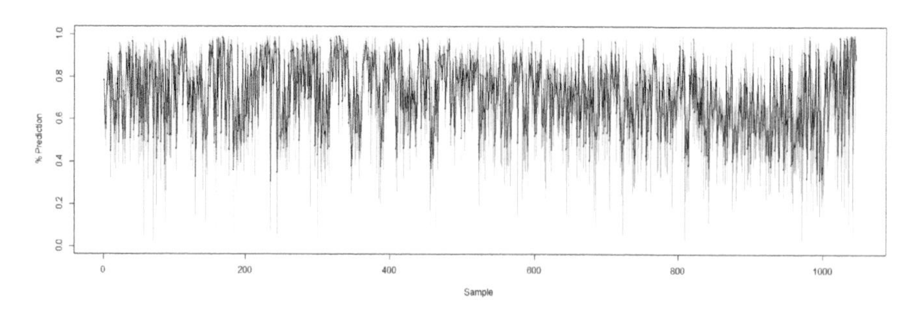

FIGURE 11.2 Uncertainty bands and prediction from rock code combined model (Bayesian approximation method).

TABLE 11.1
Confusion Matrix for Bayesian Approximation of Rock Code Model

Prediction	A0	A101	A102	A103	A202	A203	A204	A205
A0	572	9	68	10	31	56	1	16
A101	2	8	0	0	0	0	0	0
A102	44	8	112	0	1	0	2	4
A103	0	0	0	0	0	0	0	0
A202	6	0	0	0	31	3	0	1
A203	11	0	0	0	1	37	0	1
A204	1	0	2	0	0	0	6	0
A205	0	0	0	0	0	0	0	1

TABLE 11.2
Confusion Matrix Metrics for Rock Code Model from Bayesian Approximation Model

Metric	Combined model
Overall Accuracy	0.73
95% Confidence Interval	(0.71, 0.76)
No Information Rate	0.61
P-Value [Acc > NIR]	0.00
Kappa	0.49
Sensitivity	0.43
Specificity	0.93
WAIC	1737.00
Looic	1745.30
Pareto K < 0.5	97.30

each class; however, it is most likely that these two rock code classifications need more work on their interpretation or data to improve the classifications and interpretation. The low prediction and high uncertainty in these rock codes point to issues in the geological model and sub-set domain. The project may require more drilling to reduce the uncertainty in these domains or for the geologist to carry out alternative interpretations to improve the uncertainty for the model.

Figure 11.3 is a cross-section showing the uncertainty presented in the boreholes. As in Figure 11.2, there are varying amounts of uncertainty, but in this section the contact boundaries between rock codes as well as certain rock codes display high to very high uncertainty. In particular, the sub-vertical borehole BHDD046 and the rock code is interpreted to belong to the other three boreholes around elevation 200, with very high uncertainty. This has been linked to the high probability that this hole is not sub-vertical and is indeed inclined at a 60° angle to the west; in effect, a measurement error causing an error in geological interpretation.

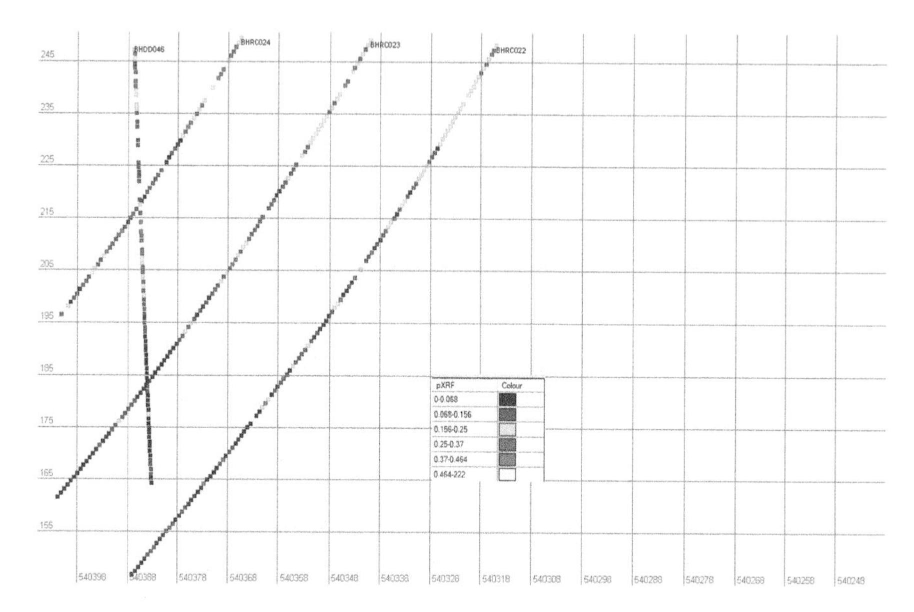

FIGURE 11.3 Bayesian interpretation uncertainty in borehole intervals.

TABLE 11.3
Variogram Parameters Used in SIS Simulation

	Nugget	Sil	Sil 2	Z	X	Y	Ranges		
A101	0.08	0.37	0.48	65.67	2.50	29.90	130	86	36
A102	0.06	0.94	0.00	62.51	9.85	28.48	129	74	48
A103	0.02	0.98	0.00	-40.00	60.00	0.00	86	86	28
A202	0.00	1.00	0.00	54.10	22.52	20.36	69	40	37
A203	0.00	1.00	0.00	65.00	15.00	90.00	67	30	15
A204	0.04	0.96	0.00	140.00	0.00	0.00	140	74	50
A205	0.04	0.96	0.00	113.53	7.64	49.57	109	50	40

11.3.2 Uncertainty Assessment Using SIS

The rock code spatial domain and associated indicator data was used to run SIS and calculate spatial uncertainty for each borehole. Variography for each rock code was modeled in order to determine parameters (the sill and range) to use in the simulation. The variogram parameters for each rock code are summarized in Table 11.3, and Table 11.4 summarizes variogram cross-validation results. Each of the domains had a low nugget effect, suggesting spatial variance was low at short distances and the ranges for the structures extended beyond the spacing of the boreholes. The boreholes were spaced approximately every 25 or 50m; in the along strike range, these extended from 67m to 130m, in the across strike range these ranged from 30 to 86 meters, and in the down dip direction, these had the lowest range from 15 to 50m. So in

TABLE 11.4
Variogram Error Metrics from Cross-Validation Checks

	Mean error	Mean Absolute Error	Mean Error Squared	Mean Kriging Variance	Mean Sample Value	Mean Estimate Value	Correlation Coefficient
A101	0.00	0.02	0.01	0.00	0.02	0.02	0.85
A102	0.00	0.03	0.01	0.02	0.17	0.17	0.96
A103	0.00	0.00	0.00	0.00	0.01	0.01	0.94
A202	0.00	0.01	0.01	0.00	0.06	0.06	0.95
A203	0.00	0.01	0.01	0.01	0.09	0.09	0.97
A204	0.00	0.00	0.00	0.00	0.01	0.01	0.94
A205	0.00	0.01	0.01	0.00	0.02	0.02	0.87

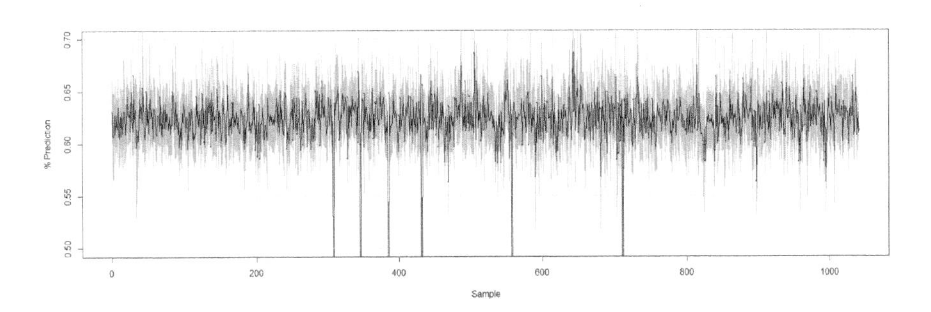

FIGURE 11.4 Uncertainty bands for SIS simulation model.

cross-sections where the along strike direction matches our point of view across the screen variability will be low whilst predicated blocks are within those ranges and closer to the sample points and there are two or more boreholes contributing samples within that range of relationship.

One hundred SIS realizations of each rock code were obtained and the simulated indicator results were converted to probabilities for classification into a rock code for each individual cell. The simulation variance, range, mean, and median were summarized for each cell. To compare spatial uncertainty with the interpretation uncertainty from the Bayesian approximation, the borehole intercepts were coded with the values of the simulated block cells that they intersected with.

Figure 11.4 shows the uncertainty band calculated from the SIS realization's variance. Whilst there are some small areas of uncertainty the effects are very subtle and hard to see, compared to the results from the Bayesian approximation in Figure 11.1. In Figure 11.4 the prediction falls to zero on the graph, which is due to there being only one realization at that block; the summary statistics to determine uncertainty require more than one realization value. The blocks were 2.5x2.5x2.5 m in size, which is double the 1m borehole sample interval. Due to this difference in block size to drill interval size, back coding the block values to the borehole intercept has created a

TABLE 11.5
Confusion Matrix for SIS Simulation Model

	A0	A101	A102	A103	A202	A203	A204	A205
A0	571	8	18	3	7	22	3	4
A101	4	16	5	0	0	0	0	0
A102	13	0	169	0	0	0	0	0
A103	1	0	0	9	0	0	0	0
A202	9	0	0	0	55	0	0	0
A203	7	0	0	0	0	89	0	0
A204	0	0	0	0	0	0	9	0
A205	9	0	0	0	0	0	0	14

TABLE 11.6
Model Metrics for SIS Geostatistical Simulation

Metric	Value
Overall Accuracy	0.89
95% Confidence Interval	(0.87, 0.91)
No Information Rate	0.59
P-Value [Acc > NIR]	0.00
Kappa	0.82
Sensitivity	0.81
Specificity	0.97

smoothing effect. An attempt was made to use a smaller block size to reduce this smoothing effect when back calculating, b tuthe increased processing time and CPU requirements for the simulations made this impractical. This smoothing effect, however, cannot fully account for overall improved uncertainty compared to the Bayesian interpretation uncertainty.

The Confusion Matrix is shown in Table 11.5 and the Confusion Matrix metrics are in Table 11.6. The spatial domain classification using SIS has high overall accuracy of 0.89, kappa of 0.82, sensitivity of 0.81, and selectivity of 0.97. All rock codes, even the two that were problematic for the Bayesian approximation method, A103 and A205, performed well individually. This begins to show the difference between the interpretation uncertainty methodology and the spatial uncertainty methodology. Where we had high uncertainty for these two rock codes assessing their interpretation uncertainty, we now have low uncertainty when assessing their spatial uncertainty.

A cross-section (Figure 11.5) that compares the borehole intercepts and the high uncertainty blocks from the simulation shows that, as expected, blocks at a distance to the borehole samples had a higher uncertainty then those close within the spatial continuity of the variograms. Issues that have come close to the boreholes match some of the higher uncertainty seen in the Bayesian approximation method, which usually

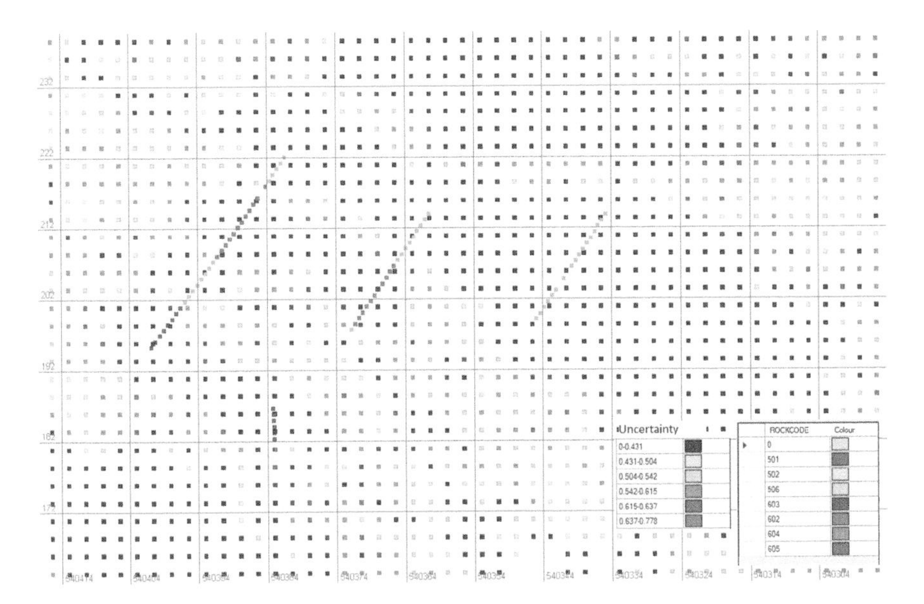

FIGURE 11.5 Cross-section comparison of borehole trace with rock codes and high uncertainty blocks from SIS simulation.

occurs at the start of a borehole, where there is higher oxidation and weathering (and there for chemical mixing over shorter distances as well as soil transportation) or at the boundary between two rock codes (where there may also be chemical mixing from alteration interactions).

A second cross-section (Figure 11.6) shows the uncertainty from the block SIS that has been back coded to the drill hole intervals so the SIS spatial uncertainty can be compared to the Bayesian interpretation uncertainty. Much of the uncertainty is limited to contacts and the upper oxidized/soil zone and the high uncertainty shown in Figure 11.3 due to the measurement error in the sub-vertical borehole are not evident. This could be due to the spatial uncertainty being low due to the blocks that have provided the data for the coding of the boreholes are close to the original data and therefore have lower spatial uncertainty due to density of data and the estimation variance being low as they are spatially within the first structure of the variogram.

11.3.3 Comparison of Interpretation and Spatial Uncertainty

When comparing the back-coded SIS uncertainty to the Bayesian uncertainty in boreholes (Figure 11.7) there is a non-significant very small positive correlation ($r = 0.002$, $p > 0.05$). Using regression analysis it was found that there is no significant linear relationship between SIS spatial uncertainty and interpretation uncertainty using Bayesian approximation for this case study ($df = 1,1040$, $p > 0.05$, $R^2 = 0.0001$).

Figure 11.8 and Figure 11.9 show a side-by-side comparison for two boreholes from top to bottom. Whilst some areas of uncertainty are highlighted as issues from each model, the SIS spatial uncertainty highlights fewer intervals than the interpretation uncertainty.

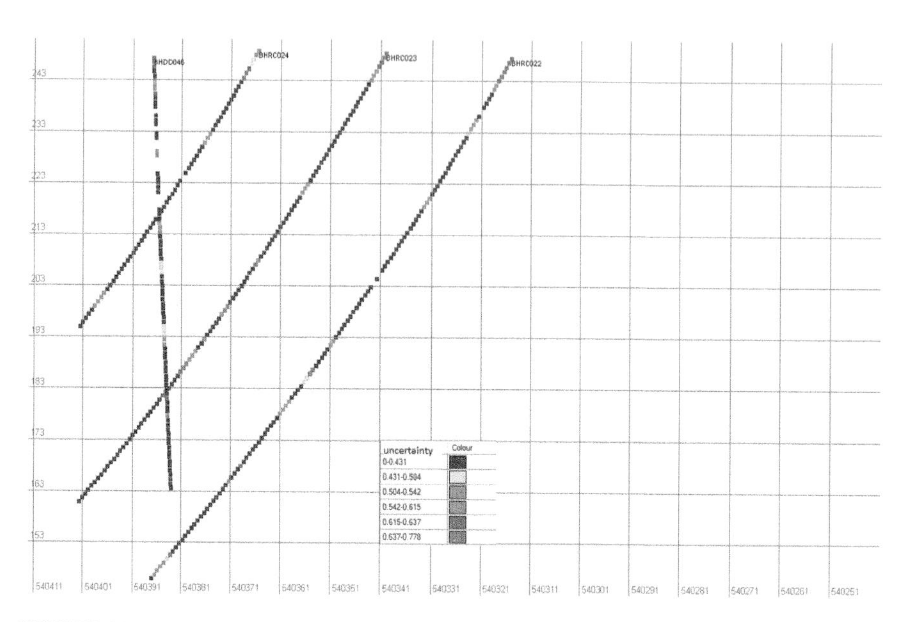

FIGURE 11.6 SIS uncertainty back calculated to borehole intervals.

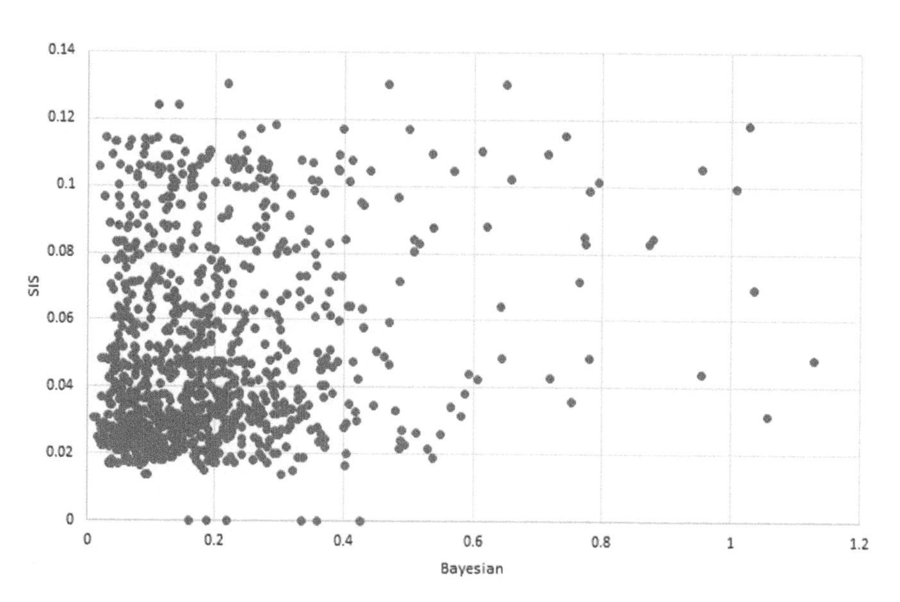

FIGURE 11.7 Scatterplot of Bayesian interpretation uncertainty vs SIS spatial uncertainty.

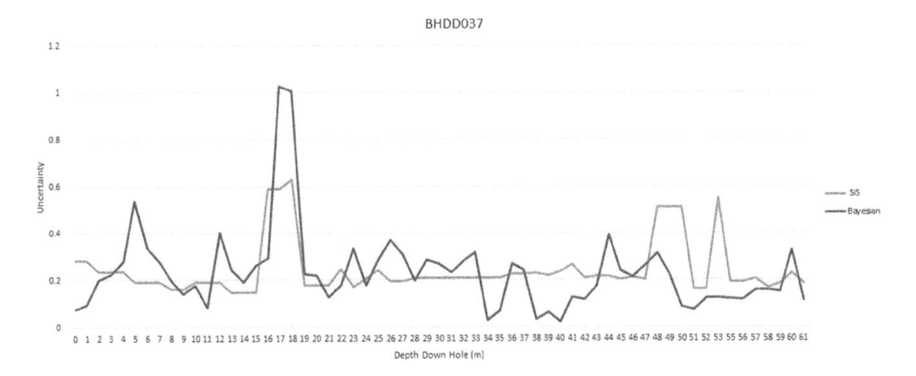

FIGURE 11.8 Comparison of interpretation uncertainty values with SIS uncertainty values in borehole037.

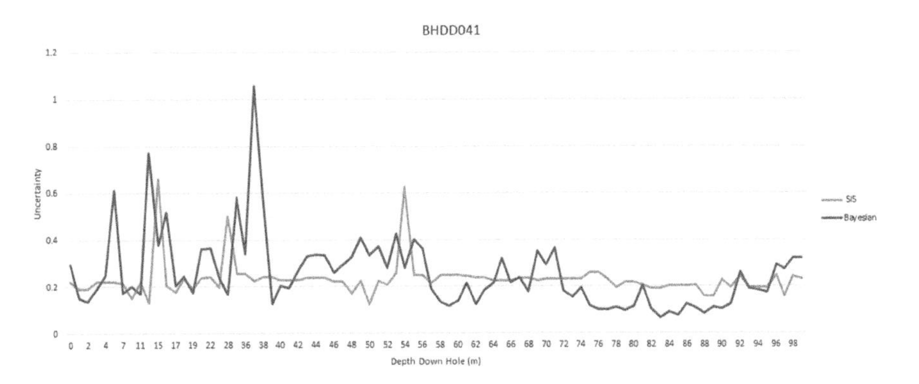

FIGURE 11.9 Comparison of interpretation uncertainty values with SIS uncertainty values in borehole041.

11.3.4 Discussion

Reviewing the uncertainty mapped in cross-section and long section from SIS simulations it is clear that blocks close to the borehole, within the range of the variogram, have a much lower variance then blocks further away (Harjule et al., 2021). This would be expected with the kriging algorithm attempting to minimize variance coupled with the conditioning effect of the simulation. The variogram parameters with lower variance at a shorter range will also minimize the variance of blocks close to boreholes due to the data density.

The method used to back assign boreholes with values from the simulation may also have had a smoothing effect, minimizing the higher variance values. The 3D grid used in the SIS simulation was 2.5x2.5x2.5 m, but boreholes had a length of 1m, which can explain smoothing of the variance with at least two blocks used to determine the value for one borehole intercept. As the borehole intercepts were only informed by the block they intercepted, with no information included from adjacent blocks, it is possible that adjacent blocks that may have had a higher uncertainty value did not contribute to the borehole coding.

In spite of these drawbacks and model assumptions that arise when comparing the two methods, there is still strong confidence that the spatial uncertainty calculated with SIS are a true measure of the error under the model assumptions.

After assigning SIS simulated values to the boreholes, a direct comparison with Bayesian interpretation uncertainty on a sample-by-sample basis using regression analysis shows that there is only a negligible correlation between the two methods.

The model metrics for the SIS were much higher than those achieved for using the Bayesian approximation method; 89% compared to 73% overall accuracy. The key difference was that sensitivity was 81% compared to a very low 30–40%. This was further highlighted in the problematic A205 and A103 rock codes, which had very high accuracy with the SIS method. These differences can be explained by the difference between the two methods.

If the variograms are robust and the nugget effect is low (suggesting the variogram model accounts for most of the variances), it is expected that spatial uncertainties are low near the borehole samples. This happens consistently except when rock codes change or provide contact boundaries down hole causing a subtle increase in variance, as the different realizations predict different rock code outcomes more often. At these contacts there is often good agreement between the two methods.

This calls into question the ability of the interpretation uncertainty method in predicting spatial uncertainty using SIS, but also the ability of geostatistical SIS to account for geological uncertainty in the interpretation of the points when assigning an interval to a spatial domain. A dual approach of using both methods, the first to describe the uncertainty around the interpretation and classifying of each borehole intercept to a spatial domain coupled with the spatial uncertainty in blocks generated from the intercepts to describe the uncertainty due to data density and distance from samples may be required to fully assess the uncertainty in a spatial domain. One without the other is only providing a part of the uncertainty story for the spatial domain. A modified workflow including both methods in the MRE workflow is presented in Figure 11.10 modified after McManus et al. (2022). The workflow progresses through the phases of the mining lifecycle from early to late-stage production. The method of assessing uncertainty in the figure is shown in the green boxes, and the corresponding phase is the earliest that that method of assessment can be used. The newer Bayesian method should be continued to be used in mid- and late-stage projects alongside the two geostatistical simulation methods, SIS and SGS, to

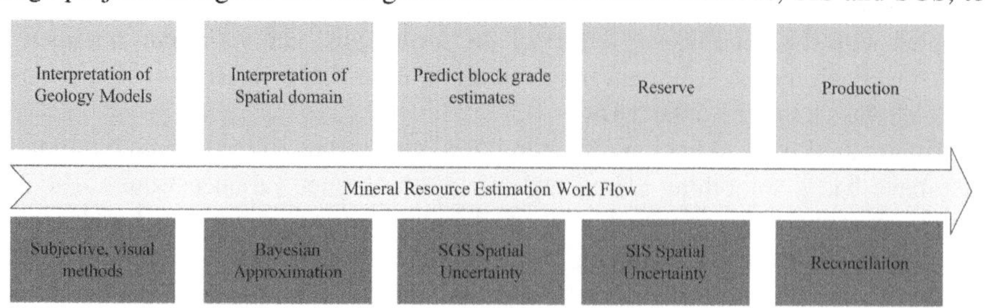

FIGURE 11.10 Modified workflow for MRE incorporating uncertainty assessment.

TABLE 11.7
Comparison of SIS and Bayesian Approximation Methods

SIS	Bayesian Approximation
Spatial uncertainty of geological model	Interpretation uncertainty of geological model
Requires late stage, robust data.	Requires only small amounts of data
Not suitable for early-stage projects	Suitable for any stage project
Does not capture Interpretation uncertainty	Does not capture spatial uncertainty

provide a full picture of uncertainty with the mineral resource estimates for public reporting. When assessing the overall uncertainty of a project, both interpretation and spatial uncertainty should be propagated alongside all other uncertainties. Prior to the application of Bayesian approximation this would be challenging to achieve in early-stage projects. Table 11.7 summarizes the benefits of both the SIS and Bayesian approximation methods of assessing uncertainty; it can be seen that they are able to complement each other in assessing the overall project geological uncertainty.

It is recommended that future work address both the suggested modified methodology and additional case studies to confirm there is no relationship between interpretation and spatial uncertainty in other mineralization styles and at other projects. Whilst it will still be problematic for early-stage projects to implement a SIS strategy in the spatial domain uncertainty workflow, it is clear that advanced projects need to incorporate an interpretation uncertainty element as well as a spatial uncertainty element in their geological or spatial domain uncertainty assessment workflow.

CONFLICT OF INTEREST

On behalf of all authors, the corresponding author states that there is no conflict of interest.

ACKNOWLEDGMENTS

The authors would like to acknowledge the support provided by the Australian Government Research Training Program Scholarship, CSU Spatial Analysis Unit, the CSU Indigenous Support Centre (Port Macquarie), CSU School of Computing and Mathematics, and the CSU School of Environmental Science. The authors are thankful to the anonymous reviewers that helped improve this application chapter.

NOTE

1 These variables are the names given by the company to describe the geological features categorially logged by professional staff after visual inspection. Veins (1–4) are the vein style if a vein is present for the 1st to 4th main vein and Minerals refers to the minerals present in the interval, which are often indicative of mineralisation style, alteration, and rock formation.

REFERENCES

Abdi, H., & Williams, L. J. (2010). Principal component analysis. *Wiley interdisciplinary reviews: computational statistics, 2*(4), 433–459.

Adeli, A., & Emery, X. (2021). Geostatistical simulation of rock physical and geochemical properties with spatial filtering and its application to predictive geological mapping. *Journal of Geochemical Exploration, 220*, 106661. https://doi.org/https://doi.org/10.1016/j.gexplo.2020.106661

Agresti, A. (2013). *Categorical data analysis* (3rd ed. ed.). Wiley.

Agarwal, B., Agarwal, A., Harjule, P., & Rahman, A. (2022a). Understanding the intent behind sharing misinformation on social media. Journal of Experimental and Theoretical Artificial Intelligence, 1–15. https://doi.org/10.1080/0952813X.2021.1960637

Agarwal, B., Rahman, A., Patnaik, S., & Poonia, R. C. (Eds.) (2022b). Proceedings of International Conference on Intelligent Cyber-Physical Systems. (Algorithms for Intelligent Systems). Springer. www.springer.com/gp/book/9789811671357.

Akaike, H. (1974). A new look at the statistical model identification. *IEEE transactions on automatic control, 19*(6), 716–723. https://doi.org/10.1109/TAC.1974.1100705

Breiman, L. (2001). Random Forests. *Machine Learning, 45*(1), 5–32. https://doi.org/10.1023/A:1010933404324

Brooks, S., Gelman, A., Jones, G., & Meng, X.-L. (2011). *Handbook of markov chain monte carlo*. CRC press.

Bürkner, P.-C. (2017). brms: An R package for Bayesian multilevel models using Stan. *Journal of statistical software, 80*(1), 1–28.

Carpenter, B., Gelman, A., Hoffman, M. D., Lee, D., Goodrich, B., Betancourt, M., ... Riddell, A. (2017). Stan: A probabilistic programming language. *Journal of statistical software, 76*(1).

Chiles, J. P., Aug, C., Guillen, A., & Lees, T. (2007). *Modelling the geometry of geological units and its uncertainty in 3D from structural data - The potential-field method* Australasian Institute of Mining and Metallurgy Publication Series,

Coombes, J. (1996). Latest Developments in Visualising Spatial Continuity from Variogram Analysis. AusIMM Conference "Diversity-the Key to Prosperity",

Coombes, J. (2008). *The Art and Science of Resource Estimation*.

Croke, B. (2009). *Representing uncertainty in objective functions: extension to include the influence of serial correlation*.

Deraisme, J., & Field, M. (2006, 21/08/2016). *Geostatistical Simulations of Kimberlite Orebodies and Application to Sampling Optimisation* 6th International Mining Geology Conference, Darwin. www.geovariances.com/en/ressources/geostatistical-simulations-kimberlite-orebodies-application-sampling-optimisation/

Deutsch, C. V. (2003). Geostatistics. In R. A. Meyers (Ed.), *Encyclopedia of Physical Science and Technology (Third Edition)* (pp. 697–707). Academic Press. https://doi.org/https://doi.org/10.1016/B0-12-227410-5/00869-3

Deutsch, C. V., & Journel, A. G. (1997). *GSLIB: Geostatistical Software Library and User's Guide*. (2nd ed.). Oxford university press.

Fisher, R. A. (1992). Statistical methods for research workers. In K. S. & J. N.L. (Eds.), *Breakthroughs in statistics* (pp. 66–70). Springer. https://doi.org/https://doi.org/10.1007/978-1-4612-4380-9_6

Gelman, A. (2013). *Bayesian data analysis* (3rd ed. ed.). CRC Press.

Gelman, A., Carlin, J. B., Stern, H. S., Dunson, D. B., Vehtari, A., & Rubin, D. B. (2013). *Bayesian data analysis*. CRC press.

Gelman, A., & Hill, J. (2007). *Data analysis using regression and multilevel/hierarchical models*. Cambridge University Press.

Gelman, A., Jakulin, A., Pittau, M. G., & Su, Y.-S. (2008). A weakly informative default prior distribution for logistic and other regression models. *Annals of applied Statistics*, *2*(4), 1360–1383.

Gelman, A., & Rubin, D. B. (1992). Inference from iterative simulation using multiple sequences. *Statistical science*, *7*(4), 457–472.

Goovaerts, P. (1997). *Geostatistics for natural resources evaluation*. Oxford University Press on Demand.

Hand, D. J., & Till, R. J. (2001). A Simple Generalisation of the Area Under the ROC Curve for Multiple Class Classification Problems. *Machine Learning*, *45*(2), 171–186. https://doi.org/10.1023/A:1010920819831.

Harjule, P., Rahman, A., & Agarwal, B. (2021). A cross-sectional study of anxiety, stress, perception and mental health towards online learning of school children in India during COVID-19. Journal of Interdisciplinary Mathematics, 24(2), 411–424. https://doi.org/10.1080/09720502.2021.1889780

Hill, E. J., Oliver, N. H., Cleverley, J. S., Nugus, M. J., Carswell, J., & Clark, F. (2014). Characterisation and 3D modelling of a nuggety, vein-hosted gold ore body, Sunrise Dam, Western Australia. *Journal of Structural Geology*, *67*, 222–234.

Jones, P., Douglas, I., & Jewbali, A. (2013). Modeling Combined Geological and Grade Uncertainty: Application of Multiple-Point Simulation at the Apensu Gold Deposit, Ghana [Article]. *Mathematical Geosciences*, *45*(8), 949–965. https://doi.org/10.1007/s11004-013-9500-3

Journel, A. G., & Bitanov, A. (2004). Uncertainty in N/ G ratio in early reservoir development. *Journal of Petroleum Science and Engineering*, *44*(1–2), 115–130. https://doi.org/10.1016/j.petrol.2004.02.009

Kruschke, J. K. (2015). *Doing Bayesian data analysis: A tutorial with R, JAGS, and Stan* (Second edition ed.). Academic Press.

Kuhn, M., & Johnson, K. (2013). *Applied predictive modeling* (Vol. 26). Springer.

Lark, R. M., Mathers, S. J., Thorpe, S., Arkley, S. L. B., Morgan, D. J., & Lawrence, D. J. D. (2013). A statistical assessment of the uncertainty in a 3-D geological framework model. *Proceedings of the Geologists' Association*, *124*(6), 946–958. https://doi.org/https://doi.org/10.1016/j.pgeola.2013.01.005

Liu, C., Frazier, P., & Kumar, L. (2007). Comparative assessment of the measures of thematic classification accuracy. *Remote Sensing of Environment*, *107*(4), 606–616. https://doi.org/https://doi.org/10.1016/j.rse.2006.10.010

Mahmudah, K. R., Purnama, B., Indriani, F., & Satou, K. (2021). Machine Learning Algorithms for Predicting Chronic Obstructive Pulmonary Disease from Gene Expression Data with Class Imbalance.

Matheron, G. (1965). *Les variables régionalisées et leur estimation: une application de la théorie de fonctions aléatoires aux sciences de la nature* (Vol. 4597). Masson et CIE.

McCarthy, P. L. (2014). Managing Risk in Feasibility Studies. In A. C. Edwards, M. Australasian Institute of, & Metallurgy (Eds.), *Mineral resource and ore reserve estimation: The AusIMM guide to good practice / edited by A. C. Edwards. Monograph / Australasian Institute of Mining and Metallurgy; 30* (pp. 13–18). Australasian Institute of Mining and Metallurgy.

McManus, S., Coombes, J., Horta, A., & Rahman, A. (2019). A workflow for assessing interpretation uncertainty in spatial domains using Bayesian approximation. *In International Future Mining Conference 2019: Incorporating the 11th Symposium on Green Mining;*.

McManus, S., Rahman, A., Coombes, J., & Horta, A. (2021a). Comparison of interpretation uncertainty in spatial domains using portable x-ray fluorescence and ICP data. *Applied Computing and Geosciences, 12*, 100067.

McManus, S., Rahman, A., Coombes, J., & Horta, A. (2021b). Uncertainty assessment of spatial domain models in early stage mining projects – a review. *Ore Geology Reviews*, 104098. https://doi.org/https://doi.org/10.1016/j.oregeorev.2021.104098

McManus, S., Rahman, A., Horta, A., & Coombes, J. (2020). Applied Bayesian Modeling for Assessment of Interpretation Uncertainty in Spatial Domains. In A. Rahman (Ed.), *Statistics for Data Science and Policy Analysis* (pp. 3–13). Springer Singapore.

McManus, S., Rahman, A., Horta, A., & Coombes, J. (2022). Measuring spatial domain models' uncertainty for mining industries. 12th International Mining Geology Confernce: Conference Proceedings,

Mery, N., Emery, X., Cáceres, A., Ribeiro, D., & Cunha, E. (2017). Geostatistical modeling of the geological uncertainty in an iron ore deposit. *Ore Geology Reviews, 88*, 336–351. https://doi.org/https://doi.org/10.1016/j.oregeorev.2017.05.011

Mustapha, H., & Dimitrakopoulos, R. (2010). High-order Stochastic Simulation of Complex Spatially Distributed Natural Phenomena [journal article]. *Mathematical Geosciences, 42*(5), 457-485. https://doi.org/10.1007/s11004-010-9291-8

Nielsen, S. H. H., Partington, G. A., Franey, D., & Dwight, T. (2019). 3D mineral potential modelling of gold distribution at the Tampia gold deposit. *Ore Geology Reviews, 109*, 276–289. https://doi.org/10.1016/j.oregeorev.2019.04.012

Oliver, M., & Webster, R. (2014). A tutorial guide to geostatistics: Computing and modelling variograms and kriging. *CATENA, 113*, 56–69.

Peters, K., Partington, G., Blevin, P., Downes, P., & Nelson, M. (2017). The Southern New England Orogen Mineral Potential Project. AusIMM New Zealand Branch Conference,

Pham, B. T., Luu, C., Phong, T. V., Trinh, P. T., Shirzadi, A., Renoud, S., ... Clague, J. J. (2021). Can deep learning algorithms outperform benchmark machine learning algorithms in flood susceptibility modeling? *Journal of Hydrology, 592*, 125615. https://doi.org/https://doi.org/10.1016/j.jhydrol.2020.125615

R Core Team. (2018). *R: A language and environment for statistical computing*. In R Foundation for Statistical Computing. www.R-project.org/.

Rahman, A. (2008). *Bayesian predictive inference for some linear models under Student-t errors*. VDM Publishing.

Rahman, A. (2019). Statistics-based data preprocessing methods and machine learning algorithms for big data analysis. *International Journal of Artificial Intelligence, 17*(2), 44.

Rahman, A. (2020). *Statistics for Data Science and Policy Analysis*. Springer.

Rahman, A., & Harding, A. (2016). *Small area estimation and microsimulation modeling*. Chapman and Hall/CRC.

Rahman, A., Nimmy, S., & Sarowar, G. (2018). Developing an Automated Machine Learning Approach to Test Discontinuity in DNA for Detecting Tuberculosis. In J. P. Davim, *Proceedings of the Twelfth International Conference on Management Science and Engineering Management*

Remy, N., Boucher, A., & Wu, J. (2009). *Applied geostatistics with SGeMS: A user's guide*. Cambridge University Press.

Ripley, B., & Lapsley, M. (2020). Package 'RODBC'.

Ripley, B., Venables, B., Bates, D. M., Hornik, K., Gebhardt, A., & Firth, D. (2018). Package 'mass'. *CRAN Repos. Httpcran R-Proj. OrgwebpackagesMASSMASS Pdf.*

Scheidt, C., & Caers, J. (2009). Representing Spatial Uncertainty Using Distances and Kernels. *Mathematical Geosciences, 41*(4), 397–419. https://doi.org/10.1007/s11004-008-9186-0

Siler, D. L., Hinz, N. H., Faulds, J. E., Ayling, B., Blake, K., Tiedeman, A., ... Rhodes, G. (2018). The geologic and structural framework of the Fallon FORGE site. Proceedings of the 43rd Workshop on Geothermal Reservoir Engineering, Stanford University,

Sokolova, M., & Lapalme, G. (2009). A systematic analysis of performance measures for classification tasks. *Information Processing & Management*, *45*(4), 427–437. https://doi.org/https://doi.org/10.1016/j.ipm.2009.03.002

Sterk, R., de Jong, K., Partington, G., Kerkvliet, S., & van de Ven, M. (2019). *Domaining in Mineral Resource Estimation: A Stock-Take of 2019 Common Practice* www.kenex.co.nz/documents/papers/Sterk%20et%20al%20-%20Domaining%20In%20Resource%20estimation.pdf

Sullivan, S. (2021). *Recognising the impact of uncertainty in resource models* 3rd AEGC: Geosciences for Sustainable World – September 2021, Brisbane, Australia, Brisbane.

Sullivan, S. (2022). Harnessing data complexity - how machine learning applies all project data for accurate resource modelling. International Mining Geology Confernce 2022,

Vallee, M. (2000). Mineral resource+ engineering, economic and legal feasibility= ore reserve. *CIM bulletin*, *93*(1038), 53–61.

Vehtari, A., Gelman, A., & Gabry, J. (2017). Practical Bayesian model evaluation using leave-one-out cross-validation and WAIC. *Statistics and computing*, *27*(5), 1413–1432. https://doi.org/10.1007/s11222-016-9696-4

Vehtari, A., Simpson, D., Gelman, A., Yao, Y., & Gabry, J. (2019). Pareto Smoothed Importance Sampling.

Wellmann, J. F., Horowitz, F. G., Schill, E., & Regenauer-Lieb, K. (2010). Towards incorporating uncertainty of structural data in 3D geological inversion. *Tectonophysics*, *490*(3), 141–151. https://doi.org/https://doi.org/10.1016/j.tecto.2010.04.022

Witter, J. B., Siler, D. L., Faulds, J. E., & Hinz, N. H. (2016). 3D geophysical inversion modeling of gravity data to test the 3D geologic model of the Bradys geothermal area, Nevada, USA. *Geothermal Energy*, *4*(1), 14.

Yarus, J., Chambers, R., Maučec, M., & Shi, G. (2012). *Facies Simulation in Practice: Lithotype proportion mapping and Plurigaussian Simulation, a powerful combination* Ninth International Geostatistics Congress, At Oslo, Norway, www.geovariances.com/en/resources/facies-simulation-practice-lithotype-proportion-mapping-plurigaussian-simulation-powerful-combination/

12 A Comparison of Geocomputational Models for Validating Geospatial Distribution of Water Quality Index

Md Galal Uddin[*1], *Stephen Nash*[1], *Mir Talas Mahammad Diganta*[1], *Azizur Rahman*[2] *and Agnieszka I. Olbert*[1]
[1]National University of Ireland Galway, Ireland
[2] Charles Sturt University, Wagga Wagga, Australia
*Corresponding author: Md Galal Uddin, Ph.D. Candidate, Civil Engineering, College of Science and Engineering, University of Galway, Ireland. Email: u.mdgalal1@nuigalway.ie

CONTENTS

DOI: 10.1201/9781003253051-16

12.1 INTRODUCTION

Water resources management is an essential task to maintain to ensure good water quality in aquatic ecosystems. It will be extremely difficult for future generations in the world to maintain "good water quality status" in all states, since the quality of the water deteriorates over time gradually due to the association of various factors, manmade intervention being one of them. Industrialization and urbanization have accelerated day by day in order to ensure a better quality of life. As consequences, freshwater consumption has significantly increased over the decades (Javed et al., 2017; Uddin et al., 2020a). Therefore, surface and ground water quality degradation has been occurring very rapidly due to the both anthropogenic activities and natural processes (Uddin et al., 2021; Uddin et al., 2020b).

In order to ensure the good water quality status, sustainable management plans, policies, and adequate resources are required. Water resources management is a complex and critical process that involves a variety of components such as institutional framework, skilled labour, legislations and financial framework, resources availability, etc. It is a challenging process not only for developing countries but also recently addressed by developed countries and agencies. Several countries have formulated various action plans to maintain good water quality status in all water bodies but have run into frequent issues when implementing and adopting management programs.

In Europe, the European Union (EU) is mainly responsible for formulating plans and policy for ensuring good water quality in the member states. The Water Framework Directive (WFD) is one of the important tools for managing quality of water and ecosystems (Carsten Von Der Ohe et al., 2007; Zotou et al., 2020). This tool provides in detail guidelines for the management of aquatic environments, but it has not defined any specific tools or techniques that should be used to assess water quality globally in EU member states. While it recommended adopting monitoring programs for surveillance of water quality in all member states, this is very costly and time consuming. Thus, many countries have been suffering these challenges and trying to overcome them.

To date, a variety of tools and techniques are used for assessing water quality and the water quality index (WQI) model is one of them. The WQI model is a simple mathematical tool whose uses have been increasing very rapidly around the world due to its ease of application compared to other traditional models. It allows converting a range of water quality information into a single unit-less numerical value. Commonly, this technique consists of four crucial components: (i) water quality parameter selection using several techniques to identify the important parameters; (ii) sub-index process to convert different dimensional variables into universal form; (iii) weighting of water quality parameters is the process to assign the parameters weight based on their relative importance; and (iv) aggregation function is the final and most important step of the WQI model to transfer sub-index and weight values into a single numerical value (Gupta and Gupta, 2021; Parween et al., 2022; Sutadian et al., 2016; Uddin et al., 2021). Details of the WQI models and their uses can be found in Uddin et. al. (2021). Recently, several studies have revealed that the WQI model produces considerable uncertainty by its own process (Abbasi and Abbasi, 2012; Uddin et al., 2022d; 2022e). As a result, accurate water quality information does not reflect its outcomes. Many researchers have developed various techniques to reduce the uncertainty in the WQI model (Juwana et al., 2016). Recently, we carried out a comprehensive study on the WQI model. In our study, we compared eight aggregation functions (five are widely used and three are newly proposed by the authors) to identify the best aggregation method in terms of uncertainty (Uddin et al., 2022b). That study revealed that the new weighted quadratic mean (WQM) aggregation functions gave excellent performance in assessing coastal water quality. For that reason, in this study, we used the WQM-WQI model to calculate WQIs for coastal water quality in Cork Harbour, Ireland.

For the limitation of surveillance programme, WQI simulation or prediction is essential to estimate WQI's value at unknown location or grid points. Many tools and techniques are widely used to predict unknown point values and spatial interpolation technique is one of the most common approaches (Farzaneh et al., 2022). Up to now, many interpolation techniques have been utilized to predict the spatial distribution of water quality (Antal et al., 2021; Borges et al., 2016; Uddin et al., 2018). But they are mainly categorized into two groups: (i) deterministic and (ii) geostatistical methods.

Deterministic geocomputational interpolation methods well known as the exact interpolator technique. Usually, this technique is used for creating surfaces (polynomial trend surface) from known points using the extent of the similarity or the degree of smoothing (radial basis function) (Adhikary and Dash, 2017). Polynomial trend surface analysis is one of the popular techniques used for global interpolation where measured data is fitted with a linear or quadratic function in order to create the interpolated surface within the geospatial extent (Verma et al., 2019). It can be classified into two groups; one is global approaches and the other one is local. Whereas the global approaches are used for the prediction of the entire geospatial extent using the entire dataset, the local methods predict unknown grid point values from the measured grid points within neighborhoods. Generally, this technique is widely used to predict unknown grid points across a smaller area within a large geospatial extent (Pellicone et al., 2018). A number of deterministic geocomputational methods are commonly used for prediction of unknown grid points, including inverse distance weighted (IDW), spline, Thiessen polygon, and linear regression (Verma et al., 2019).

Usually, the geostatistical interpolation techniques applies for predicting unknown grid points data using the statistical properties of the known points within given dataset (Adhikary and Dash, 2017). This technique is a well-known kriging method for analysing spatial distribution of measured data. In terms of statistical approaches, it is a very effective interpolation technique because it allows spatial correlation among grid points of an entire area (Adhikary and Dash, 2017; Al-Mamoori et al., 2021). Several kringing/geostatistical methods are widely used to interpolate or predict unknown grid values; there are simple kriging (SK), ordinary kriging (OK), universal kriging (UK), co-kriging, regression kriging, indicator kriging, etc. (Bronowicka-Mielniczuk et al., 2019; Han et al., 2020; Jalili Pirani and Modarres, 2020; Wu et al., 2019).

Recently, several studies have been carried out for interpolating various environmental factors such as climatic variables (Amini et al., 2019; Attorre et al., 2007; Verma et al., 2019; Yan et al., 2005); wind speed prediction (Luo et al., 2008); air temperature; snowmelt prediction (Hock, 2003); surface water quality variable predictions (Khouni et al., 2021; Murphy et al., 2010; Uddin et al., 2020a; Wu et al., 2019); soil pH and salinity in coastal water (Emadi and Baghernejad, 2014); heavy metal distribution analysis in rivers (Madhloom et al., 2018); and rainfall/precipitation interpolation (Antal et al., 2021; Bárdossy and Pegram, 2013; Borges et al., 2016; Jalili Pirani and Modarres, 2020; Liu et al., 2020; Pellicone et al., 2018; Wang et al., 2014; Zhang and Srinivasan, 2009). A number of studies have utilized various interpolation techniques for spatial distribution of water quality index for assessing surface water quality (Kawo and Karuppannan, 2018; Kumar et al., 2022; Muzenda et al., 2019; Nikitin et al., 2021). Moreover, for the purposes of performance analysis and to identify the best interpolation/prediction techniques, recently, a few studies have been carried out comparing various geostatistical interpolation methods (Adhikary and Dash, 2017; Antal et al., 2021; Attorre et al., 2007; Emadi and Baghernejad, 2014; Farzaneh et al., 2022; Liu et al., 2020; Luo et al., 2008; Meng et al., 2013; Murphy et al., 2010; Uddin et al., 2020a; Verma et al., 2019; Wang et al., 2014; Wu et al., 2019). The present study applied eight interpolation techniques including four deterministic: local polynomial interpolation (LPI), global polynomial interpolation (GPI), inverse distance weighted interpolation (IDW), and radial basis function (RBF) and four geostatistical models: simple kriging (SK), universal kringing (UK), disjunctive kriging (DK), and empirical bayesian kriging (EBK) to determine the best interpolation/prediction model for predicting WQM-WQIs value for coastal water. Finally, the model performance was evaluated using the cross-validation (CV) approach, a technique widely used to evaluate prediction model performance (Belete and Huchaiah, 2021; Agarwal et al. 2022a; Agarwal et al. 2022b).

Therefore, the aim of this study was to identify the appropriate geocomputational-interpolation method for the spatial distribution of WQM-WQIs properly at each unknown grid point in Cork Harbour, Ireland. This study can directly support the water quality-monitoring program by providing an insightful application of the WQI model. For the purposes of achieving the goals of this research, a few specific objectives have been carried out as follows:

(i) To calculate WQIs using WQM-WQI model for coastal water quality in Cork Harbour.
(ii) To utilize eight geocomputational-interpolating techniques to predict the spatial distribution of WQIs.
(iii) To identify the best geocomputational model by comparing eight interpolation techniques in order to predict WQIs accurately at each monitoring site.

The chapter is structured as follows: Section 12.2 describes the study domain, Section 12.3 provides the descriptions of various geostatistical and other statistical methods for assessing the models (Harjule et al. 2021), Section 12.4 presents results and discusses the output of the prediction models, and Section 12.5 summarizes the main conclusions of this research.

12.2 APPLICATION DOMAIN: A CASE STUDY IN CORK HARBOUR

The present study was conducted in Cork Harbour, a Special Protection Area (SPA) that is relatively the deepest and longest (17.72 km) surface waterbody in Ireland (Hartnett and Nash, 2015; Nash et al., 2011). The harbour is a large surface area (85.85 km^2) and brackish estuary on the south coast of Ireland (Nash et al., 2011; Uddin et al., 2022c). It is a macro-tidal harbour with a typical spring tide range of 4.2 m at the entrance to the harbour (Hartnett and Nash, 2015). The Cork city is the industrial hub of the Irish southwest region and the surrounding hinterlands are subject to relatively intense agricultural activities, which impact water quality in the region (EPA, 2017). Recently, several annual environmental reports from the EPA revealed that Cork and Donegal received the highest raw discharge wastewater directly without any treatment from the various sources (EPA, 2017). Moreover, the Cork Harbour geological patterns are vital for harbour area ecosystem and freshwater quality. It has been identified as a SPA under the 1979 Wild birds Directive (79/409/EEC).

12.3 METHODS AND MATERIALS

12.3.1 Data Obtaining Process

For the purposes of this study, the water-quality data was obtained from 29 monitoring sites out of 32 across the Cork Harbour monitoring data in the year 2020. Details of the monitoring sites are illustrated in Figure 12.1. Typically, the Irish Environmental Protection Agency (EPA) monitors the water quality of Cork Habour frequently. In this study, water-quality data was considered at one-meter below water surface at approximately high and low tides over the year. In order to calculate WQI values, in total 11 water-quality parameters were obtained from the EPA water quality-monitoring database of Cork Harbour. These were temperature (TEMP), pH, dissolved oxygen (DOX), salinity (SAL), ammonia (AMN), total organic nitrogen (TON), ammoniacal nitrogen (AMN), molybdate reactive phosphorus (MRP), biological oxygen demand (BOD), transparency (TRAN), and *Chlorophyll a* (CHL). Details of the data can be found at the following web resource: www.catchments.ie/data. These parameters were selected for this study based on the availability of variables in the monitoring database and taking account of the spatial distribution of monitoring sites.

12.3.2 WQI Calculation

To date, a range of WQI models have been used to calculate the WQI values. Its application has increased sequentially due to its simple mathematical functions and ease of

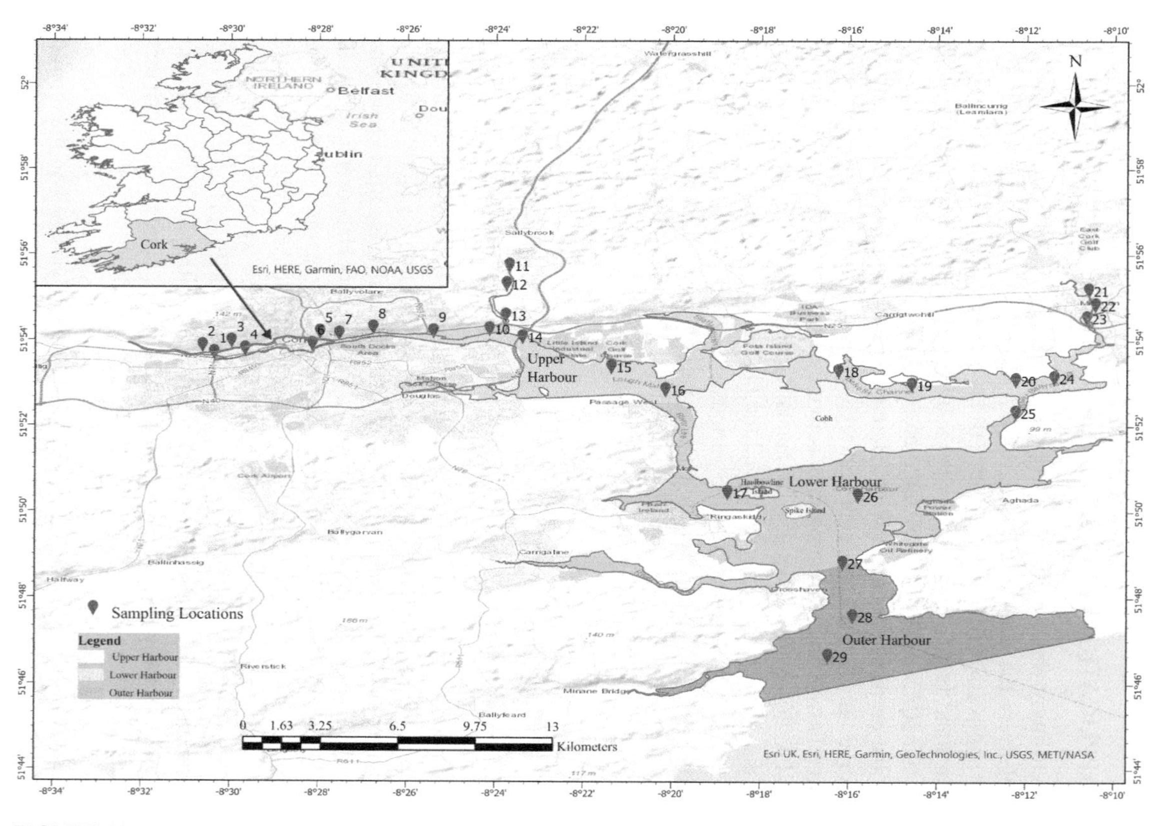

FIGURE 12.1 Study domain: Cork Harbour, Ireland.

application. Commonly, an ideal WQI model comprises four components: (i) water indicators selection; (ii) sub-index function; (iii) indicator weight value generation; and (iv) aggregation function. The existing literature on WQI models is extensive and focuses particularly on details of WQI models and their uses (Gupta and Gupta, 2021; Sutadian et al., 2016; Uddin et al., 2021, 2022a). In this research, WQI values were calculated using the weighted quadratic mean (WQM) WQI model, a technique performed based on the conceptual framework proposed by Uddin et al. (2022b). The WQM-WQI function can be expressed as follows:

$$WQM - WQI = \sqrt{\sum_{i=1}^{n} w_i s_i^2} \tag{12.1}$$

where w_i is the weight value and s_i is refers to the sub-index value of water quality ith indicators.

12.3.3 Prediction Techniques

Thus far, several geocomputational tools and techniques have been used to analyse the spatial distribution of water quality. Geocomputational-interpolation techniques allow to calculate unknown grid point values using the weighted means of sampled data by applying a set of mathematical, statistical, and spatial approaches (Rahman, 2020; Das et al., 2018; McManus et al., 2020, 2021; Rahman et al., 2013; Rahman & Harding, 2016; Uddin et al., 2018; Uddin et al., 2020a; Verma et al., 2019; Xie et al., 2011; Zandi et al., 2011). In this study, eight geocomputational prediction models were utilized to compare their performances to find the best model to properly predict WQI values.

12.3.3.1 Spatial Computational Methods

Available geocomputational-interpolation techniques can be classified into two groups: deterministic and geostatistical (Rahman and Harding, 2016). In this research, in total eight interpolation techniques were selected from two groups, with four techniques included in each group. These techniques were selected for this study based on their application and literature resources. A short description of various interpolation techniques and details of the methodological framework are presented in Figure 12.2.

12.3.3.1.1 Deterministic Interpolation Methods

(a) Deterministic Technique

In general, the geometric properties of the samples are the main attributes of the deterministic techniques, whereas the spatial autocorrelations of the target variables are the main attributes of the geocomputational techniques (Verma et al., 2019). The four techniques used in this research are as follows.

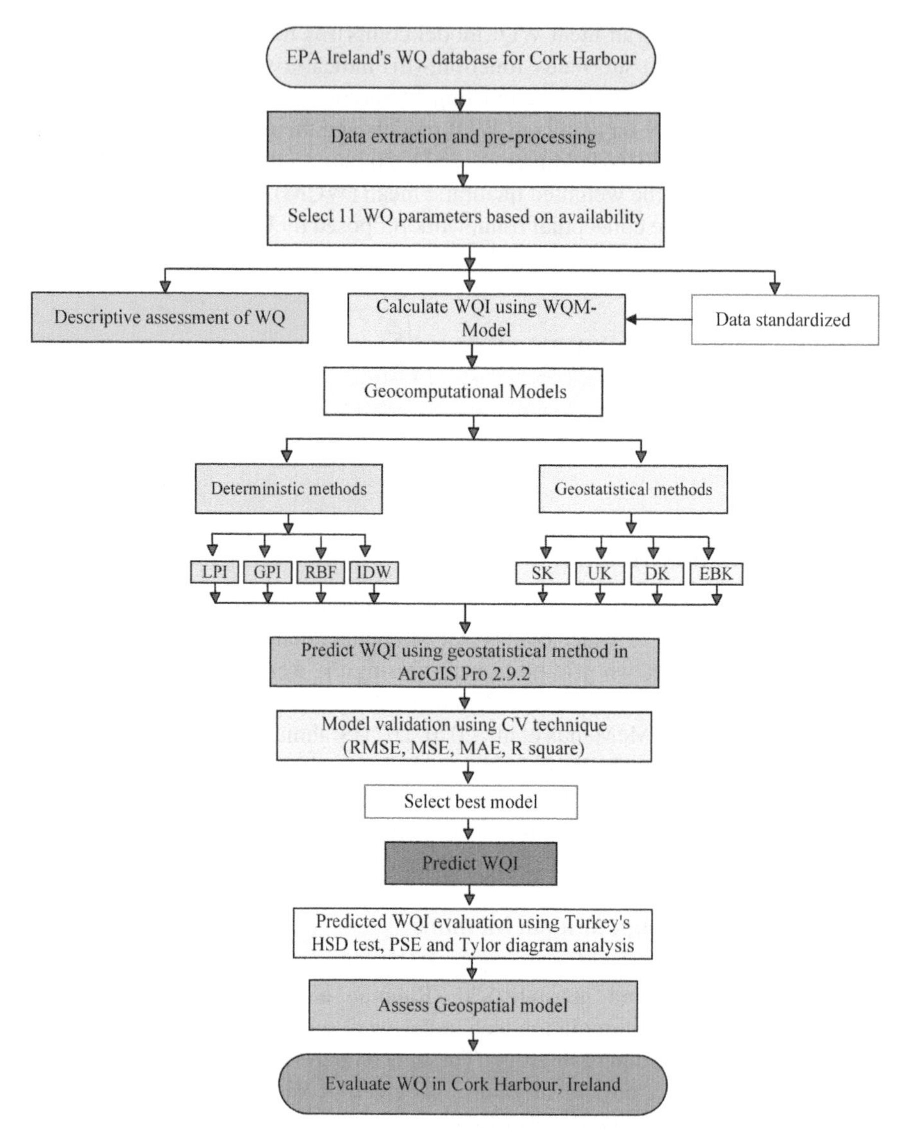

FIGURE 12.2 A comprehensive framework for the assessment of geospatial distribution tools for predicting WQIs.

(i) Global polynomial interpolation (GPI)

GPI is an effective interpolation technique to predict the lowest data variability space. It fits a smooth surface that is defined by a polynomial mathematical function to the unknown sample point (Wang et al., 2014). This technique enables to capture coarse-scale pattern in the data where the global polynomial surface changes gradually (Antal et al., 2021; ESRI, n.d.; Verma et al., 2019). The polynomial of degree n is fitted for the surface changes and is defined as follows:

$$z = ax + by + c \tag{12.2}$$

where z is the value of target variable, a, b, and c are input parameters, and x and y are the coordinates of sample.

(ii) Local polynomial interpolation (LPI)

LPI is completely different than the GPI approach. This technique is effective to predict short-range surface area whereas GPI is performed for long-range data variation (Luo et al., 2008; Wang et al., 2014). LPI fits the local polynomial using known points only within a specific neighbourhood instead of all the data. The neighbourhood is able to overlap after the polynomial has been fitted using any kind of kernel, and the surface value in the neighborhood's centre predicts an unknown value (Antal et al., 2021). The LPI can be defined as:

$$l = \left(1 - \frac{d_i}{N}\right)p \tag{12.3}$$

where d_i is the range between actual and predicted WQI values, N refers to the neighbourhood area, and p is the order of the polynomial function that is defined by the operators.

(iii) Radial basis function (RBF)

Unlike GPI and LPI, RBF requires the surface to pass through all actual data points (Wang et al., 2014). It is a sequence of actual interpolation techniques that utilizes the basic mathematical equation depending on the detachment between the known point and unknown grid point (Uddin et al., 2018). This technique enables to predict the unknown value of any roughness surface effectively whereas other interpolation methods are suitable only for the plain surface (Amini et al., 2019). The RBF can be defined as (Uddin et al., 2018):

$$K(y) = \sum_{i=1}^{p} C_i f_i(y) + \sum_{i=1}^{q} E_j \varnothing(d_j) \tag{12.4}$$

where (dj) expresses the radial basis functions and E_j the distance from the observation point to projection point y trend function is $f_i(y)$, C is member of a basis for the space of polynomials of degree $<p$. E_j are calculated and the coefficients C_i through means of the determination of the subsequent coordination of $q + p$ linear equations; and q is the whole number of selected points used in the interpolation.

(iv) Inverse distance weighted interpolation (IDW)

IDW is a widely used interpolation technique that calculates an unknown grid value using a linear combination of known grid points. It allows predicting the value of any unknown grid point using actual values surrounding the prediction location. Likely, the closest predicted point is highly influenced by the nearest known sample point, and its farthest points gradually decrease (ESRI, n.d.). Commonly, this technique predicts the value at each point using the weighted average of the nearest known points. Whereas the weight values are estimated using the inverse distance between known and predicted sample points (Uddin et al., 2018; Xiao et al., 2016). Measurements of distances can be defined as:

$$D = \frac{\sum_{1}^{n} \frac{z_i}{d_i^p}}{\sum_{1}^{n} \frac{1}{d_i^p}} \tag{12.5}$$

where D refers to the predicted point, n is the number of neighbours nominated for the prediction, p is the power of distance, z_i is the known point of ith observation, and d_i denotes to the distance between unknown and known points.

12.3.3.1.2 Geostatistical Methods

(i) Simple or Ordinary Kriging (SK)

Compared to the deterministic methods, SK is the most powerful interpolation method to precisely predict unknown grid values more. It is considered a spatial and statistical correlation between known and unknown grid values (Wang et al., 2014). This technique assumes that the mean of the known points is constant, whereas the unknown focuses on spatial attributes, and it uses only local known points within a neighbourhood to predict the unknown point value. Recently, several studies have revealed that the SK methods provide the smallest standard error in order to predict unknown values of any grid point within a surface (Verma et al., 2019; Wang et al., 2014). SK can be expressed as:

$$K_s = \sum_{i=1}^{n} w_i M(X_i) \tag{12.6}$$

where K is the predicted unknown value, w_i is the unknown values of the known point, and ad $M(X_i)$ is the known vale of ith grid points.

(ii) Universal Kriging (UK)

The UK technique is an extension of the SK and is applied to predict unknown point values more accurately. This technique was developed by the mathematician Georges Matheron in 1963 (Borges et al., 2016). The UK assumes the spatial interpolation model as:

$$Z(s) = \mu(s) + \varepsilon(s) \tag{12.7}$$

where $\mu(s)$is a few deterministic functions that are obtained using SK concepts, and $\varepsilon(s)$is the errors of the samples obtained from the randomly modelled data (ESRI, n.d.).

(iii) Disjunctive Kriging (DK)

The DK predicts unknown values using the regionalized variable theory. It calculates the conditional probabilities; that is, the known values of samples of interest equal or exceed the defined thresholds. This technique involves defining a new disjoint parameter from the original continuous variable where the parameters values equal to or exceed the threshold, then it accepts 1, and 0 otherwise (Oliver, 1991). DK assumes a model that can be defined as:

$$f(Z(s)) = \mu_1 + \varepsilon(s) \tag{12.8}$$

where μ_1 is the unknown constant and $f(Z(s))$ is an arbitrary function of Z(s) and $\varepsilon(s)$ is the errors of the samples that estimates from the predictive errors randomly.

(iv) Empirical Bayesian kriging (EBK)

EBK is a robust geostatistical interpolation technique that interpolates unknown point values automatically within most difficult aspects and optimizes the model parameters with an instinctive process of subsetting and simulation (Krivoruchko, 2012; Konstantin Krivoruchko, 2012; Mainuri and Owino, 2017; Pellicone et al., 2018; Uddin et al., 2020a). A few researchers have revealed that the EBK technique is more reliable for predicting unknown grid values with minimum errors (Antal et al., 2021; Gupta et al., 2017; K Krivoruchko, 2012; Mainuri and Owino, 2017; Payblas,2018;; Uddin et al., 2020a). Unlike other kriging approaches, the EBK generates the unknown points prediction values considering model uncertainty while other kriging techniques predict unknown value using the semivariogram of known data points (ESRI, 2016, 2007; Gupta et al., 2017; Kumari et al., 2018). Consequently, the EBK estimates prediction outputs more accurately than other kriging methods (Gupta et al., 2017; Mainuri and Owino, 2017; Payblas,2018. Several studies have successfully applied the EBK technique for predicting unknown locations data and analysis of prediction data uncertainty (Acharya and Panigrahi, 2016; Çaldırak and Kurtuluş, 2018; Gunarathna et al., 2016; Hussain et al., 2016; Kumari et al., 2018; Pandey et al., 2015; Payblas, 2018; Pellicone et al., 2018). The empirical semivariogram was calculated using the following equation:

$$\gamma(h \pm \delta) = \frac{1}{2|N(h \pm \delta|} \sum_{(i,j)\in N(h\pm\delta)} \left|z_i - z_j\right|^2 \tag{12.9}$$

where h is the distance between unkown points, zδ is the tolerance range between points, and N(h $\pm$ δ) is a set of points $N(h \pm \delta) \equiv \{(S_{i,} S_j) : |S_i S_j| = h \pm \delta; i, j = 1, 2, \ldots, N\}$. The $\left|z_i - z_j\right|^2$ are the squared variances between observations. The squared variances are added and normalized by the natural number N (h $\pm$ δ). The empirical transformation function was employed to predict the probability distribution of the aggregation value of the WQI model.

In this research, all interpolation techniques were performed using the Geostatistical tool from ArcGIS Pro 2.9.2 and other relevant statistical analysis and graphical presentations have been carried out using the R programming language on R Studio.

12.3.4 Model Performance Analysis

12.3.4.1 Cross-validation (CV) Approaches

To evaluate the performance of the geocomputational model, the present study used four evaluation criteria of the CV technique: (i) root mean square error (RMSE); (ii) mean absolute error (MAE); (iii) mean square error (MSE); and (iv) coefficient of determination (R^2). Recently, several studies have utilized this technique to compare performance among geocomputational interpolation methods (Adhikary and Dash, 2017; Amini et al., 2019; Antal et al., 2021; Bronowicka-Mielniczuk et al., 2019; Farzaneh et al., 2022; McManus et al., 2020, 2021; Rahman et al., 2013; Rahman & Harding, 2016; Xie et al., 2011). In this research, a 10-fold cross-validation (CV) technique was applied to calculate the evaluation criteria. Details technique can be found in Xiong et al. (2020). Except for R^2, the performance criteria expect a predictive model's performance to be as small as possible (Al-Mamoori et al., 2021). In general, the R^2 value refers to how well the models fit with the model inputs and predicted data (Uddin et al., 2022c; Wu et al., 2019). It is expected to be close to 1 (He et al., 2015). Model evaluation criteria are measured as follows:

$$RMSE = \sqrt{\frac{1}{n}\sum_{i=1}^{n} \left(y_i - \hat{y}_i\right)^2} \tag{12.10}$$

$$MAE = \frac{1}{n}\sum_{i=1}^{n} \left|y_i - \hat{y}_i\right| \tag{12.11}$$

$$MSE = \frac{1}{n}\sum_{i=1}^{n} \left(y_i - \hat{y}_i\right)^2 \tag{12.12}$$

$$R^2 = 1 - \frac{\sum_{i=1}^{n} \left(y_i - \hat{y}_i\right)^2}{\sum_{i=1}^{n} \left(y_i - \hat{y}\right)^2} \tag{12.13}$$

where y_i and $\hat{y}$ are the ith observed and mean of the predicted values, respectively. n is the number of observations.

12.3.4.2 Prediction Uncertainty Analysis

For the purposes of uncertainty analysis in water quality models, several techniques are used, such as Monte Carlo simulation, ML algorithms, etc. In approaching a geospatial model, several studies have used the prediction standard errors (PSEs) technique by performing the geocomputational-interpolation method (Antal et al., 2021). In this study, the UK model was used to calculate the PSE in various interpolation models, and excellent prediction performance was found for the UK interpolation model. For the analysis of prediction uncertainties, predicted WQI scores of each interpolation technique were used to calculate the PSE by implementing the optimized hyper-parameter settings of the UK model. The details of the hyper-parameters of the UK can be found in Table 12.3. Figure 12.11 presents the uncertainty results of the WQI scores obtained from the various interpolation models.

12.3.4.3 Model Suitability Analysis

The suitable model was identified by comparing all interpolation models using the Tylor diagram. This technique is commonly used to compare various methods, techniques, or models in terms of data deviation. It is effective to identify an appropriate model because it allows three statistical measures, including the correlation between observations and predictions, the root-mean-square deviation (RMSD), and their standard deviations (SD), which help in understanding model reliability (Calim et al., 2018). Recently, a few studies have utilized this technique to select the best performer (Seifi et al., 2020; Xu et al., 2016). Figure 12.12 compares the summary statistics for various geospatial prediction techniques.

12.4 RESULTS AND DISCUSSION

12.4.1 Descriptive Assessment of Water Quality

The general characteristics of the studied physicochemical parameters and the statistical relationship among them are summarized and visualized in Table 12.1 and in Figure 12.3. The surface water temperature varied between 13.90°C and 16.90°C with a mean value of 15.58 ± 0.74°C (Table 12.1). The maximum and minimum water pH was observed to be 8.33 and 7.6 with a mean value of 8.00 ± 0.23 while the range of salinity varied from 0.83 to 30.10 PSU with a mean value of 16.87 ± 12.00 PSU (Table 12.1). Water transparency was found within the guideline value of greater than 1 m/depth (Table 12.1) and showed significant positive association with water pH ($r = 0.60$, $p < 0.01$) (Figure 12.3). The highest concentration of DOX (136.17 % sat) exceeded the upper standard threshold value while minimum (69.33 % sat) and mean (107.90 % sat) concentration of DOX were found within the prescribed limit (Table 12.1) and the DOX were significantly correlated with water pH ($r = 0.86$, $p < 0.01$) (Figure 12.3), which was previously well documented by Hoque et al. (2015).

Concentration of BOD in the harbour was found with a mean value of 1.74 ± 0.87 mg/l, which was below that of upper threshold level. Except DIN, mean concentration of TON (1.47 ± 1.78 mg/l), MRP (0.02 ± 0.01 mg/l), and AMN (0.07±0.05 mg/l) were found within the upper standard limit (Table 12.1). However, maximum concentration of TON (6.00 mg/l) and MRP (0.06 mg/l) exceeded the upper threshold level (Table 12.1). The DIN correlated positively with TON ($r = 1.00$, $p < 0.01$) whereas both parameters displayed significant negative correlation with water pH ($r = -0.77$, $p < 0.01$) and with TRAN ($r = -0.70$, $p < 0.01$) and significant moderate positive correlation with MRP ($r = 0.59$, $p < 0.01$) (Figure 12.3). Concentration of CHL was found ranging from 1.00 mg/m^3 to 10.43 mg/m^3 with a mean value of 5.32 ± 3.22 mg/m^3, which implied the concentration of CHL was within the guideline value (Table 12.1).

TABLE 12.1
Descriptive Statistics of Water Quality Parameter in Cork Harbour

Parameter	Unit	Min.	Max.	Mean	SD	Variance	Skewness	Kurtosis	Standard threshold*	
									Lower	Upper
Temp.	°C	13.9	16.90	15.58	0.74	0.54	-0.36	0.01	-	25
pH	-	7.60	8.33	8.00	0.23	0.05	-0.22	-1.55	5	9
TRAN	m/depth	0.00	5.60	1.57	1.47	2.17	1.00	1.05	> 1	-
DOX	% sat	69.33	136.17	107.90	15.18	230.52	-0.18	0.18	72	128
BOD	mg/l	0.00	3.55	1.74	0.87	0.76	-0.27	0.17	0	7
DIN	mg/l	0.04	6.02	1.54	1.78	3.18	1.18	0.39	0.0	1.20
TON	mg/l as N	0.00	6.00	1.47	1.78	3.19	1.21	0.49	0.0	2
MRP	mg/l as P	0.01	0.06	0.02	0.01	0.00	2.13	5.91	0.0	0.05
AMN	mg/l	0.02	0.26	0.07	0.05	0.003	1.89	3.92	0	1.5
CHL	mg/m^3	1.00	10.43	5.32	3.22	10.36	0.29	-1.53	0.0	14.2

* EPA (2001)

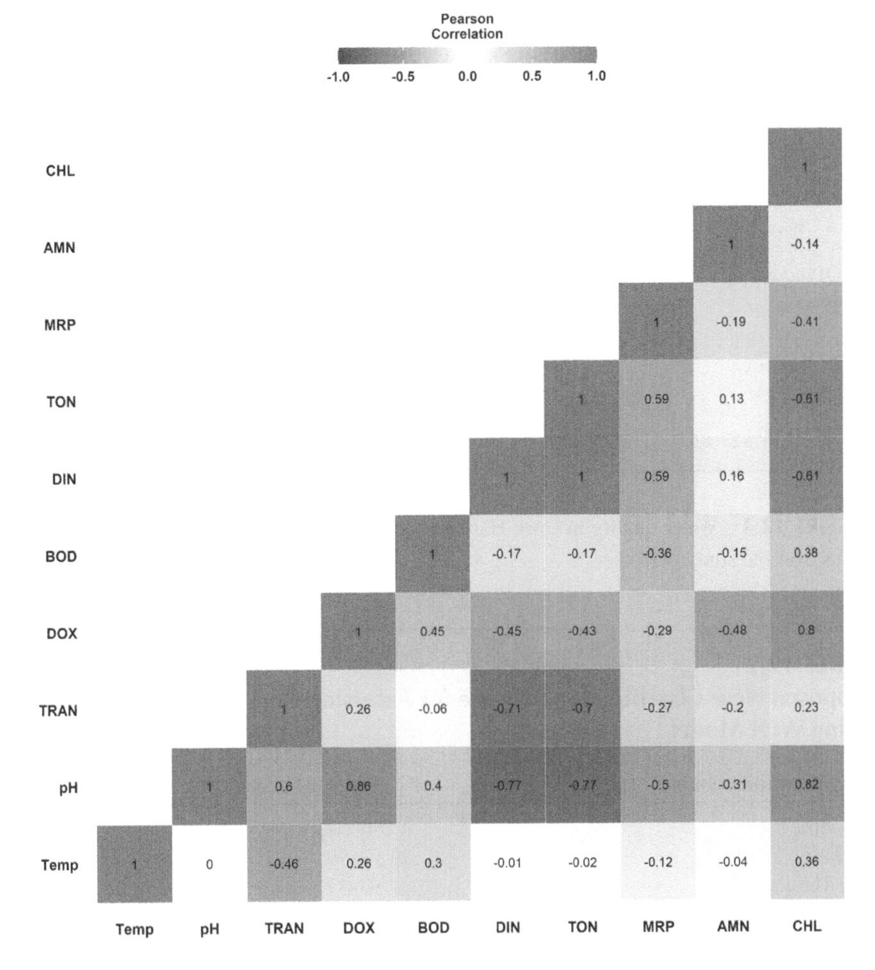

FIGURE 12.3 Correlation between water quality parameters in Cork Harbour.

The concentration of chlorophyll-*a* showed significant positive association with water pH ($r = 0.82$, $p < 0.01$) and with DOX ($r = 0.80$, $p < 0.01$) where significant negative correlation was found with DIN and TON ($r = -0.61$, $p < 0.01$) (Figure 12.3).

12.4.2 Assessing Water Quality Using WQI Models

The WQI is a widely used tool to assess water quality by the simple mathematical transformation of water quality data into numerical values. In this research, the WQI values were obtained from the WQM-WQI model. The WQI values ranged from 33 to 73, with an average of 56.19. Figure 12.4 shows the WQI and water quality status in Cork Harbour during the study period, respectively. As can be seen from Figure 12.4a, higher WQI values were found at the lower and outer monitoring sites in Cork harbour, while the lowest WQI values were obtained for the upper monitoring sites in the

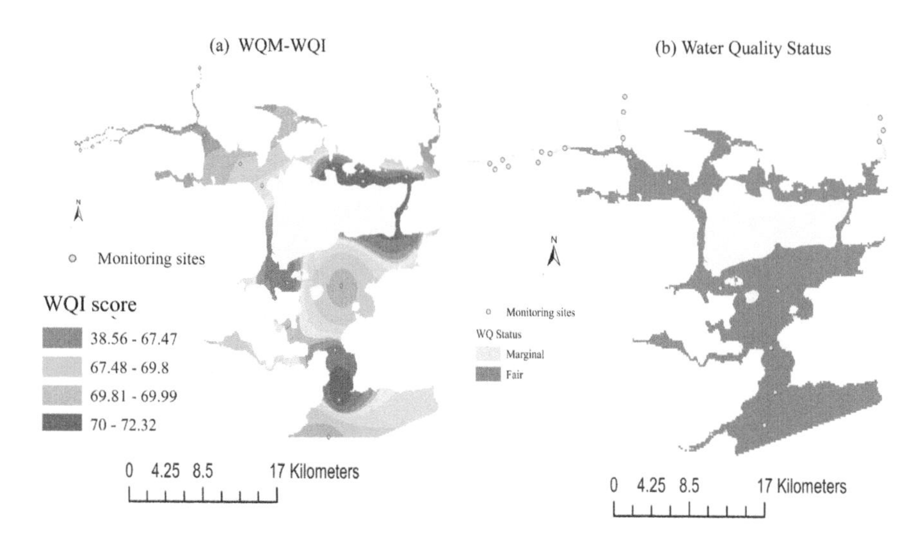

FIGURE 12.4 Water quality in Cork Harbour: (a) WQM-WQI values and (b) water quality status over the study period.

TABLE 12.2
Proposed New Classification Scheme for Assessing Coastal Water Quality Using WQI Model

Classification scheme	Range of score
(i) Good	80-100
(ii) Fair	50–79
(iii) Marginal	30–49
(iv) Poor	0-29

Source: Uddin et al., 2022.

Harbour (Figure 12.4a). Once the WQI values were obtained, the water quality was classified using the WQM water quality classification scheme (Table 12.2). As seen in Figure 12.4b, the WQM evaluated two water quality classes. These varied from "marginal" to "fair". The marginal class water quality was evaluated in the upper Lee estuary and the upper part of the river Owenacurra (Midleton). The past decade has seen an increase in the use of the effluent treatment plants (ETPs) in this area. Consequently, the ETPs can contribute to raw wastewater discharges into the estuary directly without treatment (Hartnett and Nash, 2015). As a result, it is expected that the water quality in the upper part of the Harbour's had depleted due to the overloaded wastewater and this might have also contributed to this trend. The remaining part of the Harbour was dominated by the "fair" water quality over the study period (Figure 12.4b). These results reflect those of Uddin et al. (2022b) and Wall et al. (2020); both studies also found similar water states in Cork Harbour.

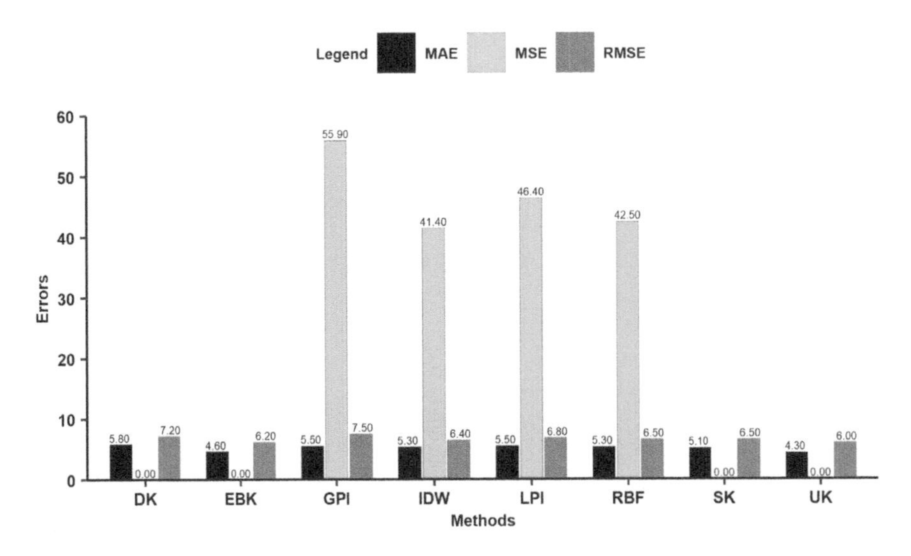

FIGURE 12.5 Cross-validation results of various geostatistical prediction model.

12.4.3 Comparison of Geostatistical Perdition Models

In this research, eight interpolation models, including four commonly used deterministic geocomputational-interpolation methods (LPI, GPI, RBF, and IDW) and four advanced geostatistical (EBK, SK, UK, and DK) models, were utilized to identify the best geocomputational technique for predicting WQI values. Figure 12.6 presents the predicted WQIs spatial distribution in Cork Habour for various geostatistical prediction models. For the purposes of model evaluation, the present study used the CV approaches to optimize the hyper-parameters of the predictive model. Table 12.3 provides the list of optimized hyper-parameters for eight geocomputational models. In order to evaluate the prediction performance, the present study utilized four evaluation metrics, including RMSE, MSE, MAE, and R^2, which indicate the model prediction (Antal et al., 2021; Uddin et al., 2018).

Figure 12.5 presents the CV results of various interpolation models. The lower prediction errors were found for the all geocomputational models. Compared to four geostatistical models, robust performance was shown by the UK (RMSE = 6.0, MSE = 0.0, and MAE = 4.3) and the EBK models (RMSE = 6.2, MSE = 0.0, and MAE = 4.6), respectively, whereas the remaining two, DK and SK, also well performed to predict WQI values. On the other hand, all deterministic models (GPI, IDW, LPI, and RBF) including the widely used IDW technique also had higher prediction errors in this study. A similar finding was also investigated by Adhikary and Dash (2017). Her finding revealed that the UK method outperformed when predicting water quality. It can be seen from Figure 12.5 that higher RMSE, MSE, and MAE were obtained for those models. Surprisingly, the IDW and RBF methods had the highest prediction errors among the most popular techniques (Figure 12.5). These findings are in line with previous research, which found similar results for the LPI, GPI, IDW, and RBF methods (Antal et al., 2021).

TABLE 12.3
Optimized Hyper-parameters of Various Geocomputational Prediction Models

	Geostatistical prediction models							
Hyper-parameters	**EBK**	**SK**	**UK**	**DK**	**LPI**	**GPI**	**IDW**	**RBF**
(i) **Model settings**								
Output	Prediction	prediction	prediction	prediction	prediction	prediction	prediction	prediction
subset size	100	100	100	-	100	-	-	-
Power	-	-	-	-	1	2	2	-
overlap factor	1	1	1	-	1	-	-	-
Number of simulation	1000	1000	1000	-	1000	-	-	-
Number of simulation	1000	1000	-	-	-	-	-	-
(ii) **Transformation**	Empirical	normal score	-	normal score		-	-	-
approximation	-	density Skew	-	density Skew		-	-	-
Kernels	Exponential	-	-	8	Exponential	-	-	CRP*
base distribution	-	Empirical	-	Student's t-distribution		-	-	-
(iii) **Neighborhood searching type**	standard circular	standard	standard		standard	-	standard	standard
Neighbors to include	15	5	5	5	1000	-	15	15
Include at least	10	2	2	2	10	-	10	10
Sector type	full	4 and 45 degree	4 and 45 degree	5 and 45 degree	full	-	full	Full
Major semiaxis	-	14374.34	9541.8	-	10046.94	-	11588	11588
Minor semiaxis	-	14374.34	9541.8	-	10046.94	-	11588	11588
radius	16510.28	-	-	-	-	-	-	-
Angle	-	0	0	-	0		0	0

(iv) Variogram	-	covariance	semivariogram	covariance	-	-	-	-
number of lags	-	12	12	12	-	-	-	-
lag size	-	1647.17	1091.93	1,766.94	-	-	-	-
Nugget	-	0	27.14	0	-	-	-	-
(v) Model type	-	stable	stable	stable	-	-	-	-
parameter	-	1.075	2	0.7941	-	-	-	0.0031
partial sill	-	1.18	112.97	0.9725	-	-	-	-
Anisotropy	-	No	No	No	-	-	-	-
range	-	14374.34	9541.8	14399.2	-	-	-	-

* Completely Regularized Spline.

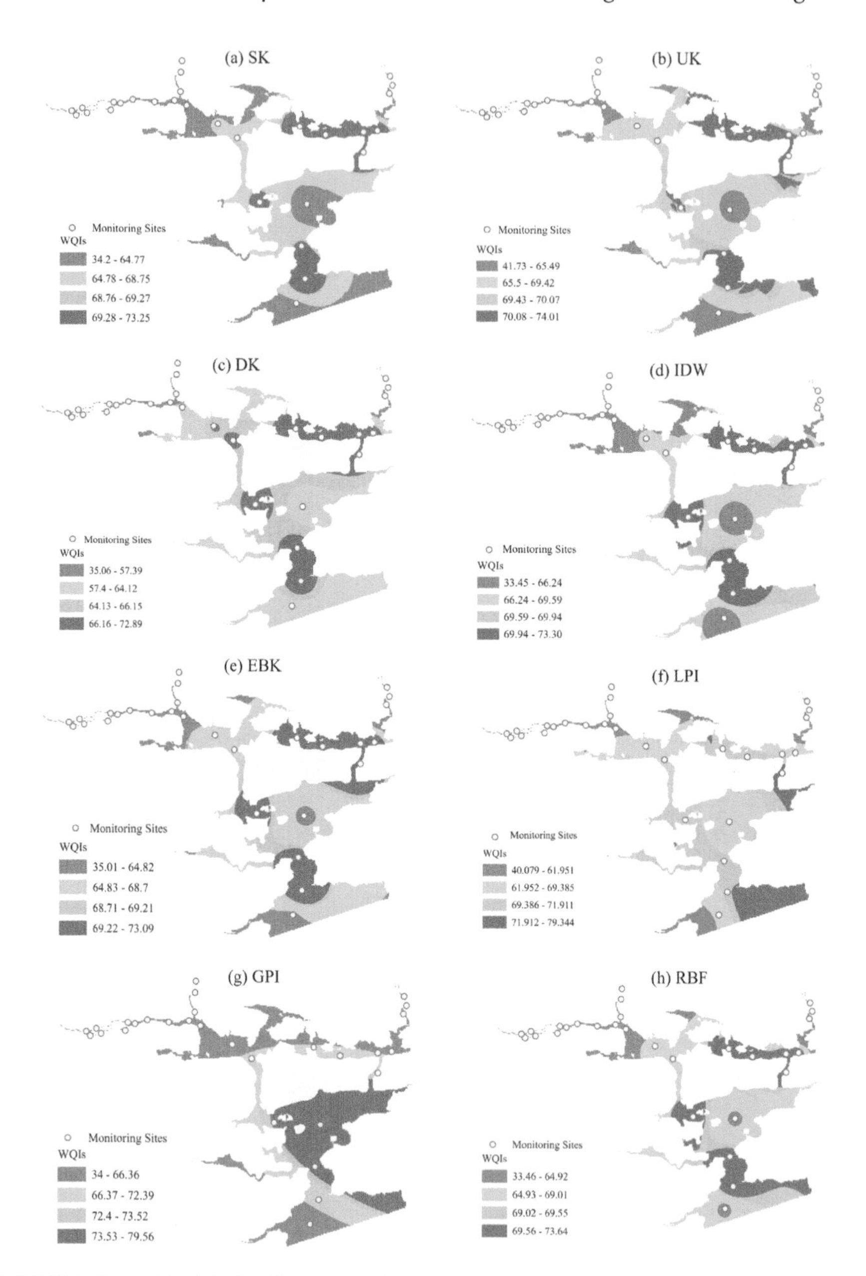

FIGURE 12.6 Spatial distribution of WQI in Cork Harbour; maps obtained from various geocomputational prediction models.

As shown in Figure 12.6, all geocomputational-interpolation models produced similar spatial distribution patterns of predicted WQIs except the LPI and GPI methods. Even though all prediction models predicted higher WQI values in the upper and outer part of the Harbour, those models (LPI and GPI) differ from others

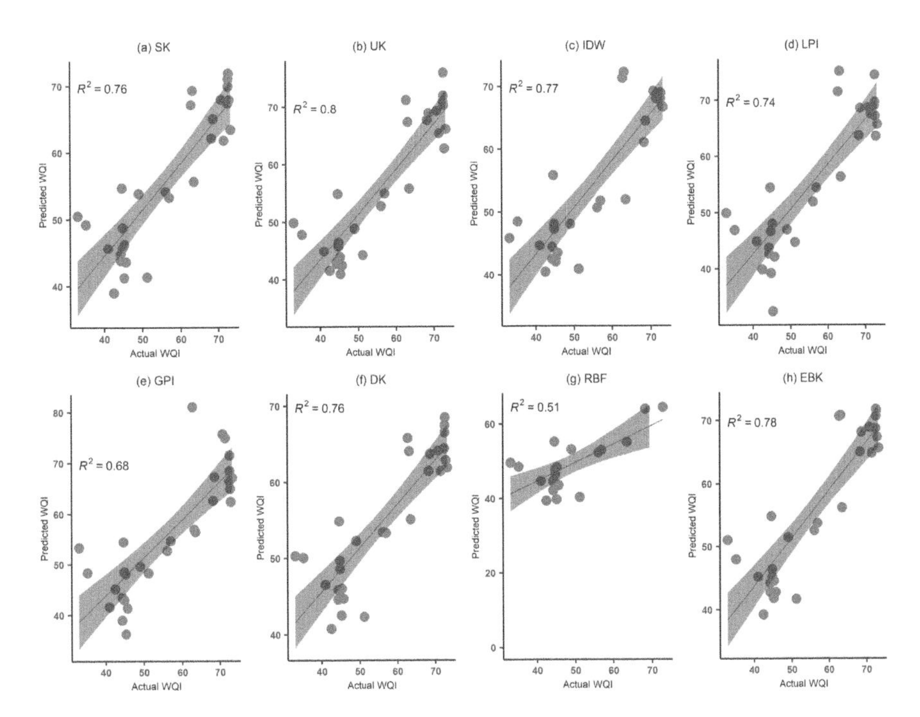

FIGURE 12.7 Scatter plots of actual vs. predicted WQI scores based on the model testing dataset of different geocomputational prediction models.

(Figure 12.6f; Figure 12.6g). The GPI model predicted the higher WQIs in the lower-eastern part of the Harbour (Figure 12.6g), whereas the LPI showed a completely different pattern of spatial distribution of WQIs in the Harbour (Figure 12.6f). Moreover, the present study utilized the coefficient of determination for assessing the relationship between model predictors and response. Figure 12.7 presents the scatter plots of actual and predicted WQI values. Based on the results of R^2, a strong relationship between the actual and predicated WQI values was observed for all models except the RBF technique ($R^2 = 0.51$). This finding is contrary to previous studies, which have suggested that the RBF model performed the best in predicting river water quality (Wu et al., 2019)

To validate the model performance, a comparison scenario was generated between actual and predicted WQI values for each deterministic and geostatistical model. Figure 12.8 provides a comparison overview between predicted and actual WQI values at each monitoring site in Cork harbour. As can be seen from Figure 12.8, all models performed well in this study, but they did not accurately predict WQI at each point. Comparison results of each site indicate that all the models followed the trend to the actual WQI values, whereas a slight difference was observed in the outer Harbour monitoring sites.

Figure 12.9 presents a comparative result of various statistical measures for the predicted and actual WQIs in boxplots. In Figure 12.9a, boxplots show the deterministic interpolation methods are different from the geostatistical techniques in several

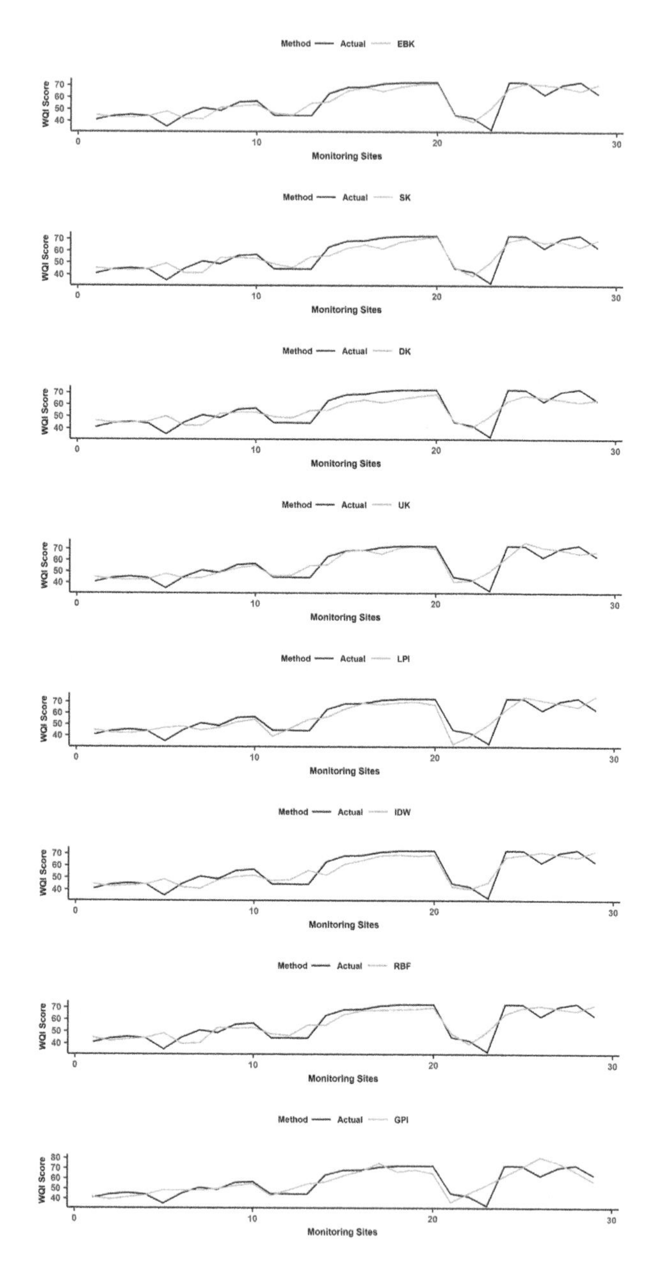

FIGURE 12.8 Comparison of the model validation performance between predicted and actual WQI values at each monitoring site.

statistical measures. It can be seen from that the predicted mean and median values differ for each group, whereas similar statistical measures were found for the UK methods when comparing actual statistics. Figure 12.9b shows the cumulative distribution function (CDF) of the eight methods. The CDF results indicate that 98% of monitoring sites were predicted correctly except the UK methods.

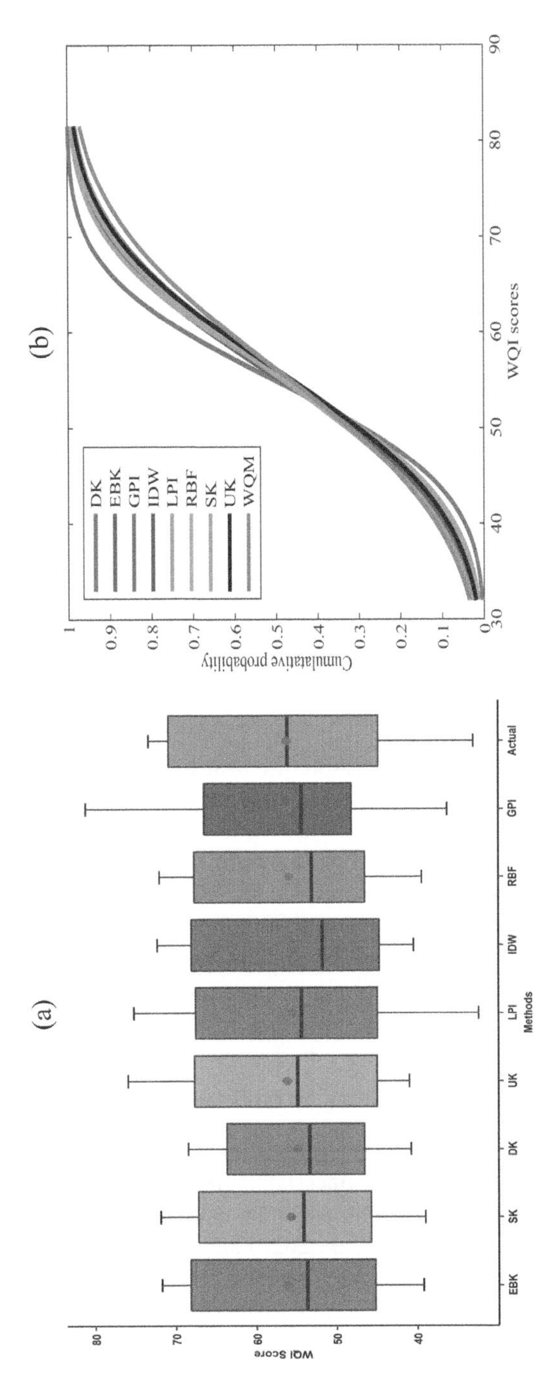

FIGURE 12.9 Comparison of predicted WQI from various geocomputational models: (a) Boxplots show a comparison between actual and predicted WQI scores and (b) CDF comparison of predicted WQI scores of various techniques.

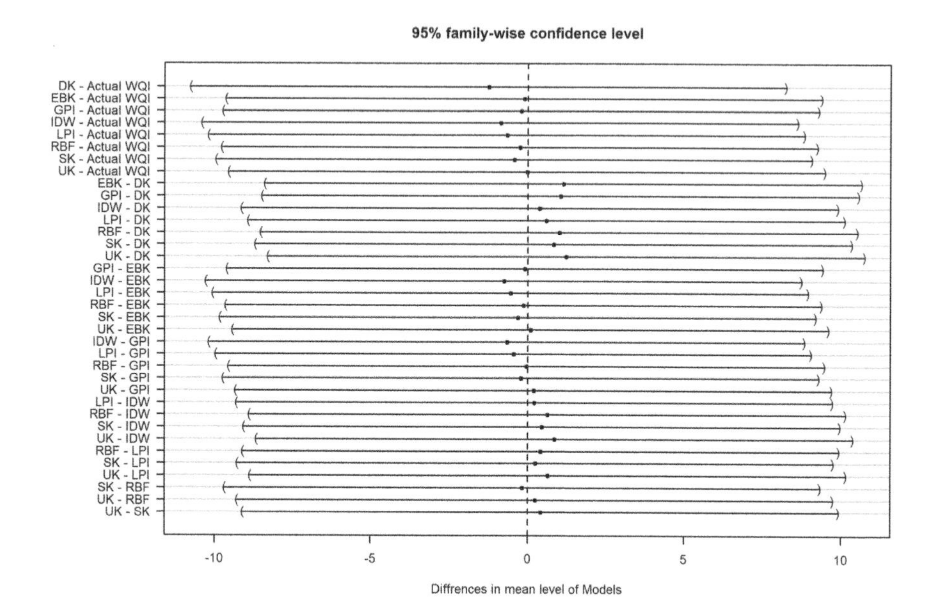

FIGURE 12.10 Multiple comparison results of pair-wise geocomputational models with 95% CI from Tukey's HSD. The vertical dashed line indicates the point where the difference between the means is equal to zero or similarity of model statistics, the filled-dot refers to the means are equal of both models.

However, it is very hard to determine the best method using CV and R^2 analysis. Hence, Turkey's HSD analysis was performed for the purposes of comparing among interpolation methods. Recently, several studies have utilized this technique to identify the unique method from groups by comparing overall and pair-wise aspects (Nanda et al., 2021; Rouder et al., 2016). Figure 12.10 presents the multiple comparison results of pair-wise various deterministic and geostatistical models with a 95% CI from Tukey's HSD. The results of the Tukey's HSD indicate that there were no statistically significant differences among methods at $p < 0.05$.

12.4.4 Evaluation of Uncertainty of Geocomputational-Interpolation Models

For real-world applications, model uncertainty is a major concern. All models, including prediction models, contain inherent uncertainties (Antal et al., 2021; Farrance and Frenkel, 2012; Seifi et al., 2020) and therefore it is critical to figure out how much data variance the model contains. It is essential to understand the level of uncertainty in the modelled spatial distribution of predicted WQIs in order to assess the reliability of prediction data for assessing water quality; otherwise, the predicted data can not reflect the actual water quality indicators. As a result, misinterpretation/wrong information can be transferred to water managers. In order to assess the prediction uncertainty of various interpolation techniques, the present study utilized the PSE value of the predicted WQIs of various interpolation techniques. As shown in Figure 12.11, all of interpolation models overestimated the WQIs in the outer part of the Harbour. In comparison of various geostatistical interpolation methods, the present study found small PSE (PSE: 0.67 to 8.64) for the EBK method (Figure 12.11e), whereas the largest PSE (PSE: 2.62 to 12.9) was produced by the DK method (Figure 12.11c). On the other hand, compared to the deterministic methods, the GPI model produced the highest interpolation PSE values; it ranged from 1.45 to 12.56 (Figure 12.11g). Surprisingly, the LPI generated the lowest PSE values (0.16 to 7.03) while this technique showed poor prediction performance (RMSE= 6.8, MSR= 46.40, MAE = 5.5, and $R^2 = 0.74$). Wang et al. (2014) reported that the LPI outperformed in estimating the spatial distribution of precipitation. The comparison results of deterministic and geostatistical methods indicate that the geostatistical model could be effective to predict WQM-WQIs for coastal water quality. Similar results are in accord with recent studies indicating that the geostatistical interpolation model outperformed deterministic methods (Antal et al., 2021; Pellicone et al., 2018). Higher PSE values were generated in the outer Harbour of monitoring locations over the study period by all interpolation methods. The higher PSE in the outer Harbour found may be due to the inadequate sampling locations (Antal et al., 2021; Borges et al., 2016). The results of uncertainties of various models reveals that the EBK model could be effective and reliable to predict spatial distribution of WQIs in Cork Harbour. The studies by Antal et al. (2021) and Al-Mamoori et al. (2021) both revealed that the EBK method is the best interpolation technique for spatial distribution analysis.

12.4.5 Comparison of Model Suitability for the Prediction of WQIs

To identify the suitable model for predicting WQIs, Tylor diagram analysis was utilized in this research. Recently, this approach has been widely used to compare various methods/datasets/models in terms of data variances (Annapoorna et al., 2016 Xu et al., 2016). This technique is more effective in identifying the best model by comparing three-dimensional statistics, including centred root-mean-square deviation (RMSD), their standard deviations (SD), and correlation between predicted and actual values (Seifi et al., 2020). Figure 12.12 provides insight into how the model performed in terms of three statistical measures, where statistics were obtained from various interpolation techniques by using actual and predicted WQI values in Cork Harbour, respectively.

As seen in Figure 12.12, the GPI showed poor performance compared to other geocomputational-interpolation methods, whereas the remaining methods showed good agreement with statistical measures. The results from Tylor statistics indicate there were no statistically significant differences between models with $p < 0.05$.

However, the present study compared various geocomputational-interpolation methods using the cross-validation results and coefficient of determination analysis. To determine the best model in order to predict WQIs appropriately, CV results were utilized in this research. According to the CV results of interpolation models, the

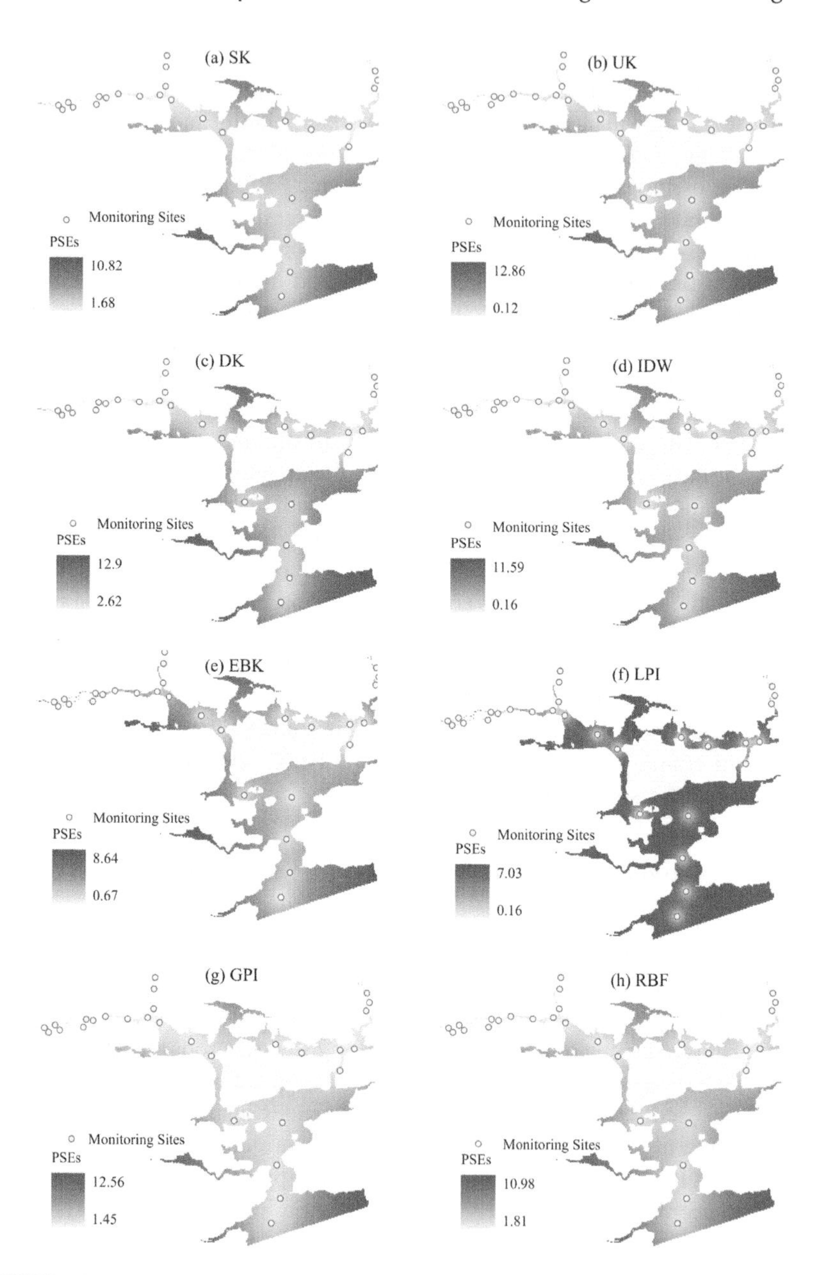

FIGURE 12.11 WQIs prediction uncertainties of various geocomputational-interpolation models.

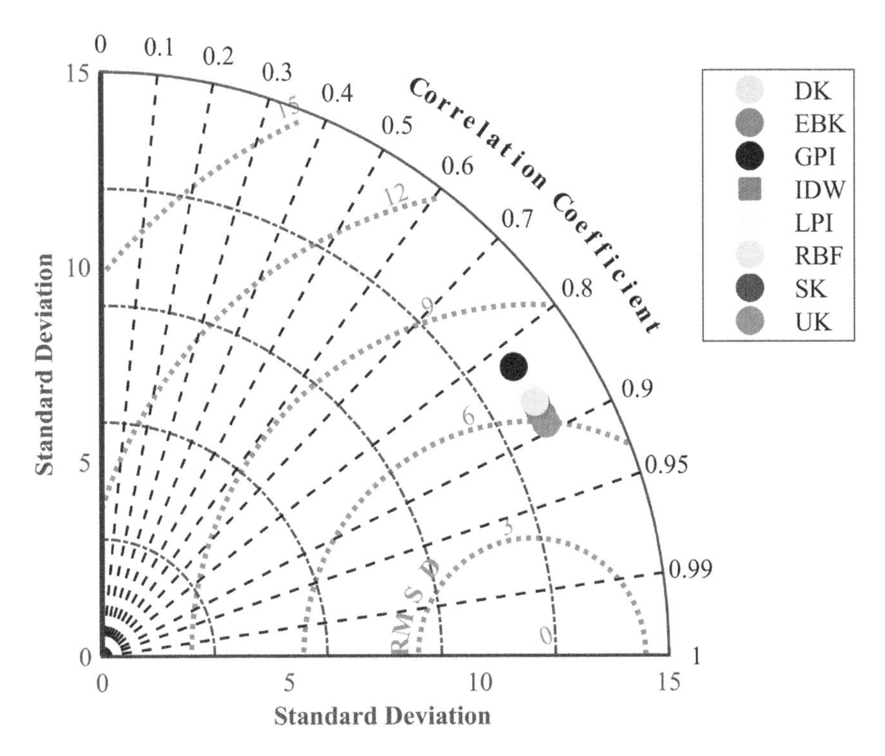

FIGURE 12.12 Comparison of various geospatial predictive models using Tylor diagram.

UK (RMSE = 6.0, MSE = 0.0, MAE = 4.30 and R^2 = 0.8) and EBK (RMSE = 6.2, MSE = 0.0, MAE = 4.60, and R^2 = 0.78) had the lowest prediction errors, respectively, compared to the other techniques. Results of the CV reveal that the UK method could be effective to interpolate the WQIs in Cork Harbour. This study also compared between actual and predicted WQIs of various geocomputational using Tukey's HSD test and Tylor diagram. The results of those tests revealed that there were no statistically significant differences between predicted and actual WQI values for all geocomputational predictive models, except for the deterministic GPI. But when comparing the CV and uncertainty results of the geocomputational-interpolation methods, the findings of this study suggest that the EBK method could be a useful technique for reducing the uncertainty in WQI spatial distribution analysis.

12.5 CONCLUSION

The aim of this research was to identify the best geocomputational-interpolation technique for predicting spatial distribution of WQIs by comparing various interpolation methods. To achieve this goal, eight widely used interpolation techniques (LPI, GPI, IDW, RBF, SK, UK, DK, and EBK) were performed in this study. Predictive models were validated using the cross-validation process. The key findings of this research are as follows:

- Geostatistical interpolation methods showed better performance than deterministic techniques.
- The CV results reveal that the UK and the EBK methods could be effective in predicting the spatial distribution of the WQIs, whereas these methods showed robust performance (RMSE = 6.0, MSE = 0.0, MAE = 4.3, and R2 = 0.8) and (RMSE = 6.2, MSE = 0.0, MAE = 4.6, and R2 = 0.78), respectively.
- Tukey's HSD family-wise multi-comparison results reveal there was no significant difference between predicted and actual WQIs among geocomputational models.
- Compared to the uncertainty of the spatial distribution of WQIs, the lowest PSE was found for the EBK, where PSE ranged from 1.01 to 13.31.

However, geospatial EBK predictive models could be effective to interpolate WQIs more precisely to reduce uncertainty in predicting WQIs. The findings of this research could be useful in predicting the geospatial distribution of WQIs more accurately in order to assess coastal water quality. Further studies should be carried out in order to validate the other geostatistical methods in terms of predicting WQIs. Although the conclusion of this study was drawn based on the findings of analysis it is critical to compare between the results of this study and findings of the literature because most of the recent studies have only focused on surface (river, lakes) and groundwater quality whereas previous studies are limited to coastal water quality.

DECLARATION OF COMPETING INTEREST

The authors state that there are no competing financial interests or personal relationships that influenced the work reported in this chapter.

ACKNOWLEDGMENTS

This research was funded by the Hardiman Research Scholarship of the University of, Galway, which funded the first author as part of his PhD program. The authors would like to thank the Environmental Protection Agency of Ireland for providing water quality data. The research was also supported by MaREI, the SFI Research Centre for Energy, Climate, and Marine and Ryan Institute, University of Galway, Ireland. The authors also sincerely acknowledge Charles Sturt University for providing all necessary supports to this PhD project through the international co-supervision. Moreover, the authors also sincerely acknowledge the EcohdroInformatics Research Group (EHIRG), School of Engineering, College of Science and Engineering, University of Galway, Ireland for providing computational laboratory facilities to complete this research. The research was also supported by the Environmental Protection Agency, Ireland for the AquaCop project [Grant Ref No: 2022-NE-1128].

FUNDING INFORMATION

There was no funding or special allocation for this study.

REFERENCES

Abbasi, T., Abbasi, S., 2012. Water Quality Indices, Water Quality Indices. https://doi.org/10.1016/C2010-0-69472-7

Acharya, S.S., Panigrahi, M.K., 2016. Evaluation of factors controlling the distribution of organic matter and phosphorus in the Eastern Arabian Shelf: A geostatistical reappraisal. Cont. Shelf Res. https://doi.org/10.1016/j.csr.2016.08.001

Adhikary, P.P., Dash, C.J., 2017. Comparison of deterministic and stochastic methods to predict spatial variation of groundwater depth. Appl. Water Sci. 7, 339–348. https://doi.org/10.1007/s13201-014-0249-8

Al-Mamoori, S.K., Al-Maliki, L.A., Al-Sulttani, A.H., El-Tawil, K., Al-Ansari, N., 2021. Statistical analysis of the best GIS interpolation method for bearing capacity estimation in An-Najaf City, Iraq. Environ. Earth Sci. 80, 1–14. https://doi.org/10.1007/s12665-021-09971-2

Agarwal, B., Agarwal, A., Harjule, P., & Rahman, A. (2022a). Understanding the intent behind sharing misinformation on social media. Journal of Experimental and Theoretical Artificial Intelligence, 1–15. https://doi.org/10.1080/0952813X.2021.1960637

Agarwal, B., Rahman, A., Patnaik, S., & Poonia, R. C. (Eds.) (2022b). Proceedings of International Conference on Intelligent Cyber-Physical Systems. (Algorithms for Intelligent Systems). Springer. www.springer.com/gp/book/9789811671357.

Amini, M.A., Torkan, G., Eslamian, S., Zareian, M.J., Adamowski, J.F., 2019. Analysis of deterministic and geostatistical interpolation techniques for mapping meteorological variables at large watershed scales. Acta Geophys. 67, 191–203. https://doi.org/10.1007/s11600-018-0226-y

Antal, A., Guerreiro, P.M.P., Cheval, S., 2021. Comparison of spatial interpolation methods for estimating the precipitation distribution in Portugal. Theor. Appl. Climatol. 145, 1193–1206. https://doi.org/10.1007/s00704-021-03675-0

Attorre, F., Alfo, M., De Sanctis, M., Francesconi, F., Bruno, F., 2007. Comparison of interpolation methods for mapping climatic and bioclimatic variables at regional scale. Int. J. Climatol. 27, 1825–1843. https://doi.org/10.1002/JOC.1495

Bárdossy, A., Pegram, G., 2013. Interpolation of precipitation under topographic influence at different time scales. Water Resour. Res. 49, 4545–4565. https://doi.org/10.1002/WRCR.20307

Belete, D.M., Huchaiah, M.D., 2021. Grid search in hyperparameter optimization of machine learning models for prediction of HIV/AIDS test results. Int. J. Comput. Appl. https://doi.org/10.1080/1206212X.2021.1974663

Borges, P. de A., Franke, J., da Anunciação, Y.M.T., Weiss, H., Bernhofer, C., 2016. Comparison of spatial interpolation methods for the estimation of precipitation distribution in Distrito Federal, Brazil. Theor. Appl. Climatol. 123, 335–348. https://doi.org/10.1007/S00704-014-1359-9/FIGURES/9

Bronowicka-Mielniczuk, U., Mielniczuk, J., Obroślak, R., Przystupa, W., 2019. A Comparison of Some Interpolation Techniques for Determining Spatial Distribution of Nitrogen Compounds in Groundwater. Int. J. Environ. Res. 13, 679–687. https://doi.org/10.1007/s41742-019-00208-6

Çaldırak, H., Kurtuluş, B., 2018. Evidence of Possible Recharge Zones for Lake Salda (Turkey). J. Indian Soc. Remote Sens. https://doi.org/10.1007/s12524-018-0779-x

Calim, M.C., Nobre, P., Oke, P., Schiller, A., Siqueira, L.S.P., Castelão, G.P., 2018. A new tool for model assessment in the frequency domain – Spectral Taylor Diagram: application to a global ocean general circulation model with tides. Geosci. Model Dev. https://doi.org/10.5194/gmd-2018-5

Carsten Von Der Ohe, P., Prüß, A., Schäfer, R.B., Liess, M., De Deckere, E., Brack, W., 2007. Water quality indices across Europe - A comparison of the good ecological status of five river basins. J. Environ. Monit. 9, 970–978. https://doi.org/10.1039/b704699p

Das, S., Rahman, A., Ahamed, A., & Rahman, S. T. (2018). Multi-level models can benefit from minimizing higher-order variations: an illustration using child malnutrition data. Https://Doi.Org/10.1080/00949655.2018.1553242, 89(6), 1090–1110. https://doi.org/10.1080/00949655.2018.1553242

Emadi, M., Baghernejad, M., 2014. Comparison of spatial interpolation techniques for mapping soil pH and salinity in agricultural coastal areas, northern Iran. Arch. Agron. Soil Sci. 60, 1315–1327. https://doi.org/10.1080/03650340.2014.880837

EPA, 2017. Urban Waste Water Treatment, Official Journal of the European Union.

ESRI, 2016. How inverse distance weighted interpolation works – Help | ArcGIS for Desktop [WWW Document]. ARCMAP.

ESRI, 2007. How Inverse Distance Weighted (IDW) interpolation works. ArcGIS 9.2 Deskt. Help.

ESRI, n.d. How global polynomial interpolation works – ArcGIS Pro | Documentation [WWW Document]. URL https://pro.arcgis.com/en/pro-app/2.7/help/analysis/geostatistical-analyst/how-global-polynomial-interpolation-works.htm (accessed 4.9.22a).

ESRI, n.d. Understanding universal kriging – ArcGIS Pro | Documentation [WWW Document]. URL https://pro.arcgis.com/en/pro-app/2.7/help/analysis/geostatistical-analyst/understanding-universal-kriging.htm (accessed 4.9.22b).

Farrance, I., Frenkel, R., 2012. Uncertainty of measurement: A review of the rules for calculating Uncertainty components through functional relationships. Clin. Biochem. Rev. 33, 49–75.

Farzaneh, G., Khorasani, N., Ghodousi, J., Panahi, M., 2022. Application of geostatistical models to identify spatial distribution of groundwater quality parameters. Environ. Sci. Pollut. Res. https://doi.org/10.1007/s11356-022-18639-8

Gunarathna, M.H.J.P., Nirmanee, K.G.S., Kumari, M.K.N., 2016. Are geostatistical interpolation methods better than deterministic interpolation methods in mapping salinity of groundwater? Int. J. Res. Innov. Earth Sci.

Gupta, A., Kamble, T., Machiwal, D., 2017. Comparison of ordinary and Bayesian kriging techniques in depicting rainfall variability in arid and semi-arid regions of north-west India. Environ. Earth Sci. https://doi.org/10.1007/s12665-017-6814-3

Gupta, S., Gupta, S.K., 2021. A critical review on water quality index tool: Genesis, evolution and future directions. Ecol. Inform. 63, 101299. https://doi.org/10.1016/j.ecoinf.2021.101299

Han, H., Lee, K.H., Jeon, Y., Yoon, J.W., 2020. Comparison of various interpolation techniques to infer localization of audio files using ENF signals. Proc. - 2020 Int. Conf. Softw. Secur. Assur. ICSSA 2020 46–51. https://doi.org/10.1109/ICSSA51305.2020.00015

Hartnett, M., Nash, S., 2015. An integrated measurement and modeling methodology for estuarine water quality management. Water Sci. Eng. 8, 9–19. https://doi.org/10.1016/j.wse.2014.10.001.

Harjule, P., Rahman, A., & Agarwal, B. (2021). A cross-sectional study of anxiety, stress, perception and mental health towards online learning of school children in India during COVID-19. Journal of Interdisciplinary Mathematics, 24(2), 411–424. https://doi.org/10.1080/09720502.2021.1889780

He, X., Gou, W., Liu, Y., Gao, Z., 2015. A Practical Method of Nonprobabilistic Reliability and Parameter Sensitivity Analysis Based on Space-Filling Design 2015.

Hock, R., 2003. Temperature index melt modelling in mountain areas. J. Hydrol. 282, 104–115. https://doi.org/10.1016/S0022-1694(03)00257-9

Hoque, M.M., Mustafa Kamal, A.H., Idris, M.H., Ahmed, O.H., Saifullah, A.S.M., Billah, M.M., 2015. Status of some fishery resources in a tropical mangrove estuary of Sarawak, Malaysia. Mar. Biol. Res. https://doi.org/10.1080/17451000.2015.1016970

Hussain, M.M., Bari, S.H., Tarif, M.E., Rahman, M.T.U., Hoque, M.A., 2016. Temporal and spatial variation of groundwater level in Mymensingh district, Bangladesh. Int. J. Hydrol. Sci. Technol. https://doi.org/10.1504/IJHST.2016.075587

Jalili Pirani, F., Modarres, R., 2020. Geostatistical and deterministic methods for rainfall interpolation in the Zayandeh Rud basin, Iran. Hydrol. Sci. J. 65, 2678–2692. https://doi.org/10.1080/02626667.2020.1833014

Javed, S., Ali, A., Ullah, S., 2017. Spatial assessment of water quality parameters in Jhelum city (Pakistan). Environ. Monit. Assess. 189. https://doi.org/10.1007/s10661-017-5822-9

Juwana, I., Muttil, N., Perera, B.J.C., 2016. Uncertainty and sensitivity analysis of West Java Water Sustainability Index - A case study on Citarum catchment in Indonesia. Ecol. Indic. 61, 170–178. https://doi.org/10.1016/j.ecolind.2015.08.034

Kärnä, T., Baptista, A.M., 2016. Evaluation of a long-term hindcast simulation for the Columbia River estuary. Ocean Model. 99, 1–14. https://doi.org/10.1016/j.ocemod.2015.12.007

Kawo, N.S., Karuppannan, S., 2018. Groundwater quality assessment using water quality index and GIS technique in Modjo River Basin, central Ethiopia. J. African Earth Sci. 147, 300–311. https://doi.org/10.1016/J.JAFREARSCI.2018.06.034

Khouni, I., Louhichi, G., Ghrabi, A., 2021. Use of GIS based Inverse Distance Weighted interpolation to assess surface water quality: Case of Wadi El Bey, Tunisia. Environ. Technol. Innov. 24, 101892. https://doi.org/10.1016/j.eti.2021.101892

Krivoruchko, K, 2012. Empirical Bayesian Kriging. ESRI Press.

Krivoruchko, Konstantin, 2012. Empirical Bayesian Kriging Implemented in ArcGIS Geostatistical Analyst. ArcUser.

Kumar, A., Bojjagani, S., Maurya, A., Kisku, G.C., 2022. Spatial distribution of physicochemical-bacteriological parametric quality and water quality index of Gomti River, India. Environ. Monit. Assess. 194. https://doi.org/10.1007/S10661-022-09814-Y

Kumari, M.K.N., Sakai, K., Kimura, S., Nakamura, S., Yuge, K., Gunarathna, M.H.J.P., Ranagalage, M., Duminda, D.M.S., 2018. Interpolation methods for groundwater quality assessment in tank cascade landscape: A study of ulagalla cascade, Sri Lanka. Appl. Ecol. Environ. Res. https://doi.org/10.15666/aeer/1605_53595380

Liu, D., Zhao, Q., Fu, D., Guo, S., Liu, P., Zeng, Y., 2020. Comparison of spatial interpolation methods for the estimation of precipitation patterns at different time scales to improve the accuracy of discharge simulations. Hydrol. Res. 51, 583–601. https://doi.org/10.2166/NH.2020.146

Luo, W., Taylor, M.C., Parker, S.R., 2008. A comparison of spatial interpolation methods to estimate continuous wind speed surfaces using irregularly distributed data from England and Wales. Int. J. Climatol. 28, 947–959. https://doi.org/10.1002/JOC.1583

Madhloom, H.M., Al-Ansari, N., Laue, J., Chabuk, A., 2018. Modeling spatial distribution of some contamination within the lower reaches of Diyala river using IDW interpolation. Sustain. 10. https://doi.org/10.3390/su10010022

Mainuri, Z.G., Owino, J.O., 2017. Spatial Variability of Soil Aggregate Stability in a Disturbed River Watershed. Eur. J. Econ. Bus. Stud. https://doi.org/10.26417/ejes.v9i1.p278-290

McManus, S., Rahman, A., Coombes, J., & Horta, A. (2021). Uncertainty assessment of spatial domain models in early stage mining projects – A review. Ore Geology Reviews, 133, 104098. https://doi.org/10.1016/J.OREGEOREV.2021.104098

McManus, S., Rahman, D. A., Coombes, D. J., & Horta, D. A. (2021). Comparison of interpretation uncertainty in spatial domains using portable x-ray fluorescence and ICP data. Applied Computing and Geosciences, 12, 100067. https://doi.org/10.1016/J.ACAGS.2021.100067

McManus, S., Rahman, A., Horta, A., & Coombes, J. (2020). Applied Bayesian Modeling for Assessment of Interpretation Uncertainty in Spatial Domains. Statistics for Data Science and Policy Analysis, 3–13. https://doi.org/10.1007/978-981-15-1735-8_1

Meng, Q., Liu, Z., Borders, B.E., 2013. Assessment of regression kriging for spatial interpolation – comparisons of seven GIS interpolation methods. Cartogr. Geogr. Inf. Sci. 40, 28–39. https://doi.org/10.1080/15230406.2013.762138

Murphy, R.R., Curriero, F.C., Ball, W.P., 2010. Comparison of Spatial Interpolation Methods for Water Quality Evaluation in the Chesapeake Bay. J. Environ. Eng. 136, 160–171. https://doi.org/10.1061/(asce)ee.1943-7870.0000121

Muzenda, F., Masocha, M., Misi, S.N., 2019. Groundwater quality assessment using a water quality index and GIS: A case of Ushewokunze Settlement, Harare, Zimbabwe. Phys. Chem. Earth, Parts A/B/C 112, 134–140. https://doi.org/10.1016/J.PCE.2019.02.011

Nanda, A., Mohapatra, D.B.B., Mahapatra, Abikesh Prasada Kumar, Mahapatra, Abiresh Prasad Kumar, Mahapatra, Abinash Prasad Kumar, 2021. Multiple comparison test by Tukey's honestly significant difference (HSD): Do the confident level control type I error. Int. J. Stat. Appl. Math. 6, 59–65. https://doi.org/10.22271/maths.2021.v6.i1a.636

Nash, S., Hartnett, M., Dabrowski, T., 2011. Modelling phytoplankton dynamics in a complex estuarine system. Proc. Inst. Civ. Eng. - Water Manag. 164, 35–54. https://doi.org/10.1680/wama.800087

Nikitin, A., Tregubova, P., Shadrin, D., Matveev, S., Oseledets, I., Pukalchik, M., 2021. Regulation-based probabilistic substance quality index and automated geo-spatial modeling for water quality assessment. Sci. Reports 2021 111 11, 1–14. https://doi.org/10.1038/s41598-021-02564-w

Oliver, M.A., 1991. Disjunctive Kriging - an Aid to Making Decisions on Environmental Matters. Area 23, 19–24.

Pandey, P.C., Kumar, P., Tomar, V., Rani, M., Katiyar, S., Nathawat, M.S., 2015. Modelling spatial variation of fluoride pollutant using geospatial approach in the surrounding environment of an aluminium industries. Environ. Earth Sci. https://doi.org/10.1007/s12665-015-4563-8

Payblas, C., 2018,. Evolution of Ground Water Quality and Source Tracking of Nitrate Contamination in the Seymour Aquifer of Texas. Boller Rev. https://doi.org/10.18776/tcu/br/3/86

Parween, S., Siddique, N.A., Mahammad Diganta, M.T., Olbert, A.I., Uddin, Md.G., 2022. Assessment of urban river water quality using modified NSF water quality index model at Siliguri city, West Bengal, India. Environmental and Sustainability Indicators 16, 100202. https://doi.org/10.1016/j.indic.2022.100202

Pellicone, G., Caloiero, T., Modica, G., Guagliardi, I., 2018. Application of several spatial interpolation techniques to monthly rainfall data in the Calabria region (southern Italy). Int. J. Climatol. https://doi.org/10.1002/joc.5525

Rahman, A., Harding, A., Tanton, R., & Liu, S. (2013). Simulating the characteristics of populations at the small area level: New validation techniques for a spatial microsimulation model in Australia. Computational Statistics & Data Analysis, 57(1), 149–165. https://doi.org/10.1016/J.CSDA.2012.06.018

Rahman, A., & Harding, A. (2016). Small area estimation and microsimulation modeling. Small Area Estimation and Microsimulation Modeling, 1–521. https://doi.org/10.1201/9781315372143

Rahman, A. (2020). Statistics for Data Science and Policy Analysis. In Statistics for Data Science and Policy Analysis. Springer Singapore. https://doi.org/10.1007/978-981-15-1735-8

Rouder, J.N., Engelhardt, C.R., McCabe, S., Morey, R.D., 2016. Model comparison in ANOVA. Psychon. Bull. Rev. 23, 1779–1786. https://doi.org/10.3758/s13423-016-1026-5

Seifi, A., Dehghani, M., Singh, V.P., 2020a. Uncertainty analysis of water quality index (WQI) for groundwater quality evaluation: Application of Monte-Carlo method for weight allocation. Ecol. Indic. 117, 106653. https://doi.org/10.1016/j.ecolind.2020.106653

Seifi, A., Dehghani, M., Singh, V.P., 2020b. Uncertainty analysis of water quality index (WQI) for groundwater quality evaluation: Application of Monte-Carlo method for weight allocation. Ecol. Indic. 117, 106653. https://doi.org/10.1016/j.ecolind.2020.106653

Sutadian, A.D. hany, Muttil, N., Yilmaz, A.G. okhan, Perera, B.J.C., 2016. Development of river water quality indices-a review. Environ. Monit. Assess. 188, 58. https://doi.org/10.1007/s10661-015-5050-0

Uddin, M.G., Moniruzzaman, M., Quader, M.A., Hasan, M.A., 2018. Spatial variability in the distribution of trace metals in groundwater around the Rooppur nuclear power plant in Ishwardi, Bangladesh. Groundw. Sustain. Dev. 7. https://doi.org/10.1016/j.gsd.2018.06.002

Uddin, M.G., Olbert, A.I., Nash, S., 2020a. Assessment of water quality using Water Quality Index (WQI) models and advanced geostatistical technique, in: Civil Engineering Research Association of Ireland (CERAI). pp. 594–599.

Uddin, M.G., Stephen Nash, Olbert, A.I., 2020b. Application of Water Quality Index Models to an Irish Estuary, in: Kieran Runae, V.J. (Ed.), Civil and Environmental Research. Civil and Environmental Research, Cork, Ireland, pp. 576–581.

Uddin, M.G., Nash, S., Olbert, A.I., 2021. A review of water quality index models and their use for assessing surface water quality. Ecol. Indic. https://doi.org/10.1016/j.ecolind.2020.107218

Uddin, M. G., Nash, S., Olbert, A. I., & Rahman, A. (2022a). Development of a water quality index model - a comparative analysis of various weighting methods. In Prof. Dr. A. Çiner (Ed.), Mediterranean Geosciences Union Annual Meeting (MedGU-21) (pp. 1–6).

Uddin, M. G., Nash, S., Rahman, A., & Olbert, A. I. (2022b). A comprehensive method for improvement of water quality index (WQI) models for coastal water quality assessment. Water Research, 118532. https://doi.org/10.1016/J.WATRES.2022.118532

Uddin, Md.G., Nash, S., Talas, M., Diganta, M., Rahman, A., Olbert, A.I., 2022c. Robust machine learning algorithms for predicting coastal water quality index, Journal of Environmental Management. https://doi.org/10.1016/j.jenvman.2022.115923

Uddin, Md.G., Nash, S., Rahman, A., Olbert, A.I., 2022d. A novel approach for estimating and predicting uncertainty in water quality index model using machine learning approaches. Manuscript submitted for publication.

Uddin, MD. Galal and Nash, Stephen and Rahman, Azizur and Olbert, Agnieszka I., 2022e. Performance Analysis of the Water Quality Index Model for Predicting Water State Using Machine Learning Techniques. Available at http://dx.doi.org/10.2139/ssrn.4270761

Verma, P.A., Shankar, H., Saran, S., 2019. Comparison of Geostatistical and Deterministic Interpolation to Derive Climatic Surfaces for Mountain Ecosystem. Remote Sens. Northwest Himal. Ecosyst. 537–547. https://doi.org/10.1007/978-981-13-2128-3_24

Wall, B., Cahalane, A., Derham, J., 2020. Ireland's Environment - An Integrated Assessment.

Wang, S., Huang, G.H., Lin, Q.G., Li, Z., Zhang, H., Fan, Y.R., 2014. Comparison of interpolation methods for estimating spatial distribution of precipitation in Ontario, Canada. Int. J. Climatol. 34, 3745–3751. https://doi.org/10.1002/JOC.3941

Wu, C.Y., Mossa, J., Mao, L., Almulla, M., 2019. Comparison of different spatial interpolation methods for historical hydrographic data of the lowermost Mississippi River. Ann. GIS 25, 133–151. https://doi.org/10.1080/19475683.2019.1588781

Xiao, Y., Gu, X., Yin, S., Shao, J., Cui, Y., Zhang, Q., Niu, Y., 2016. Geostatistical interpolation model selection based on ArcGIS and spatio-temporal variability analysis of groundwater level in piedmont plains, northwest China. Springerplus 5. https://doi.org/10.1186/s40064-016-2073-0

Xie, Y., Chen, T.-B. Bin, Lei, M., Yang, J., Guo, Q.-J.J., Song, B., Zhou, X.-Y.Y., 2011. Spatial distribution of soil heavy metal pollution estimated by different interpolation methods: Accuracy and uncertainty analysis. Chemosphere 82, 468–476. https://doi.org/10.1016/j.chemosphere.2010.09.053

Xiong, Z., Cui, Y., Liu, Z., Zhao, Y., Hu, M., Hu, J., 2020. Evaluating explorative prediction power of machine learning algorithms for materials discovery using k-fold forward cross-validation. Comput. Mater. Sci. 171, 109203. https://doi.org/10.1016/j.commatsci.2019.109203

Xu, Z., Hou, Z., Han, Y., Guo, W., 2016. A diagram for evaluating multiple aspects of model performance in simulating vector fields. Geosci. Model Dev. 9, 4365–4380. https://doi.org/10.5194/gmd-9-4365-2016

Yan, H., Nix, H.A., Hutchinson, M.F., Booth, T.H., 2005. Spatial interpolation of monthly mean climate data for China. Int. J. Climatol. 25, 1369–1379. https://doi.org/10.1002/JOC.1187

Zandi, S., Ghobakhlou, a, Sallis, P., 2011. Evaluation of Spatial Interpolation Techniques for Mapping Soil pH. 19th Int. Congr. Model. Simul. 12–16.

Zhang, X., Srinivasan, R., 2009. Gis-based spatial precipitation estimation: A comparison of geostatistical approaches. J. Am. Water Resour. Assoc. 45, 894–906. https://doi.org/10.1111/J.1752-1688.2009.00335.X

Zotou, I., Tsihrintzis, V.A., Gikas, G.D., 2020. Water quality evaluation of a lacustrine water body in the Mediterranean based on different water quality index (WQI) methodologies. J. Environ. Sci. Heal. - Part A Toxic/Hazardous Subst. Environ. Eng. 55, 537–548. https://doi.org/10.1080/10934529.2019.1710956

13 Mathematical Modeling for Socio-economic Development

A Case from Palestine

Nour Jamal
Palestine Technical University – Kadoorie, Tulkarm, Palestinian Territories
Email: n.j.hajali1@student.ptuk.edu.ps

CONTENTS

13.1 INTRODUCTION: BACKGROUND AND DRIVING FORCES

Mathematical modeling was not an important research tool until the second decade of the twentieth century (Rahman, 2020). However, it began to appear in various scientific and practical fields in the middle of the nineteenth century. The language of models or comparisons appeared in the late nineteenth century among physicists, and then biologists who needed to represent many scientific and cosmological phenomena using mathematics (Schank & Twardy, 2009). A mathematical model is a representation of the real world using mathematics (Squires & Tappenden, 2011; Rahman et al. 2013).

The use of mathematical modeling exceeded the previous scientific aspects. It is used in the literature, guiding policy decisions, and the decision-making process. The development of regional economic models that imitate the macroeconomic model began in the 1950s by the Nobel Prize winner Pan Tinbergen, and the aim was to translate spatial economic mechanisms in a quantitative way in order to make an appropriate decision. In the 1960s, two types of economic models were developed. The first model known as the traditional Keynesian model, and the second as the regional input and output model. The expression of these models was largely using matrix algebra in order to obtain a simple economic model that is easy to study

DOI: 10.1201/9781003253051-17

(Nijkamp & Poot, 2008). However, there are many examples of economic growth studies and development models that used mathematical modeling as the basis for analyzing and obtaining accurate and correct results such as in Lin & Wang (2021) and Yadav & Wadia (2021).

Van (2020) aimed to determine the long-term economic growth in developing countries such as Cote d'Ivoire, Bangladesh, or Uganda. The study relied on the use of the Public Choice Growth Model (PCGM). The model seeks to combine two basic elements in economic theory, and microeconomics summarizes the principles of the theory of public choice. Macroeconomics includes the factors of the slow growth model. Furthermore, in Boldyrev et al. (2019) socio-economic growth in the Russian Arctic Zone was described. As such, a mathematical model consists of three nonlinear differential equations. The proposed model here is based on three basic variables: the population, jobs, and energy. A control vector was used in Boldyrev et al. (2019) to connect variables and explains the relationship with nine components, each of which has specific significance. In addition, the researchers in Boldyrev et al. (2019) used the stochastic model to test the components precisely, and reached to need more parameters to make the model more accurate.

The nature of the relationship between macroeconomic indicators and growth through capital accumulation in India in the period between 1983–2007 was described in Kumar & Rai (2020). The pattern of market capitalization and gross domestic product (GDP) growth, and the provision of local labor to understand the future direction of the stock market was verified. This research group used the Gompertz model, which is a growth equation that applies the Gompertz curve, and the equation predicts the growth of any phenomenon. Montemayor at al. (2018) presented three mathematical models for forecasting the GDP in Mexico for the period 1935–2016 by using the regression. Three models of regression were used: linear regression, exponential regression, and parabolic regression. It was found that the parabolic regression is the most appropriate and accurate model.

Abu Eideh (2015) aimed to study the causal relationship between public expenditure and GDP growth in Palestine in the period 1994–2013 using Wagner's Law and formulas. Data stability was checked using Augmented-Dickey Fuller (ADF) test. Engle-Granger cointegration test used to test the long-term relationship between public expenditure and GDP growth. Eventually, the study illustrated proportional relation between public development expenditure and GDP in Palestine. There are several measures to study the economic and social development in a country. Whereas, the ecconomic and social development may be based on the measure of the GDP and public spending of the region, number of populations, jobs and energy, government lending, inflation, employment, the rate of general income for population, labor, technology, stock, and others. However, the common factor between the cited studies was mathematical modeling as a basic tool in representing the problem, analyzing it, and obtaining the results.

Economic development is generally defined as "the level of diversity of economic resources in a country and the equitable distribution of these resources, whereas they cover the needs of most of the population" (Zakour & Swager, 2018), it can be read as "changes in income, savings and investment along with gradual changes in the social and economic structure of the country (institutional and technological changes)" (Economic growth and Economic development, n.d.). Economic growth in the State of Palestine as in many other countries is heavily influence by political and demographical and natural resources of the region. The most influential is the political portion, in particular the occupation in Palestine. Since the Oslo Accords of 1993, the Palestinian economy has faced a series of ongoing shocks including the division of the West Bank, Paris Protocol of the same year, the first and second Intifadas, and the siege of Gaza resultant in three wars. This combined with freedom of movement restrictions enforced by the occupation security apparatus limited access to basic services and resources, restrictions on investment and trade, and poor access to finance has created an unsustainable economy artificially propped up by donor aid (Government of Palestine and the United Nations Development Program, 2016; The World Bank, 2019).

The main characteristics of the Palestinian economy depend on the occupation and donors' community and lack of self-sufficiency. The fragmentation of the Palestinian territory has resulted in three main areas: the West Bank, Gaza strip, and east Jerusalem. As a result, differing levels of governance and autonomy have led to an unviable economy, each suffering from economic dependence, limited growth potential, and subject to the occupation policies of economic strangulation (Government of Palestine and the United Nations Development Program, 2016; The World Bank, 2019).

In this study, we present a description of the socio-economic development in Palestine considering four main basic variables that have the greatest impact on the Palestinian economic and social reality. In this chapter, we mainly relied on mathematical modeling in representing the problem, analyzing it, and obtaining results. The model consists of four nonlinear differential equations containing four main variables performed by a vector function such as $n(t) = [n_1(t), n_2(t), n_3(t), n_4(t)]$, where $n_1(t)$ is the number of populations, $n_2(t)$ is the number of jobs, $n_3(t)$ the available of energy in the area, and $n_4(t)$ is the occupation. In addition, we have control vectors such as $\alpha(t) = [\alpha_1(t), \alpha_2(t)..., \alpha_{16}(t)]$ with sixteen components that connect main variables to each other and explain those relationships.

13.2 METHODOLOGY

Socio-economic development in Palestine is described by a dynamic model with four nonlinear differential equations. The proposed model introduces four main variables:

- $n_1(t)$: the number of population,
- $n_2(t)$: the number of jobs,
- $n_3(t)$: the available of energy in the area, and
- $n_4(t)$: the occupation.

The coefficients that connect between variables can be defined as $\alpha(t) = [\alpha_1(t), \alpha_2(t), ..., \alpha_{16}(t)]$. We built and analyzed the model based on the basic concepts in economic and social development. In addition, we considered local Palestinian social and economic reality as observed from Palestinian data and statistics. The following equations show the four-dimensional model:

$$\frac{dn_1}{dt} = \alpha_1 n_1 - \alpha_2 n_1 n_2 + \alpha_3 n_1 n_3 + \alpha_4 n_1 n_4 \tag{13.1}$$

$$\frac{dn_2}{dt} = \alpha_5 n_2 + \alpha_6 n_1 n_2 + \alpha_7 n_2 n_3 - \alpha_8 n_2 n_4 \tag{13.2}$$

$$\frac{dn_3}{dt} = \alpha_9 n_3 - \alpha_{10} n_1 n_3 - \alpha_{11} n_2 n_3 - \alpha_{12} n_3 n_4 \tag{13.3}$$

$$\frac{dn_4}{dt} = \alpha_{13} n_4 + \alpha_{14} n_1 n_4 - \alpha_{15} n_2 n_4 - \alpha_{16} n_3 n_4 \tag{13.4}$$

The above model illustrates the interactions between the four variables and the effect of each variable on the growth of others by α's factor. In the first equation, dn_1/dt is the rate of change in population over time. Clearly, the rate is proportional to the regional population n_1 itself. The higher the population, the greater the growth rate by α_1, which is the demographic activity coefficient. In addition, we note the negative impact of job opportunities n_2 on the population growth rate. As there are a large number of holders of bachelor and postgraduate degrees greater than number of job opportunities available in the region. As such, there is an increase in the demand for workplaces on experiences, skills, and competencies, leading to an increase in the unemployment rate in the young and working age community. As a result, population growth rate decreases under the effect of number of jobs by α_2 which is the coefficient of academic and professional level for the population (Palestinian Central Bureau of Statistics, 2018; Courbage et al., 2016). In the third term of the equation, we noticed a proportional relation between energy supply n_3 and population growth rate. More energy entering the region encourages population increase and growth, therefore energy supply indicator increases the population growth rate by α_3 is the energy supply coefficient (West Bank & Gaza Energy Efficiency Action Plan 2020–2030, n.d.). The fourth equation term explains the proportional relationship between population and occupation. Hence, when occupation is increasing, population will also increase by α_4, the coefficient of people's motivation for childbearing. This increase results in the number of residential buildings increasing. Moreover, there are many causes for this increase. First, to resist the occupiers as well as to compensate for the shortfall in the number of people who pass away because of the war (this is very clear in the Gaza Strip). Second, to recompense the absence due to imprisonment (Palestinian Central Bureau Statistics, 2020).

In the second equation, dn_2/dt is the rate of change of the economic development over time. The rate of change of economic development is proportional to the number of jobs in real sector n_2. Such that, when the number of jobs increases in a region the rate of economic development will rise by α_5, which is the coefficient of the real sector economic development. It is measured through the contribution of different sectors such as agricultural, tourism, transportation, and commercial to the gross domestic product (GDP) (Palestinian Central Bureau of Statistics, 2014). Moreover, there is a positive relation between the rate of change of the economic development and regional population n_1. Hence, when the population of the region rises, the people become more interested in economic development by increasing their contribution to the workforce by factor α_6 which is the coefficient of the people's interest in economic development (The World Bank, 2019).

In third part, the energy supply n_3 will increase the economic growth rate by the factor α_7 the coefficient of energy supply per internal workplace. That is, when the total energy supplied to the region increases at a reasonable price and at a low cost, it leads to an increase and expansion in the economic sector in various fields and thus leads to economic growth (Pcbs, n.d.; Palestine Resilience conference, 2016, n.d.). In the fourth term, there is an inverse relation between jobs and occupation, since when the occupation increases the number of jobs will decrease by $\alpha_{8,}$ which is the coefficient of occupation per internal workplace. However, the occupation costs Palestinian heavy economic losses. It prevents exports abroad. In addition, it demolishes the economic infrastructure of many economic institutions, establishments, places of business, and employment opportunities for many laborers. Furthermore, the occupation imposes heavy taxes on imports for Palestinians into works, and the resulting debts accumulated on the Palestinian National Authority (Arkadie, 1977; Secretariat, U. N. C. T. A. D. 2019).

In the third equation, dn_3/dt is the rate of change in the region energy supply. There is a positive relation between the rate of change in energy supply and the energy consumption n_3. In other words, when the amount of energy consumed decreases the energy supply rate will also decline by $\alpha_{9,}$ the energy supply coefficient (World Bank, 2007). On the other hand, the effect of population n_1 on the rate of change in energy supply is negative. Hence, when the population increases the rate of change in energy supply will decrease by $\alpha_{10,}$ which is the conformity ratio of the population with energy supply. In addition, there is an inverse relation between the number of jobs n_2 and the rate of change in energy supply. That is, when the number of jobs decreases the rate of change in energy supply will decrease by α_{11} which is the conformity ratio of the economic development with the energy supply (Wang et al., 2018; Pcbs, n.d.; Zabel, 2009). In the fourth term, there is an inverse relation between energy and occupation. That is, when the occupation increases the energy supply will decrease by α_{12} which is the conformity ratio of the occupation with energy supply. As such, the occupation take control of the natural resources, natural gas, oil, and petroleum wells in Palestine by occupying the lands in 1967 and prevent Palestinians to access own resources. Moreover, it aeffects the electricity sector and the percentage of electricity in the Palestinian lands. In addition, Palestinians are importing and buying electricity through the occupation imposing consequences such as the accumulation of debts and taxes on Palestinians that has a negative effect on the energy sector (United Nations, 2019).

In the fourth equation, we present rate of change of occupation dn_4/dt during time and the factors that effect on it. However, there is a proportional relation between occupation n_4 itself and rate of change of occupation. Naturally, when occupation population increases the rate of change of occupation will increase by $\alpha_{13,}$ which is the occupation demographic activity coefficient. Meanwhile, at the beginning of occupation in Palestine, the number of Jewish populations increases, either with the increasing immigration to Palestine or with the increase in the number of births. As such, their number becomes more than the number of Palestinians themselves. Consequently, there is an increase in the construction of settlements, major expansion at the expense of the Palestinian lands, and the confiscation of their lands (Pcbs, n.d.). The second term of the equation illustrates the proportional relationship between number of populations in Palestine n_1 and occupation n_1 by $\alpha_{14,}$ which is the coefficient of forced displacement of residents from their lands. At the beginning of occupation, large number of Palestinians were forcibly to displaced from their lands, which were confiscated for the benefit of settlements and the Jewish population in Palestine. On the other hand, the increase in the number of Palestinians will be limited for many reasons such as migration and travel to obtain alternatives on the standard of living and to reduce stress (Abu-Lughod, 1983; Forced displacement – reliefweb, n.d.). In the third term of the equation, we notice an inverse relation between number of jobs n_2 and occupation by $\alpha_{15,}$ which is the coefficient of relationship between Palestinian government, private sector, and civil society. However, when the partnership between the private and government sector increases, economic development and the rate of employment will increase. On the other hand, support of the government by the private sector will reduce their debts and taxes on the Israeli side. That is, reducing the impact of the occupation and its extortion over time (Palestinian Non-Governmental Organizations (ngos) and the Private Sector: Potentials for Cooperation and Partnerships, n.d.). In the fourth term of the equation, the interaction happens between energy n_3 and occupation. Clearly, there is an inverse relation too, since when the use of renewable energy sources increases the effect of occupation will decrease by α_{16}, which is the coefficient of creating new energy alternatives for Palestinians. Whereas, when the use of renewable energy alternatives increases, the outcome motivate to activate and exploit more projects as well as prepare the available manpower in this field, which lead to negative impact on occupation. Eventually, most needs of Palestinians will be met reducing dependance on occupation resources (Abu Hamed et al., 2012). A summary of the coefficients (α) in the system is as follows:

α_1: demographic activity coefficient.
α_2: coefficient of academic and professional level for population.
α_2: energy supply coefficient.
α_4: coefficient of people's motivation to childbearing.
α_5: coefficient of the real sector economic development.
α_6: coefficient of people's interest in economic development.
α_7: coefficient of energy supply per internal workplace.
α_8: coefficient of occupation per internal workplace.
α_9: energy supply coefficient.
α_{10}: conformity ratio of population with energy supply.
α_{11}: conformity ratio of economic development with energy supply.
α_{12}: conformity ratio of the occupation with energy supply.
α_{13}: occupation demographic activity coefficient.

α_{14}: coefficient of forced displacement of residents from Palestinian lands.
α_{15}: coefficient of relationship between government, private sector, and civil society.
α_{16}: coefficient of creating new energy alternatives for Palestinians.

13.3 FIXED POINTS AND STABILITY FOR THE SYSTEM

In order to find equilibrium points for the system, we need to solve the proposed system of differential equations at certain point that is by making the derivatives equal to zero such as in ($n_1^{`} = 0, n_2^{`} = 0, n_3^{`} = 0, n_4^{`} = 0$).

The set of equilibrium points will be $(n_1^*, n_2^*, n_3^*, n_4^*)$.

We have 12 fixed points of the system given as follows:

$$F_{p1} = (0,0,0,0) \tag{13.5}$$

$$F_{p2} = \left(\frac{-\alpha_5 + \alpha_7 \eta - \alpha_8 \beta}{\alpha_6}, \frac{\alpha_1 - \alpha_3\ \eta - \alpha_4 \beta}{\alpha_2}, -\eta, -\beta \right) \tag{13.6}$$

$$\text{Whereas } \eta = \frac{(\alpha_{14}\alpha_8 - \alpha_4\alpha_{15})(\alpha_2\alpha_9\alpha_6 - \alpha_{11}\alpha_1 + \alpha_{10}\alpha_5) + (\alpha_{10}\alpha_8 + \alpha_{11}\alpha_4 + \alpha_2\alpha_6\alpha_{12})(\alpha_2\alpha_6\alpha_{13} - \alpha_5\alpha_4 - \alpha_1\alpha_5)}{(\alpha_{10}\alpha_7 - \alpha_{11}\alpha_3)(\alpha_{14}\alpha_8 - \alpha_4\alpha_{15}) - (\alpha_{10}\alpha_8 + \alpha_{11}\alpha_4 + \alpha_2\alpha_6\alpha_{12})(\alpha_7\alpha_{14} + \alpha_3\alpha_{15} + \alpha_2\alpha_6\alpha_{16})} \tag{13.7}$$

$$\text{and } \beta = \frac{-\alpha_2\alpha_9\alpha_6 + \alpha_{11}\alpha_1 - \alpha_{10}\alpha_5 - (\alpha_{10}\alpha_7 - \alpha_{11}\alpha_3)\eta}{(\alpha_{10}\alpha_8 + \alpha_{11}\alpha_4 + \alpha_2\alpha_6\alpha_{12})} \tag{13.8}$$

$$F_{p3} = \left(0, \frac{\alpha_1 - \alpha_3\ \eta - \alpha_4 \beta}{\alpha_2}, -\eta, -\beta \right) \tag{13.9}$$

$$F_{p4} = \left(\frac{-\alpha_5 + \alpha_7 \eta - \alpha_8 \beta}{\alpha_6}, 0, -\eta, -\beta \right) \tag{13.10}$$

$$F_{p5} = \left(\frac{-\alpha_5 - \alpha_8 \beta}{\alpha_6}, \frac{\alpha_1 - \alpha_4 \beta}{\alpha_2}, 0, -\beta \right) \tag{13.11}$$

$$F_{p6} = \left(\frac{-\alpha_5 + \alpha_7 \eta}{\alpha_6}, \frac{\alpha_1 - \alpha_3 \eta}{\alpha_2}, -\eta, 0 \right) \tag{13.12}$$

$$F_{p7} = (0, 0, -\eta, -\beta) \tag{13.13}$$

$$F_{p8} = \left(0, \frac{\alpha - \alpha\beta}{\alpha}, 0, -\beta\right) \tag{13.14}$$

$$F_{p9} = \left(0, \frac{\alpha_1 - \alpha_4\eta}{\alpha_2}, -\eta, 0\right) \tag{13.15}$$

$$F_{p10} = \left(\frac{-\alpha_5 - \alpha_8\beta}{\alpha_6}, 0, 0, -\beta\right) \tag{13.16}$$

$$F_{p11} = \left(\frac{-\alpha_5 + \alpha_7\eta}{\alpha_6}, 0, -\eta, 0\right) \tag{13.17}$$

$$F_{p12} = \left(\frac{-\alpha_5}{\alpha_6}, \frac{\alpha}{\alpha}, 0, 0\right) \tag{13.18}$$

To analyze the stability for the system at each fixed point, we need to find the general Jacobin matrix and explain the eigenvalues (λs) to determine the condition of stability. The general Jacobin matrix at any fixed point is illustrated as the following.

$$J = \begin{bmatrix} \alpha_1 - \alpha_2 n_2^* + \alpha_3 n_3^* + \alpha_4 n_4^* - \alpha_2 n_1^* \alpha_6 n_2^* \ \alpha_5 + \alpha_6 n_1^* + \alpha_7 n_3^* - \alpha_8 n_4^* - \alpha_{10} \\ n_3^* - \alpha_{11} n_3^* \alpha_{14} n_4^* - \alpha_{15} n_4^* \ \alpha_3 n_1^* \alpha_4 n_1^* \alpha_7 n_2^* - \alpha_8 n_2^* \alpha_9 - \alpha_{10} - \alpha_{11} n_2^* - \\ \alpha_{12} n_4^* - \alpha_{12} n_3^* - \alpha_{16} n_4^* \alpha_{13} + \alpha_{14} n_1^* - \alpha_{15} n_2^* - \alpha_{16} n_3^* \end{bmatrix} \tag{13.19}$$

To check the stability at general fixed point, $F_{p1} = (0, 0, 0, 0)$ the Jacobian matrix of F_{p1} is illustrated as the following.

$$J = \left[\alpha_1 0\ 0\ 0\ 0\ \alpha_5\ 0\ 0\ 0\ 0\ \alpha_9\ 0\ 0\ 0\ 0\ \alpha_{13}\right] \tag{13.20}$$

The above matrix is triangular and the eigenvalues are the diagonal values. If these eigenvalues are real and positive (negative), then the point is unstable (stable). Nevertheless, if at least one of these eigenvalues has a different sign, the point is an unstable saddle node (Amon, 2007). On the other hand, we want to check the stability at general fixed point F_{p2}. Hence, the other ten fixed points are branching from this main point so we can apply the same process and characteristic equation to others. To demonstrate the stability of this fixed point, we need to find their eigenvalues by using the characteristic polynomial equation for the Jacobin matrix:

$$\lambda^4 - tr(J)\lambda^3 + A\lambda^2 - A\lambda + \det(J) = 0 \qquad (13.21)$$

Numerically, to find the value of fixed points, the value of coefficients is computed based on the Palestinian Central Bureau of Statistics (Pcbs) by using the percentage rate method (Cooper & John, 2013). The value of coefficients is computed as follows:

$$\alpha_1 = 0.7127,\ \alpha_2 = 1.909,\ \alpha_3 = 0.5242,\ \alpha_4 = -0.2483,\ \alpha_5 = 0.0457,\ \alpha_6 = 2.739,$$
$$\alpha_7 = 1.422,\ \alpha_8 = 0.85,\ \alpha_9 = 0.0506,\ \alpha_{10} = 0.1012,\ \alpha_{11} = 0.01325,\ \alpha_{12} = 1.1014,$$
$$\alpha_{13} = 0.493,\ \alpha_{14} = 0.67,\ \alpha_{15} = 1.826,\ \alpha_{16} = 1.0162.$$

The stability state of the equilibrium points and bifurcation nodes determines the trajectory of the system and the points of change. The Newton method (Remani, 2013) is used to obtain the best values of the variables that will increase population, jobs, and energy and decrease the occupation, which on the other hand match the system objectives. Table 13.1 shows the stability of numerical fixed points. Figure 13.1 and Figure 13.2 demonstrate the system numerical solution.

Table 13.1 illustrates the stability of the numerical fixed points.

13.4 NUMERICAL SOLUTION AND BIFURCATION

The sensitivity of the system with change in initial conditions is calculated. The following Figures 13.1 and 13.2 illustrate the system behavior and variable correlation and their attitude toward the stability. The initial conditions are (0.205, 0.7, -0.205, -0.7) and (0.23, 0.7, -0.23, -0.7) with time period $t \in [0,100]$.

Figure 13.3 presents bifurcations of main variable n_4 vs. bifurcation factor $m(\alpha_{15})$, it may be more than one local bifurcation coalesce to each other which lead to global bifurcation and local chaos.

The fluctuation in population repeats itself in the second half of the 100-year period. Jobs and energy fluctuate along the period with deep increases in jobs after 45 years. Occupation increases and maintains their values. Figure 13.4 demonstrates a decrease at half time to a limit point. In the second part of the defined period, there is a high increase in jobs and population, but constant decrease in energy and occupation.

13.5 CONCLUSION

The literature has set the foundation for using and applying mathematical modeling for regional socio-economic development in Palestine. A number of modeling equations were proposed and formed for the problem. The equilibrium points and bifurcation were calculated and measured for stability options by analyzing the Jacobin matrix. The numerical calculations verified that the system looks like butterfly models (i.e., the system is chaotic). In future works, we suggest to focus

TABLE 13.1
Stability of Numerical Fixed Points

Fixed points	Eigenvalues	Stability
F_{p1} (–3.0194, 0, –1.5056, –0.3234)	$\lambda_1 = 0.0802+.7297i$, $\lambda_2 = -0.1603$ $\lambda_3 = 0.0802-.7297i$, $\lambda_4 = -10.6403$	Unstable
F_{p2} (0, 0.2737, –0.0066, –0.0426)	$\lambda_1 = -0.1583$, $\lambda_2 = 0.0792+0.052i$, $\lambda_3 = 0.0792-0.052i$, $\lambda_4 = 0.1968$	Unstable
F_{p3} (0, 0, 0, 0)	$\lambda_1 = 0.7127$, $\lambda_2 = 0.0457$, $\lambda_3 = 0.0506$, $\lambda_4 = 0.4930$	Unstable
F_{P4} (–0.7358, 0, 0, 3.0110)	$\lambda_1 = -0.3418$, $\lambda_2 = 1.8916 + 2.7406i$, $\lambda_3 = 1.8916-2.7406i$, $\lambda_4 = 0.5897$	Unstable
F_{p5} (0.5, 0, –1.35016, 0)	$\lambda_1 = 0.1917$, $\lambda_2 = -0.1868$, $\lambda_3 = -0.5047$, $\lambda_4 = 2.200$	Unstable
F_{p6} (0, 0.2700, 0, -0.0538)	$\lambda_1 = -0.18$, $\lambda_2 = 0.0839+0.0621i$, $\lambda_3 = 0.0839-0.0621i$, $\lambda_4 = 0.2100$	Unstable
F_{p7} (0, 0, 0.4851, –0.0459)	$\lambda_1 = 0.0486i$, $\lambda_2 = -0.0486i$, $\lambda_3 = 0.9779$, $\lambda_4 = 0.6965$.	Unstable
F_{p8} (0, 3.8189, –0.0321, 0)	$\lambda_1 = -0.0466$, $\lambda_2 = 0.0467$, $\lambda_3 = -6.5944$, $\lambda_4 = -6.4477$	Unstable
F_{p9} (–0.0167, 0.3733, 0, 0)	$\lambda_1 = 0.1807$, $\lambda_2 = -0.1806$, $\lambda_3 = 0.0473$, $\lambda_4 = -0.1998$	Unstable
F_{p10} (0.0506, 0.3484, –0.1075, –0.0371)	$\lambda_1 = 0.0144+0.2625i$, $\lambda_2 = 0.0144-0.2625i$, $\lambda_3 = -0.0145+0.0624i$, $\lambda_4 = -0.0145-.0624i$	Unstable
F_{p11} (–3.3723, –0.9674, 0, 10.8131)	$\lambda_1 = -1.3016+5.4081i$ $\lambda_2 = -1.3016-5.4081i$ $\lambda_3 = 7.4587+8.8410i$ $\lambda_4 = 7.4587-8.8410i$	Unstable
F_{p12} (0.4859, 0.1075, -0.9681, 0)	$\lambda_1 = 0.0259+0.4982i$, $\lambda_2 = 0.0259-0.4982i$, $\lambda_3 = -0.0520$, $\lambda_4 = 1.6060$	Unstable

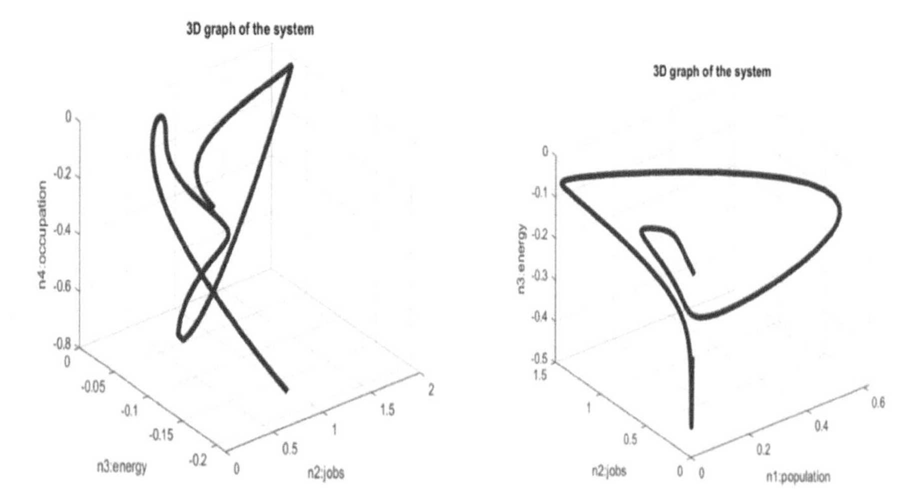

FIGURE 13.1 3D view space. (a) 3D view in the n_2-n_3-n_4 space with initial condition (0.205, 0.7, -0.205, -0.7). (b) 3D view in the n_1-n_2-n_3 with initial condition (0.23, 0.7, -0.23, -0.7).

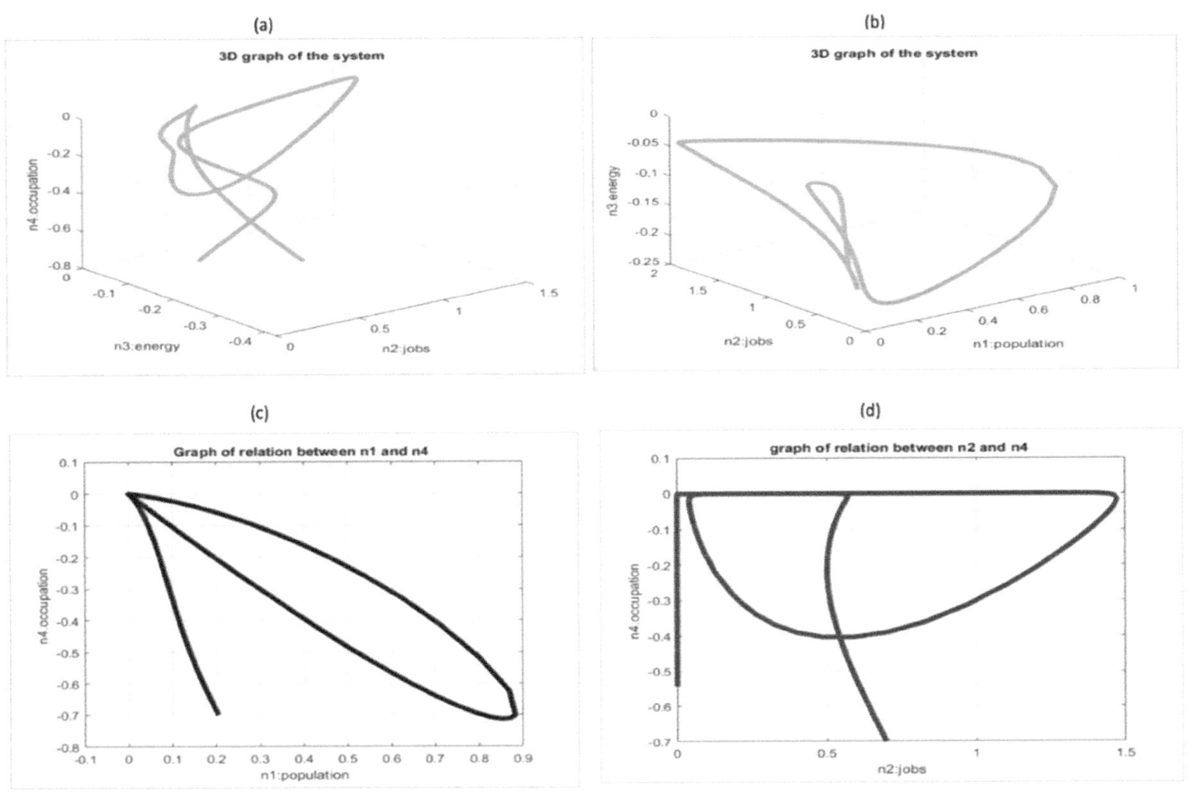

FIGURE 13.2 2D and 3D view space. (a) 3D view in the n_2-n_3-n_4 space with initial condition (0.23, 0.7, -0.23, -0.7). (b) 3D view in the n_1-n_2-n_3 space with initial condition (0.205, 0.7, -0.205, -0.7). (c) 2D view in the n_1-n_4 space with initial condition (0.205, 0.7, -0.205, -0.7). (d) 2D view in the n_2-n_4 space with initial condition (0.23, 0.7, -0.23, -0.7).

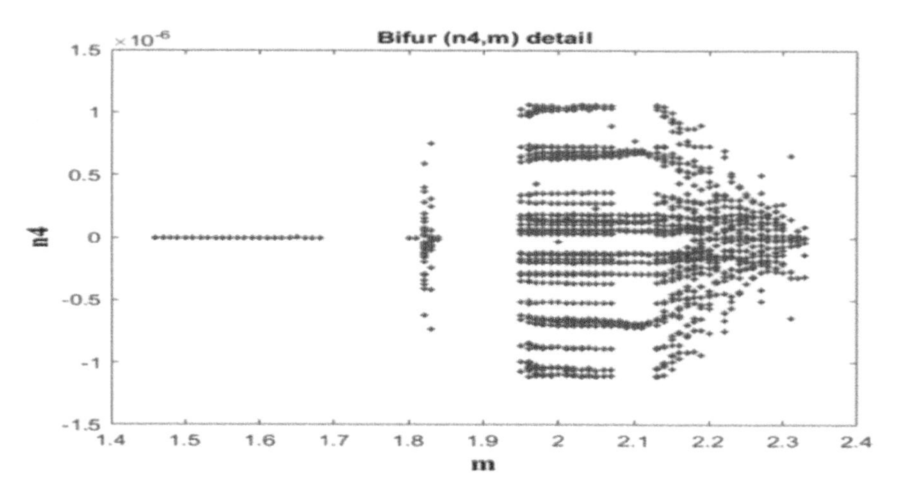

FIGURE 13.3 Direction field of the system n_4 versus m.

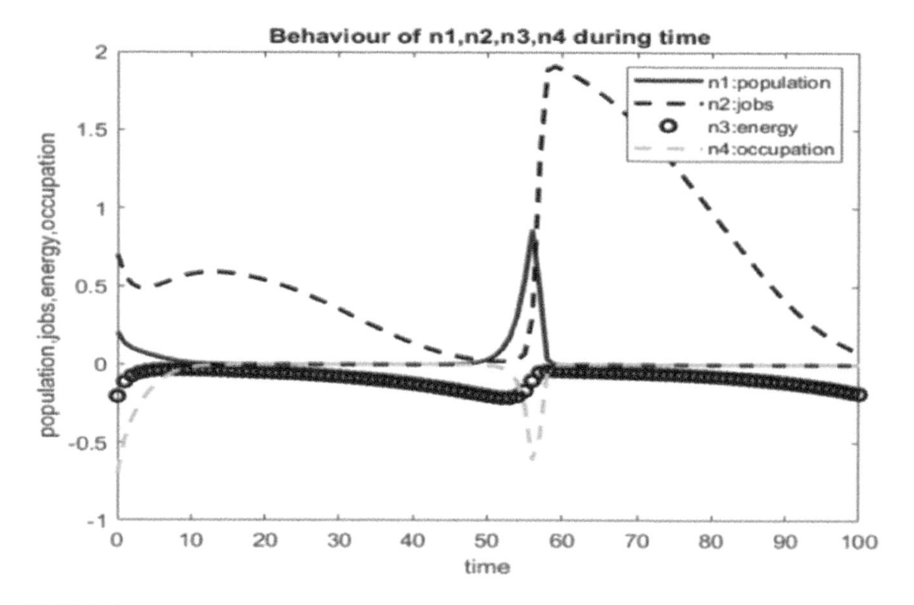

FIGURE 13.4 Behavior of n_1, n_2, n_3, n_4 during t∈ [0,100] with (0.205, 0.7, -0.205, -0.7).

the system stability on specific variables such as jobs or energy. In addition, specific parameters can be defined and calculated based on specific years for a precise and harmonic model.

REFERENCES

Abu-Eideh, O. M. (2015). Causality between public expenditure and GDP growth in Palestine: An econometric analysis of Wagner's law. *Journal of Economics and Sustainable Development*, *6*(2), 189–199.

Abu-Lughod, J. L. (1983). Demographic consequences of the occupation. *MERIP Reports* (115), 13–17.
Amon, A. (2007). *Nonlinear dynamics* (Doctoral dissertation, Université Rennes 1).
Arkadie, B. V. (1977). The impact of the Israeli occupation on the economies of the West Bank and Gaza. *Journal of Palestine Studies, 6(2),* 103–129.
Boldyrev, Y., Chernogorskiy, S., Shvetsov, K., Zherelo, A., & Kostin, K. (2019). A mathematical model of regional socio-economic development of the Russian Arctic zone. *Resources*, *8*(1), 45.
Cooper, R., & John, A. A. (2013). *Macroeconomics: Theory through Applications.*
Courbage, Y., Abu Hamad, B., & Zagha, A. (2016). Palestine 2030-demographic change: opportunities for development. *State of Palestine: United Nations Population Fund Palestine.*
Economic growth and Economic development. (n.d.). Retrieved February 1, 2022, from https://nios.ac.in/media/documents/SrSec318NEW/318_Economics_Eng/318_Economics_Eng_Lesson3.pdf
Forced displacement - reliefweb. (n.d.). Retrieved November 29, 2021, from https://reliefweb.int/sites/reliefweb.int/files/resources/Forced%20displacement%20in%20Palestine%20and%20Israel.pdf.
Government of Palestine and the United Nations Development Program (2016). Building Economic Resilience in the State of Palestine. *Palestine resilience conference*, Amman, Jordan, 24–25 November 2016.
Hamed, T. A., Flamm, H., & Azraq, M. (2012). Renewable energy in the Palestinian Territories: Opportunities and challenges. *Renewable and Sustainable Energy Reviews*, *16*(1), 1082–1088.
Kumar, N., & Rai, L. P. (2020). Use of Mathematical Models to Forecast Market Capitalization and Economic Growth in India.
Lin, Z., & Wang, H. (2021). Modeling and application of fractional-order economic growth model with time delay. *Fractal and Fractional*, *5*(3), 74.
Montemayor, O. M. F., Rojas, A. L., Chavarría, S. L., Elizondo, M. M., Vargas, I. R., & Hernandez, J. F. G. (2018). Mathematical modeling for forecasting the gross domestic product of Mexico. *International Journal of Innovative Computing, Information and Control*, *14*(2), 423–36.
Nijkamp, P., & Poot, J. (2008). Mathematical models in regional economics.
Palestinian Central Bureau of Statistics (2014). National Accounts at Current and Constant Prices 1994–2012. Ramallah – Palestine.
Palestinian Central Bureau of Statistics (2018). Population, Housing and Establishments Census 2017. Ramallah – Palestine.
Palestinian Central Bureau Statistics (2020). Sustainable Housing Development in Palestine 2007–2017. Ramallah-Palestine.
Palestinian Non-Governmental Organizations (ngos) and the Private Sector: Potentials for Cooperation and Partnerships. (n.d.). Retrieved November 29, 2021, from www.ndc.ps/sites/default/files/Potentials%20for%20Cooperation-Partnerships%20final-English.pdf.
Palestine Resilience conference 2016. UNDP in Programme of Assistance to the Palestinian People. (n.d.). Retrieved October 12, 2021, from file:///C:/Users/User/Downloads/UNDPpappresearchPRC_Building%20Economic%20Resilience%20%20(9).pdf.
Pcbs. (n.d.). Annual Energy Tables and Energy Balance. PCBS. Retrieved October 12, 2021, from www.pcbs.gov.ps/site/886/Default.aspx
Pcbs. (n.d.). Dr. Awad, reviews the conditions of the Palestinian people via statistical figures and findings, on the 72nd annual commemoration of the Palestinian. PCBS. Retrieved

November 29, 2021, from www.pcbs.gov.ps/site/512/default.aspx?lang=en&ItemID=3734

Pcbs. (n.d.). Total energy supply by year and type of energy, 2009–2019. PCBS. Retrieved October 12, 2021, from www.pcbs.gov.ps/statisticsIndicatorsTables.aspx?lang=en&table_id=531.

Rahman, A. (2020). Statistics for Data Science and Policy Analysis. Springer.

Rahman, A., Harding, A., Tanton, R., & Liu, S. (2013). Simulating the characteristics of populations at the small area level: new validation techniques for a spatial microsimulation model in Australia. Computational Statistics and Data Analysis, 57(1), 149–165. https://doi.org/10.1016/j.csda.2012.06.018

Remani, C. (2013). Numerical methods for solving systems of nonlinear equations. *Lakehead University Thunder Bay, Ontario, Canada.*

Schank, J. C., & Twardy, C. (2009). Mathematical Models. *The Modern Biological and Earth Sciences.*

Secretariat, U. N. C. T. A. D. (2019). Economic costs of the Israeli occupation for the Palestinian people: fiscal aspects note/by the Secretary-General.

Squires, H., & Tappenden, P. (2011). Mathematical modelling and its application to social care.

The World Bank. (2019). (rep.). Enhancing Job Opportunities for Palestinians. Retrieved January 26, 2022, from https://documents1.worldbank.org/curated/en/523241562095688030/pdf/

United Nations. (2019). In The economic costs of the Israeli occupation for the Palestinian people: The unrealized oil and natural gas potential. Geneva.

Van, G. (2020). A Mathematical Theory of Economic Growth: The Public Choice Growth Model.

Wang, L., Chen, Y., & Zhu, W. (2018, May). Correlation analysis of the population and energy--the coexistence of promotion and inhibition. In *2018 7th International Conference on Energy, Environment and Sustainable Development (ICEESD 2018)* (pp. 1–7). Atlantis Press.

West Bank & Gaza Energy Efficiency Action Plan 2020–2030. (n.d.). Retrieved November 20, 2021, from https://documents.worldbank.org/curated/en/851371475046203328/pdf

World Bank. (2007, May 1). West Bank and Gaza Energy Sector Review. Washington, DC. © World Bank. Retrieved November 20, 2021, from https://openknowledge.worldbank.org/handle/10986/19226

Yadav, R., & Wadia, S. (2021). Analyze the Effects of Indian Population on Economy of India. *Turkish Journal of Computer and Mathematics Education (TURCOMAT), 12*(6), 4681–4686.

Zabel, G., & Economics, E. (2009). Peak people: the interrelationship between population growth and energy resources. *Energy Bulletin, 20.*

Zakour, M. J., & Swager, C. M. (2018). Vulnerability-plus theory: the integration of community disaster vulnerability and resiliency theories. In *Creating Katrina, Rebuilding Resilience*(pp. 45–78). Butterworth-Heinemann.

Theme 5

Healthcare and Mental Disorder Detection with AIs

14 A Computational Study Based on Tensor Decomposition Models Applied to Screen Autistic Children

High-order SVD, Orthogonal Iteration and Discriminant Analysis Algorithms

Jamal Amani Rad[*1], *Hamidreza Pouretemad*[1], *Kourosh Parand*[2] *and Negar Sammaknejad*[1]
[1]Institute for Cognitive and Brain Sciences, Shahid Beheshti University, Tehran, Iran
[2]Department of Computer Sciences, Faculty of Mathematical Sciences, Shahid Beheshti University, Tehran, Iran
*Corresponding Author: j_amanirad@sbu.ac.ir; j.amanirad@gmail.com

CONTENTS

DOI: 10.1201/9781003253051-19

14.1 INTRODUCTION

Autism Spectrum Disorder (ASD) is a broad definition for cognitive and neurodevelopmental disorders that cause cognitive impairment and disability in social interaction and communication with other people [2] and also manifest in restricted and repetitive patterns of behavior [2]. These disorders are lifelong, believed to be congenital [15]. The rate of ASD has increased sharply in the past decades. There is no clear-cut agreement on the prevalence of autism, and estimates vary from up to 100 out of a population of 10,000 [42]. The most recent statistics indicate that approximately 1 out of every 88 children born in the United States is autistic [1]. It is expected that more than 30,000 Iranians younger than 19 years old suffer from ASD [42]. Partial and complete impairment in social interaction, language learning, impaired verbal and non-verbal communications, impaired social perception especially of human faces, deficits in executive function, and repetitive behavior have been well documented in ASD [2].

ASD presents itself in a wide range of disorders with common core symptoms, from very severe forms to mild impairments with lesser impact on socialization and communication. The term spectrum and the varying degrees of severity in which the disorder presents play an important role in the formulation of its diagnostic criteria, as these should be common to autism in all its manifestations. Thus, ASD comprises the following disorders [21]:

1. *Autistic Disorder* is the well-known type of autism and associated with impairments in verbal and non-verbal communication skills and pretend play and probability of low IQ.
2. *Childhood Disintegrative Disorder (CDD)* is a regressive form of autism that manifests after two years of typical development.
3. *Rett Syndrome* is a low-functioning form of autism essentially limited to girls and manifests itself after 6 to 8 months of typical development.
4. *Atypical Autism or Pervasive Developmental Disorder – Not Otherwise Specified (PDD-NOS)* is an exclusive diagnosis, in which significant autistic symptoms are present but are not sufficient to specify one of the other alternative diagnoses.
5. *Asperger Syndrome* is a form of high-functioning autism with no language delay and often normal IQ. Therefore, it is often diagnosed later than other forms of autism.

Although symptoms and their severity vary widely across these areas, individuals with ASD are characterized by three core impairments [2, 48]: (1) Social communication deficits such as delay in language development without non-verbal compensation and repetitive and stereotyped use of language or echolalia; (2) Impairments in social interaction (i.e., reduced use of non-verbal behaviors and lack of sensitivity to social stimuli such as poor eye contact, delayed onset of gaze-following, lack of face-orienting, and lack of social smile and emotional reciprocity); (3) Restricted repertoire of interests, behaviors and activities such as abnormal over-focus on particular topics, and and a preoccupation with specific parts of objects rather than the object as a whole.

Some basic deficits such as lack of attention and failure to name objects often appear within the first year of life [54] while other malfunctions such as deficits in social interaction, imitation, responding to the others' emotional cues, and face recognition are often manifested later on, within age 2 to 3 [16, 17, 18, 35, 46, 47]. Many of the social deficits in autism occur due to the inability to perceive and process information from faces [45]. These deficits are not typically diagnosed until about the age 3 [21].

For example, people with autism show a abnormal gaze pattern when looking at faces. As compared with individuals without ASD, they spend more time looking at the mouth and body as well as objects and often look less into the eyes of others [39, 33]. In Ref. [36], it is noted that if the information received from the eyes is critical for eye-blink entrainment, it is likely that eye-blink entrainments are lacking or reduced in adults with ASD because of the reduced time spent looking into the eyes of others [36]. Despite this impairment, high-functioning adults with autism can usually recognize people from their faces, and can identify basic emotional facial expressions [4]. A normal face inversion effect and a whole-over-parts advantage suggest intact domain-specific processing of facial information [29]. However, people with autism are impaired on more complex tasks that require understanding of social intentions and mental states based on expressions in the eyes [9]. Children with autism have deficits using eye gaze as a cue for visual attention, even though they can detect gaze direction normally [34]. On the other hand, twin studies commencing in 1977 by Folstein and Rutter with proceeding studies provide evidence on genetic components of ASD [22, 44]. Other studies estimated that as much as 90% of variance in the etiology of ASD is genetic [5]. Later behavioral genetic studies of ASD have shown that liability to its phenotype may be broader than the clinical picture of the syndrome. Studies comparing parents and siblings of children with ASD with control indicate that autistic traits manifest themselves in milder forms in first-degree relatives as mild phenotype variants of the disorder. Moreover, these relatives are at higher risk of social and communicative difficulties [5]. Researchers in several electrophysiological and neuroimaging studies of individuals with ASD have proposed that ASD subjects have abnormal connectivity in the brain regions as found using functional Magnetic Resonance Imaging (fMRI), Positron Emission Topography (PET), Diffusion Tensor Imaging (DTI), and Electroencephalography (EEG). For some examples of these studies, see [4, 23, 52, 55] and other references therein. The interested reader of neuroimaging methods can also see Ref. [32].

There are many challenges of early autism screening. In fact, several factors render the identification of early symptoms problematic. A number of symptoms are not apparent in the first two years of life. The skill set of very young infants is limited, and so are the opportunities to sense if something is amiss. As the development progresses, and specific milestones are expected, it becomes more likely that impairments in communication and social skills may surface. As already pointed out, the eyes play an important role in social communication because the way an individual observes others is an essential cue for the understanding of how he relates to them. Looking at the other is the first step towards imitation and joint attention. Observing and imitating the other creates a sense of interpersonal connection, which is an essential element of social interaction [50].

In contrast, the effect of eye-blinks on social communication has been given minor attention, even though spontaneous blinks occur 15–20 times per min on average [41, 30]. Blink rates drastically change with internal states, such as arousal, emotion, and cognitive load [41, 26, 49]. Given that the blinks of others can be easily recognized due to their relatively long duration (150–400 ms, [51]), eye-blinks can provide an additional window to the mind. This leads us to hypothesize that blinks are involved in social communication. In particular, when we blink the flow of visual information between the world and one's retina is temporarily interrupted. In that instant of blinking, visual stimulation from the external world is lost for 150–400 ms and the visual information will be missed. Then if new visual information is presented in that instant of blinking, it will be missed. Hence, blinking sets a physical limit on visual attention [31]. Studies have also shown that the timing of blinks is related to both explicit [38, 19] and implicit [37, 27] attentional pauses in task content. It should be noted that up to now the blink data are commonly discarded as artifacts or noise in most experimental studies of eye movements and visual scanning [56, 25]. If the timing of blink inhibition is an adaptive reaction to minimize the loss of critical information, then discarding these data may mean losing a measure of not simply what a person is looking at but of how engaged that person is with what is being looked at.

Using the proposed algorithm in this chapter, we can try to program a computer to detect whether children (around eight to fourteen months of age) display signs of autism. This very early detection enables doctors to train these children (when their brain plasticity is high) to behave in ways to counter the behavioral limitations autism imposes, thus allowing these babies to act more normally as they grow up.

Multilinear algebra is the algebra of higher-order tensors (multiway arrays) and can intuitively be imagined as the multidimensional equivalent of matrices (second order) or vectors (first order); i.e., as blocks of numbers in three or more dimensions ($N>2$). The entries of an Nth-order tensor are defined with respect to the basis chosen in N reference vector spaces, each of which has its own coordinate system. Hence, $a_{i,j,k}$ are elements of a third-order tensor A. Multimodal or multiway datasets of several interrelated variables with high dimensionality are often used in modern applications from complex phenomena. Tensors provide a natural way to respect multidimensional data objects whose entries are indexed by several continuous or discrete variables. Hence, tensor decompositions are appropriate mathematical tools for analysis of large and complex multidimensional datasets to reduce the dimensionality of their high-order tensor by using compressing the space to a lower dimensional tensor space and extract hidden information or core datasets which are smaller but contain more valuable information.

Tensor decomposition, dimensionally reduction, and feature extraction schemes are recently gaining more and more interest because of the structure of real-world data (e.g., images, videos, and brain signals as tensors). It should be noted that the classic two-way decomposition techniques such as principal component analysis and independent component analysis have exhibited high effectiveness in extracting the features of datasets, but they treat the data as matrices or vectors (i.e.< the number of the modes of the extracted components) is limited to two. So they are often not efficient.

Factoring tensors have several advantages over two-way matrix factorization such as they can explicitly take into account the multiway structure of the data that would otherwise be lost when analyzing the data by two-way factorization methods by collapsing some of the modes. Tensor decomposition such as Tucker, Canonical Decomposition (CANDECOMP)- Parallel Factor Analysis (PARAFAC) also known as CP, Nonnegative tensor factorization (NTF) with some constraints such as orthogonality, nonnegativity, and statistical independence have been recently proposed as meaningful and efficient representations of images, videos and signals [60].

One of the most common tensor decomposition is the Tucker decomposition (also known as higher-order singular value decomposition when orthogonality constraints on factor matrices are imposed) which was first introduced in psychometrics by Tucker in a 1966 Psychometrika paper that described the decomposition of the third-order datasets to obtain a "method for searching for relations in a body of data", for the case "when individuals are measured by a battery of measures on a number of occasions" or "each person in a group of individuals is measured on each trait in a group of traits by each of a number of methods".

The Tucker decomposition and its variation have already found many applications for example tensors have become extremely popular in the chemometrics field, see a book in 2004 [10]. In neuroimaging, a tradition has been to average data across trials or groups of subjects for the extraction of the most reproducible neural activation. Here, tensor decomposition can efficiently extract the consistent patterns of activation, whereas noisy trials/subjects can be downweighted in the averaging process. For example, functional magnetic resonance imaging (fMRI) in neurosciences is frequently organized as a third-order tensor time series comprising a three-dimensional voxel evolving time frame [59].

On the other hand, tensor decompositions are very attractive in signal processing field such as multichannel EEG and magnetoencephalography (MEG) data because they take into account spatial, temporal, and spectral information and provide links among various data and extract features. For example, the multichannel EEG can be preserved as a third-order tensor time series that is modeled as channel×frequency×time frame [14,57,58].

Also, in the bioinformatics domain, a multiple time series is assembled into a high-order time series form gene×sample×time [3].

Tensor-based dimensionality reduction, feature extraction, and classification have recently been extensively studied for computer vision applications such that these applications require particular dimensionality reduction in spatio-temporal space. As known, images and videos are naturally high-order tensors. For example, color images are often stored as a sequence of RGB triplets (i.e., as separate red, green, and blue overlays). An $m \times n$ pixel RGB image is represented by on $m \times n \times 3$ array and a collection of p such images is an $m \times n \times 3 \times p$ array and can be compressed by a low-rank approximation. Different from traditional algorithms, which unfold the image and video data to vector form before classification, tensor works on high-order forms of input data because of destroying the structure using vectorization.

The chapter is organized as follows: Section 14.3 describes the framework of the experimental design and data acquisition system. Section 14.4 is devoted to introducing the notations and giving the basic definitions of tensors, presenting our algorithm and the application of such a numerical technique to the pattern recognition system for the dynamic screening of ASD. The numerical results obtained are presented and discussed in Section 14.5 and finally, in Section 14.6, some conclusions are drawn.

14.2 EXPERIMENTAL DESIGN

In this study we made use of the data collected during a previous study by the second and fourth authors of the chapter on assessing the relationship between severity of ASD core symptoms and the average fixation duration, the gaze shifts, and the eye movement patterns when processing faces. We re-analyzed the data and focused on eye-blink pattern in children with autism spectrum disorder when they are compared to typically developed children (TD).

Thirty children with a mean chronological age of 5.12 years (SD=0.89) participated in the study, all with the written informed consent of their parents and/or legal guardians so that families were free to withdraw from the study at any time. In particular, the study involved 15 children with ASD (11 males and 4 females, age: 5.47 ± 1.29 years; mean±s.d.) previously diagnosed with ASD using DSM-IV criteria and Gilliam Autism Rating scale questionnaire (GARS) that measure three scopes of symptoms such as stereotyped behaviors, communication skills, and social interaction, and was completed by parents; and 15 TD children (11 males and 4 females, age: 4.77 ± 0.49 years) with no history of psychiatric or neurological disease and no family history of autism. As shown in Table 14.1, the TD children were matched to those with autism by chronological age and there was no significant difference between them.

The autistic children were recruited from Center for the Treatment of Autistic Disorder and Autistic Beh-ara Center in Shahid Beheshti University (SBU). All participants were medically screened in our laboratory at SBU for visual acuity and were found to have normal or corrected-to-normal vision. As cultural issues matter, all participants in both groups were Iranian with Persian as their mother tongue. All 30 children completed the experimental procedures.

Each participant sat in front of a 19-inch computer monitor from a viewing distance of 60 cm. The documentary viewing was conducted individually in an isolated

TABLE 14.1
Participant Characterization

Group	Average age	s.d.	SE	F value	Sig. F	t value	df	Sig. t
ASD	5.47	1.29	0.35	8.98	0.006	1.94	28	0.067
TD	4.77	0.49						

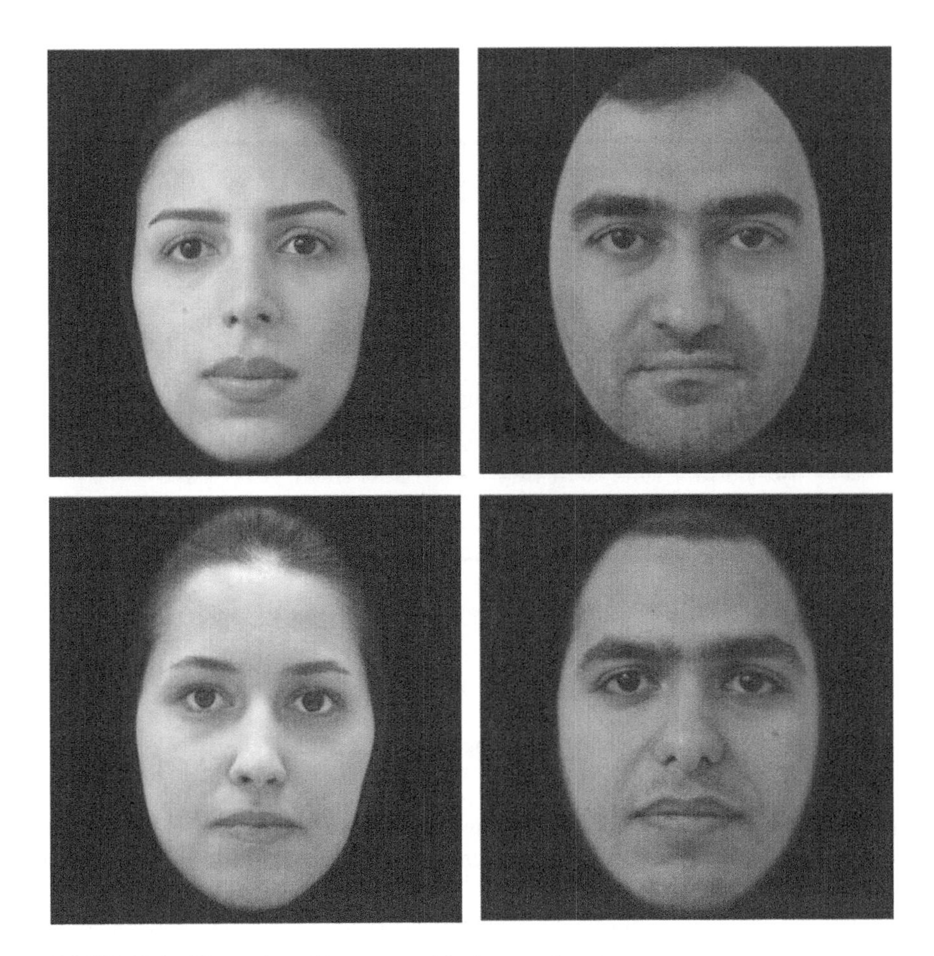

FIGURE 14.1 Illustration of images used in the experiment.

room with the light on. Temperature and humidity were not controlled but were within a comfortable range. Images of 40 males and females (age between 20 and 35) that were standardized for size (1500 × 1500 pixels), lighting, and color (71 × 71 dpi), were used in the experiment. Illustration of images can be seen in Figure 14.1. None of the participants had previously seen the images.

Photos were shown to ASD and TD children and each photo was shown for 3000 ms followed by a fixation cross that stayed on the screen for 1000 ms to draw the attention of viewers to a common fixation location. Then, the test photos were shown for 10000 ms followed by a fixation cross for 1000 ms. The fixation cross was located in four main parts corresponding to the left eye, right eye, nose, and mouth. See Figures 14.2 and 14.3for the structure and timing of the trial and test for each subject, respectively.

A Senso Motoric Instrument (SMI) eye tracking system was used to collect the eye movements at a rate of 120 Hz when the participants were looking at images of

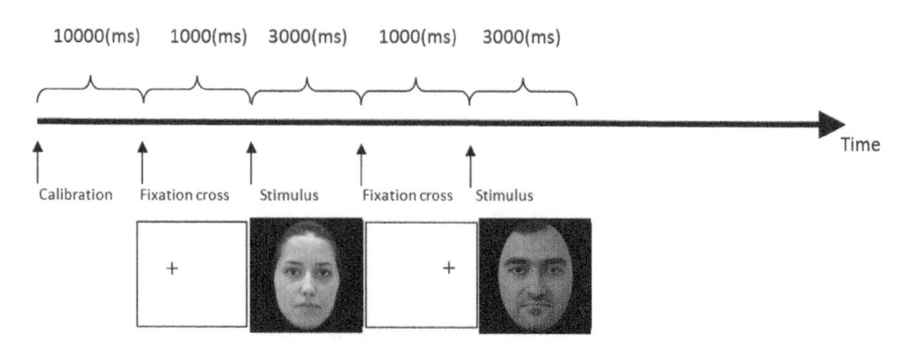

FIGURE 14.2 Structure of trial data capturing.

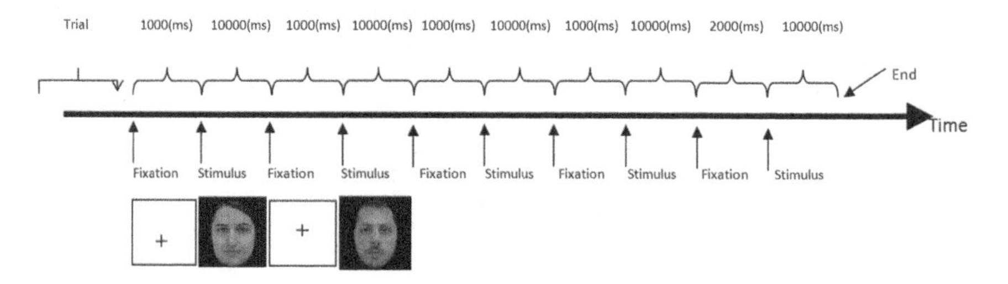

FIGURE 14.3 Structure of test data capturing.

human faces. This system reported the average fixation duration (average duration of looking at stimuli). Five regions of interest (ROIs) were defined for each face and fixations were assigned to each region. These regions were right eye and under right eye, left eye, and under left eye, nose, mouth, and others in the face. Here one of the most important parameters is order of gaze shifts from one ROIs area to another. These paths included any shifts between any two zones of five areas.

Blinks were recorded as events with a measurable duration, identified by an automated algorithm. In fact, the algorithm measured occlusion of the pupil by rate of change in pupil diameter and by vertical displacement of the measured pupil center. The algorithm accurately detected 95.0% of all blinks.

The data were analyzed off-line using MATLAB. After detecting eye-blinks, we calculated the number of eye-blinks per second in order to reduce data size. In this study, we focused on the temporal pattern of inter-eye blink interval (IEBI). Therefore, the number of eye-blinks per second was converted to IEBI over blink number and aligned to the center time point in order to depict a representative sampling of IEBIs. In fact, the approach was to construct a sequence of blinks as an IEBI time series rather than a simple blink rate, thus providing more information for distinguishing between populations beyond mean blink levels. For data analysis, each subject's IEBI was normalized to the average of their IEBI. For example, a transformation and the resulting distribution of IEBI is illustrated in Figures 14.4(a) (blink sequence over time in seconds) and 9.4(b) (IBI sequence over blink number).

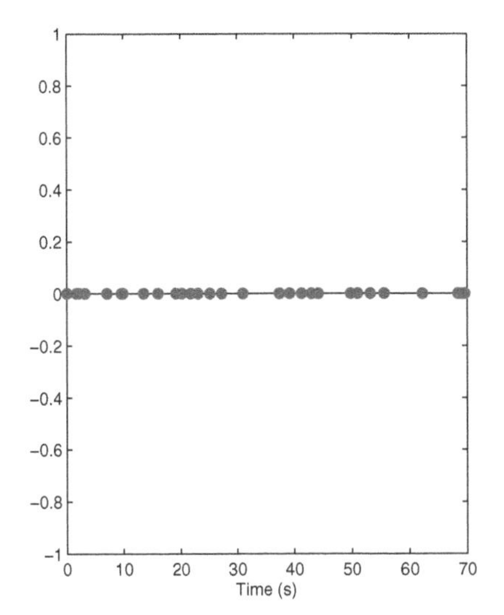

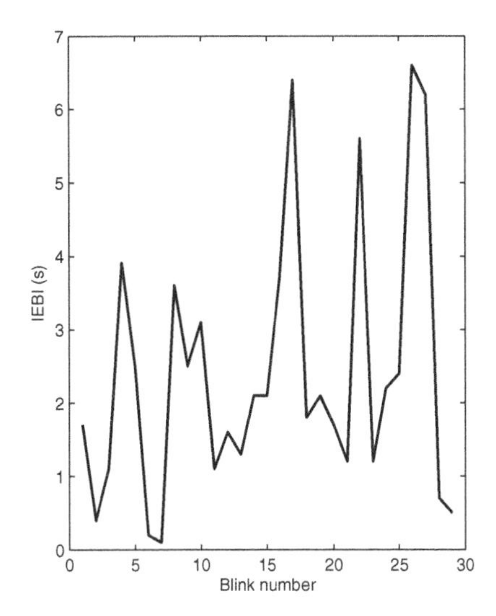

FIGURE 14.4 Illustration of the transformation of blinks over seconds to IEBIs over blink number to the distribution of IEBI based on the first 70 seconds of recorded time for a normal subject in the calibration state. (left) Blink sequence over time in seconds and (right) IEBI sequence over blink number.

14.3 METHODOLOGY

In this section we introduce the notations and give the basic definitions used in the rest of the chapter. The following notations were adopted from Ref. [14]. We shall denote a tensor by underlined bold capital letters (e.g., $\underline{X}$); matrices by bold capital letters (e.g., $A = \left[a_1, a_2, \ldots, a_J\right]$); vectors by bold italic lowercase letters (e.g., a_j); and lowercase letters for the entries (e.g., x_{ij}). Tensors are multilinear mappings on a set of vector spaces. Therefore, an Nth-order tensor of size $\prod_{n=1}^{N} I_n$ can be defined as $\underline{X} \in R^{I_1 \times I_2 \times \ldots \times I_N}$, where each I_i is the dimension of its mode i. We also denote $x_{i_1,\ldots,i_N}$ as the $\left(i_1,\ldots,i_N\right)$th entry $\underline{X}\left(i_1,\ldots,i_N\right)$ of higher-order tensor $\underline{X}$. For a three-dimensional tensor $\underline{X} \in R^{I \times J \times K}$, its frontal slice, lateral slice, and horizontal slice are denoted, respectively, by $X_k = X_{\bar{i}\bar{j}k}$, $X_{:j:}$ and $X_{i\bar{j}}$; so in general $X_{i_n=\alpha}$ is a slice obtained by fixing the nth index to α when $\underline{X} \in R^{I_1 \times I_2 \times \ldots \times I_N}$. A tube at a position (i, j) along mode-3 is denoted by $x_{ij:}$ and the corresponding tubes along mode-2 and mode-1 are $x_{i:k}$ and $x_{:jk}$, respectively (see Figure 14.5).

Mode-n matricization of $\underline{X}$ is the process of unfolding $\underline{X}$ into a matrix $X_{(n)} \in R^{I_n \times \left(\prod_{i \neq n} I_i\right)}$, such that $X_{(n)}\left(i_n, j\right) = X\left(i_1, \ldots i_n, \ldots, i_N\right)$, for $j = 1 + \sum_{k=1, k \neq n}^{N} \left(i_k - 1\right) \prod_{m=1, m \neq n}^{k-1} I_m$. Outer, Khatri-Rao, Kronecker, and Hardamard products and elementwise division are denoted by $\circ$, $\odot$, $\otimes$, $\circledast$, and $\oslash$, respectively.

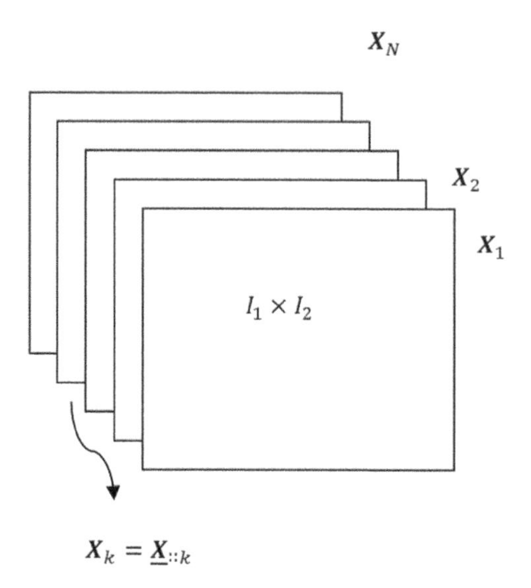

FIGURE 14.5 Set of N 2-d images represented as an $I_1 \times I_2 \times I_3$ tensor X.

The mode-n product of the tensor $\underline{X} \in R^{I_1 \times I_2 \times \ldots \times I_N}$ with a matrix $A \in R^{J_n \times I_n}$ is denoted as $\underline{Y} = \underline{X} \times_n A$, which is a Nth-order tensor of size $I_1 \times \ldots \times I_{n-1} \times J_n \times I_{n+1} \times \ldots \times I_N$ such that

$$Y_{(n)} = A\, X_{(n)}. \tag{14.1}$$

Moreover, the product of a tensor $\underline{X}$ and a set of matrices $A^{(n)}$ is denoted by

$$\left[\underline{X} \times_1 A^{(1)} \times_2 A^{(2)} \ldots \times_N A^{(N)}\right]_{(n)} = A^{(n)} X_{(n)} \left[A^{(N)} \otimes \ldots \otimes A^{(n+1)} \otimes A^{(n-1)} \otimes \ldots \otimes A^{(1)}\right]^T. \tag{14.2}$$

Multiplication of a tensor with all but one mode-n is denoted as

$$\underline{X} \times_{-n} \{A\} = \underline{X} \times_1 A^{(1)} \times_2 A^{(2)} \ldots \times_{n-1} A^{(n-1)} \times_{n+1} A^{(n+1)} \ldots \times_N A^{(N)}$$

Similarly, multiplication of a tensor and matrices when two modes are excluded is denoted as (for $n<m$)

$$\underline{X} \times_{-(n,m)} \{A\} = \underline{X} \times_1 A^{(1)} \times_2 A^{(2)} \ldots \times_{n-1} A^{(n-1)} \times_{n+1} A^{(n+1)} \ldots \times_{m-1} A^{(m-1)} \times_{m+1} A^{(m+1)} \ldots \times_N A^{(N)}$$

The inner product of two tensors $\underline{X}, \underline{Y} \in R^{I_1 \times \ldots \times I_N}$ is defined as

$$< \underline{X}, \underline{Y} \geq \sum_{i_1=1, i_2=1, \ldots, i_N=1}^{I_1, I_2, \ldots, I_N} x_{i_1, i_2, \ldots, i_N} \cdot y_{i_1, i_2, \ldots, i_N}$$

Accordingly, the Frobenius norm of $\underline{X}$ is defined as $\| \underline{X} \|_F =< \underline{X}, \underline{X} >^{1/2}$. Also, the tensor distance of two1 N tensors, $\underline{X}$ and $\underline{Y}$, is given by $d(\underline{X}, \underline{Y}) = \| \underline{X} - \underline{Y} \|_F$. The n-rank of a tensor $X \in R^{I_1 \times I_2 \times \ldots \times I_N}$ is the dimension of the vector space generating the n-mode vectors (i.e., $rank_n(\underline{X}) = rank\left(X_{(n)}\right)$). Also, $\underline{X}$ has rank 1 when it equals the outer product of N vectors $u^{(1)}, u^{(2)}, \ldots, u^{(N)}$ such that

$$\underline{X} = u^{(1)} \circ u^{(2)} \circ \ldots \circ u^{(N)}. \tag{14.3}$$

14.3.1 Feature Extraction, Feature Ranking, and Classification

Tensor decompositions are important tools for feature extraction and dimension reduction by capturing multilinear structures in large-scale high-order datasets with applications in image, video, signal analysis, neuroscience, and so on. Tucker decomposition is a method to extract features that interact within each modality, formulated as a decomposition of a given Nth-order tensor $\underline{X} \in R^{I_1 \times I_2 \times \ldots \times I_N}$ into an unknown core tensor $\underline{G} \in R^{J_1 \times \ldots \times J_N}$ multiplied by a set of N unknown factors $A^{(n)} \in R^{I_n \times J_n}, (n = 1, 2, \ldots, N)$, which allows to effectively perform model reduction and feature extraction:

$$\underline{Y} = \underline{G} \times_1 A^{(1)} \times_2 A^{(2)} \ldots \times_N A^{(N)} + \underline{E} = \hat{\underline{Y}} + \underline{E} \tag{14.4}$$

where $\underline{Y}$ is an approximation of $\underline{Y}$ and $\underline{E}$ denotes error. To solve this problem we can obtain the following nonlinear optimization problem, which is quite complicated:

$$argmin_{\underline{G}, A^{(1)}, \ldots, A^{(N)}} \| \underline{Y} - \underline{G} \times_1 A^{(1)} \times_2 A^{(2)} \ldots \times_N A^{(N)} \|_F^2$$

For example, in three-dimensional data the model with three factors is

$$\underline{Y} \approx \underline{G} \times_1 A \times_2 B \times_3 C. \tag{14.5}$$

This model can be reduced to the following model with two factors:

$$\underline{Y} \approx \underline{F} \times_1 A \times_2 B, \tag{14.6}$$

where $\underline{F} = \underline{G} \times_3 C$. It should be noted that parallel factor analysis is a special case of the Tucker model where the core tensor has nonzero elements on the superdiagonal. Furthermore, by imposing orthogonality conditions on the factor matrices we obtain higher-order singular value decomposition.

At first, we focus on the feature extracting concept on a set of two-dimensional samples. Thus, we consider a set of K data matrices n corresponding to D categories such that their matrix factorizations are $X^{(k)} \approx A^{(1)}F^{(k)}A^{(2)T}$ where $A^{(1)} \in R^{I_1 \times J_1}, A^{(2)} \in R^{I_2 \times J_2}, (J_n \ll I_n)$, and $F^{(k)} \in R^{J_1 \times J_2}$ are the basis matrix along the horizontal dimension, the basis matrix along the vertical dimension, and the extracted features, respectively.

Here, we aim to extend the tensor concept for the problem. Therefore, to achieve the goal, the K training sample matrices are first concatenated along mode 3 so that the training set is represented by an 3th-order tensor $\underline{X}$ defined as

$$X_{(1)} = \left(X^{(1)}X^{(2)}\ldots X^{(K)}\right) \approx A^{(1)}F_{(1)}\left(I_k \otimes A^{(2)}\right)^T,$$

or

$$X_{(2)} = \left(X^{(1)T}X^{(2)T}\ldots X^{(K)T}\right) \approx A^{(2)}F_{(2)}\left(I_k \otimes A^{(1)}\right)^T,$$

where $F_{(1)}$ and $F_{(2)}$ are mode-1 and mode-2 matricized versions of the concatenation core tensor $F_{(1)} = \left(F^{(1)}F^{(2)}\ldots F^{(k)}\right)$. Therefore, the 3-order tensor $\underline{X} \in R^{I_1 \times I_2 \times K}$ can be written as

$$\underline{X} \approx \underline{F} \times_1 A^{(1)} \times_2 A^{(2)}$$

As already pointed out, $A^{(1)}$, $A^{(2)}$ and $F^{(k)}$ can be obtained using the Alternating Least Squares (ALS) algorithm; i.e.,

$$A^{(1)} = argmin_{A^{(1)}} \left\| X_{(1)} - A^{(1)}F_{(1)}\left(I_k \otimes A^{(2)}\right)^T \right\|_F,$$

$$A^{(2)} = argmin_{A^{(2)}} \left\| X_{(2)} - A^{(2)}F_{(2)}\left(I_k \otimes A^{(1)}\right)^T \right\|_F,$$

$$F_{(3)} = argmin_{F_{(3)}} \left\| X_{(3)} - F_{(3)}\left(A^{(2)} \otimes A^{(1)}\right)^T \right\|_F.$$

Here, to obtain a unique decomposition it is clear that we should impose some additional constraints and obtain well-known models such as PARAFAC, HOSVD, and so on. In a particular case where $F^{(k)}$ are $J \times J$ diagonal matrices, the following PARAFAC model is approximated:

$$\underline{X} \approx \underline{I} \times_1 A^{(1)} \times_2 A^{(2)} \times_3 F,$$

where $\underline{I} \in R^{J \times J \times J}$is an identity tensor, $F = \left[f^{(1)}, f^{(2)}, \ldots, f^{(K)}\right]^T \in R^{K \times J}$, and $F^{(k)} = diag\left(f^{(k)}\right)$ for $J = J_1 = J_2$.

Similar to the above statements, we can formulate the following classification problem on a set on N dimensional samples. Let us consider a set of K training samples $\underline{X}^{(k)} \in R^{I_1 \times I_2 \ldots \times I_N}, (k = 1, \ldots, K)$ corresponding to C categories, and a set of test data $\dot{\underline{X}}^{(t)} \in R^{I_1 \times I_2 \ldots \times I_N}, (t = 1, \ldots, T)$. Again, we should note that the main challenge is to give suitable labels for the test data. Therefore, first the main features in the compressed core tensor $\underline{G}^{(k)} \in R^{J_1 \times J_2 \ldots \times J_N}$ and N factors $A^{(n)} \in R^{I_n \times J_n}$ for the training data should be obtained, then using the basis factors of the training set the feature extraction for test samples should be performed. At end, the key role of the algorithm (i.e., classification) is applied by comparing the test and the training features.

Instead of solving the nonlinear optimization problem $argmin_{A^{(1)}, \ldots, A^{(N)}} \sum_{k=1}^{K} \| \underline{X}^{(k)} - \underline{G}^{(k)} \times_1 A^{(1)} \ldots \times_N A^{(N)} \|_F^2$,which is quite complicated, we turn back to the reminder of the concatenating concept; that is, we rewrite all the training data $\underline{X}^{(k)}$ and convert the problem into that of a single tensor decomposition as follows:

$$\underline{X} \approx \underline{G} \times_1 A^{(1)} \times_2 A^{(2)} \ldots \times_N A^{(N)},$$

where

$$\underline{X} = cat\left(N+1, \underline{X}^{(1)}, \underline{X}^{(2)}, \ldots, \underline{X}^{(K)}\right) \in R^{I_1 \times \ldots \times I_N \times K},$$

or, equivalently,

$$X_{(N+1)} = \left[vec\left(\underline{X}^{(1)}\right), vec\left(\underline{X}^{(2)}\right), \ldots, vec\left(\underline{X}^{(K)}\right) \right]^T,$$

where $X_{(N+1)}$ is the mode-$(N+1)$ matricized version of an $(N+1)$-way tensor $\underline{X}$. Also, the core tensor $\underline{G} \in R^{J_1 \times J_2 \ldots \times J_N \times K}$ represents extracted features for the training samples; i.e.,

$$\underline{G}\left(:_1, :_2, \ldots, :_N, k_{N+1}\right) = \underline{G}^{(k)}.$$

It should be emphasized that such tensor decomposition and equivalent simultaneous matrix factorizations are quite flexible, and we can impose various constraints and obtain various methods. In fact, an N-order tensor $\underline{X} \in R^{I_1 \times \ldots \times I_N}$ can be written as

$$\underline{\boldsymbol{X}} = \underline{\boldsymbol{G}} \times_1 \boldsymbol{U}^{(1)} \times_2 \boldsymbol{U}^{(2)} \ldots \times_N \boldsymbol{U}^{(N)}, \tag{14.7}$$

where $U^{(n)} \in R^{I_n \times I_n}$ is an orthogonal matrix, and $\underline{G}$ is a tensor of the same dimensions as $\underline{X}$ such that $\left\langle G_{i_n = \alpha}, G_{i_n = \beta} \right\rangle = 0$ when $\alpha \neq \beta$ and also the norms of the slices along every mode are ordered; i.e., for every mode we have

$$\left| G_{i_n = 1} \right\| \geq \left\| G_{i_n = 2} \right\| \geq \cdots \geq \left\| G_{i_n = I_n} \right| \geq 0, \tag{14.8}$$

where the norms || || are the Frobenius norms and in fact are the n-mode singular values $\sigma_i^{(n)}$ of $\underline{X}$. Indeed, the vector $u_i^{(n)}$ in $U^{(n)} = \left(u_1^{(n)}, u_2^{(n)}, \ldots, u_{I_n}^{(n)}\right)$ is an ith n-mode singular vector.

The pseudo-code of the algorithm is described in detail in Algorithm 1. In this algorithm, `svd` refers to the MATLAB SVD function, which computes a few leading singular values and vectors.

```
Algorithm 1: HOSVD algorithm
  Input: tensor X ∈ R^(I_1×...×T_N)
    Output: N orthogonal factors U^(n) ∈ R^(I_n×I_n) and a core tensor G ∈ R^(I_1×...×T_N)
  begin
        for r = 1,2...,N do
            Find the i-mode matricization X_(i) of X
            Compute [U^(i), S^(i), V^(i)] = svd(X_(i))
      end for
      Compute the core G = X ×_1 (U^(1))^T ×_2 (U^(2))^T ... ×_N (U^(N))^T
end
```

Here, tensor approximation can be done by using Algorithm 1 and Tucker decomposition as follows:

$$\underline{X} \approx \underline{\tilde{G}} \times_1 \tilde{U}^{(1)} \times_2 \tilde{U}^{(2)} \times_3 \ldots \times_N \tilde{U}^{(N)} \tag{14.9}$$

where $\underline{\tilde{G}} \in R^{J_1 \times \ldots \times J_N}, \tilde{U}^{(n)} \in R^{I_n \times J_n}, (n = 1,2,\ldots,N)$. Then to obtain the unknown factors $\tilde{U}^{(n)}, (n = 1,\ldots,N)$, we can solve the following formula:

$$max\ J\left(\tilde{U}^{(1)},\ldots,\tilde{U}^{(N)}\right) = \| \underline{X} \times_1 \tilde{U}^{(1)T} \times_2 \tilde{U}^{(2)T} \ldots \times_N \tilde{U}^{(N)T} \|_F^2 .$$

Here again we emphasize that only the $\tilde{U}^{(i)}$ are unknown. Using fixed $\tilde{U}^{(1)},\ldots,\tilde{U}^{(n-1)},\tilde{U}^{(n+1)},\ldots,\tilde{U}^{(N)}$, we can project tensor $\underline{X}$ onto the subspace defined as

$$\underline{W}^{(-n)} = \underline{X} \times_1 \tilde{U}^{(1)T} \ldots \times_{n-1} \tilde{U}^{(n-1)T} \times_{n+1} \tilde{U}^{(n+1)T} \ldots \times_N \tilde{U}^{(N)T} = \underline{X} \times_{-(n,N+1)} \left\{\tilde{U}^T\right\},$$

and then the orthogonal matrix $\tilde{U}^{(n)}$ can be estimated as J_n leading left singular vectors of the mode-n matricized version $W_{(n)}^{(-n)}$. The pseudo-code of this algorithm, called the Higher Order Orthogonal Iteratione (HOOI) algorithm, is described in detail in Algorithm 2.

```
Algorithm 2: HOOI algorithm (orthogonal TUCKER)
  Input: concatenation tensor of all training samples i.e.
  X ∈ R^(I_1×...×I_N×K) and number of basis components for factors i.e. J_1, J_2, ..., J_N
  Output: N orthogonal factors U^(n) ∈ R^(I_n×J_n) and a core tensor G ∈ R^(I_1×...×I_N×K)
  begin
```

Perform Algorithm 1 (HOSVD algorithm) or random initialization for all factors $U^{(n)}$

```
for n = 1,2...,N do
```

$$\underline{W}^{(-n)} = \underline{X} \times_{-(n,N+1)} \{\tilde{U}^T\}$$

$$Compute\left[\tilde{U}^{(n)}, \tilde{S}^{(n)}, \tilde{V}^{(n)}\right] = svds\left(W_{(n)}^{(-n)}, J_n, 'LM'\right)$$

```
end for
```

Compute the core tensor $\underline{G} = \underline{W}^{(-N)} \times_N U^{(N)T}$

```
end
```

Remark: Although the training features obtained by the proposed method can be directly used for classification, they do not contain any label of a class, which is often useful to model the difference between classes of data. To exploit such information, we should find discriminant bases to project the training features onto the discriminant subspaces. Hence, we can use the following algorithm called the High Order Discriminant Analysis algorithm (HODA).

```
Algorithm 3: HODA algorithm
```

`Input:` *concatenation tensor of all training samples i.e.* $\underline{X} \in R^{I_1 \times \ldots \times I_N \times K}$ *and number of basis components for factors i.e.* $J_1, J_2, \ldots, J_N$

`Output:` N *orthogonal factors* $U^{(n)} \in R^{I_n \times J_n}$ *and a core tensor* $\underline{G} \in R^{J_1 \times \ldots \times J_N \times K}$

```
begin
```

Perform Algorithm 1 (HOSVD algorithm) or random initialization for all factors $U^{(n)}$

Calculate $\underline{\tilde{X}}$ *and* $\underline{\breve{X}}$ *according to Eq. (3.9) and Eq. (3.10)*

```
for n = 1,2,...,N do
```

$$\underline{\tilde{Z}} = \underline{\tilde{X}} \times_{-(n,N+1)} \{U^T\}$$

$$S_w^{-n} = < \underline{\tilde{Z}}, \underline{\tilde{Z}} >_{-n}$$

$$\underline{\breve{Z}} = \underline{\breve{X}} \times_{-(n,N+1)} \{U^T\}$$

$$S_b^{-n} = < \underline{\breve{Z}}, \underline{\breve{Z}} >_{-n}$$

$$\varphi = \frac{trace\left(U^{(n)T} S_b^{-n} U^{(n)}\right)}{trace\left(U^{(n)T} S_w^{-n} U^{(n)}\right)}$$

$$\left[U^{(n)}, \Lambda\right] = eigs\left(S_b^{-n} - \varphi S_w^{-n}, J_n, 'LM'\right)$$

$$\left[U^{(n)},\Lambda\right]=eigs\left(U^{(n)}U^{(n)T}<\underline{X},\underline{X}>_{(-n)}U^{(n)}U^{(n)T},J_n,'LM'\right)$$

```
    end for
```
Compute the core tensor $\underline{G}=\underline{X}\times_{-(N+1)}\left\{U^T\right\}$
```
end
```

It should be noted that in this algorithm, we have

$$\underline{\tilde{X}}^{(k)}=\underline{X}^{(k)}-\underline{\underline{X}}^{(c_k)} \tag{14.10}$$

where $\underline{\underline{X}}^{(c)}$ is the mean tensor of the c-th class consisting of K_c training samples; i.e.,

$$\underline{\underline{X}}^{(c)}=\frac{1}{K_c}\sum_{k\in I_c}\underline{X}^{(k)},\ c=1,2,\ldots,C$$

where I_c is the subset of indices k, which indicates the samples k belong to class c and K_c is the number of training samples in the c-th class. Indeed, c_k denotes the class to which the k-th training sample $\underline{X}^{(k)}$ belongs. It should be noted that the concatenation of all the core tensors $\underline{\tilde{X}}^{(k)}$ forms an $(N+1)$-dimensional tensor $\underline{\tilde{X}}$ so that $\underline{\tilde{X}}_k=\underline{\tilde{X}}^{(k)}$. On the other hand, in this algorithm we assume that the tensors $\underline{\breve{X}}$ are defined as

$$\underline{\breve{X}}^{(c)}=\sqrt{K_c}\left(\underline{\underline{X}}^{(c)}-\overline{\overline{\underline{X}}}\right) \tag{14.11}$$

which are parts of the concatenation tensor $\underline{\breve{X}}$ (i.e., $\underline{\breve{X}}_c=\underline{\breve{X}}^{(c)}$). Also in Eq. (3.10), $\overline{\overline{\underline{X}}}$ is the average tensor for all the data tensors; i.e.,

$$\overline{\overline{\underline{X}}}=\frac{1}{K}\sum_{k=1}^{K}\underline{X}^{(k)},$$

In addition, here the contracted product of tensors $\underline{A}\in R^{I_1\times\ldots\times I_N}$ and $\underline{B}\in R^{J_1\times\ldots\times J_N}$ along all modes except the mode-n is denoted as

$$<\underline{A},\underline{B}>_{-n}=A_{(n)}B_{(n)}^{\ T}\in R^{I_n\times J_N},\left(I_k=J_k,\ \forall k\neq n\right)$$

For a detailed description of this algorithm the reader is referred to Ref. [14]. The decomposition of the extracted feature is $J_1\times\ldots\times J_N$, which is dependent on the dimension of the factors $\tilde{U}^{(n)}$. Thus, it is clear that J_n can be determined using the n-mode singular values $\sigma_i^{(n)}$ of $\underline{X}$. The optimal dimensione J_n can be found using the following optimizing problem:

$$arg\ min_{J_n}\frac{\sum_{i=1}^{J_n}\left(\sigma_i^{(n)}\right)^2}{\sum_{i=1}^{I_n}\left(\sigma_i^{(n)}\right)^2}>\theta, \tag{14.12}$$

which means that the factors should explain the whole training data at least θ(typical value is 95%) (i.e., a large percentage of the variation of the training data is accounted by the first J_n singular values). So we observe that the number of features is reduced, but it seems that its dimension is still large and one cannot guarantee that the proposed scheme will obtain the solution with good accuracy. Our main struggle is to present the appropriate classification algorithm with the minimum number of features that have the most significant information. To achieve the goal, it is better to consider some significant features from the core tensor instead of using all the extracted features because we believe that some of them may have a negligible effect or in some cases perhaps a negative effect on the accuracy. Hence, another issue under discussion is removing redundant features. Therefore, we can apply a ranking algorithm on the extracted features, then sort them in descending order of their scores and finally choose the significant features by using an acceptance tolerance. Therefore, to combat the mentioned problem, designing the proper feature ranking algorithm plays a vital role in this step; such techniques include correlation score, minimum-redundancy-maximumrelevance selection, Fisher score, Laplacian score, and entropy [28, 40]. Overall, we bring this matter to the reader's mind that the major ranking procedure in the context of classification algorithms uses the Fisher ranking algorithm as defined in Ref. [40], which is also employed in other works.

Finally, classification based on the reduced features of training and test sets using well-known supervised binary classifier algorithms such as k-nearest neighbors, algorithm, support vector machines, and so on is performed. To satisfy our curiosity and due to the fact that it is almost impossible to select the best binary classifier between the algorithms in practice, it is better to consider some of the well-known classification algorithms. To clarify even more, what is worthy to note is the performance of the classifiers in terms of accuracy, specificity, and sensitivity. In many practical problems, the main contribution is devoted to finding the best classifier with optimal parameters (i.e., accuracy, specificity, and sensitivity), but it is an open problem that is still under intensive investigation. Thus, in this work we arbitrarily applied knearest neighbor (kNN), linear support vector machine (linSVM), and nonlinear support vector machine with polynomial kernel (polySVM), and Gaussian radial basis function kernel (rbfSVM). svmtrain, svmclassify, knnclassify, and classperf MATLAB functions were used to classify the selected features of training sets.

In this MATLAB function, the selection of parameters of the kNN and nonlinear SVM classifiers are very important for convergence issue. Therefore, the performance was optimized in a statistically rigorous manner by choosing the following parameters: width of Gaussian kernel from 0.8 to 20, with the step size of 0.2, and order of polynomial kernel from 2 to 7.

The overall architecture of the proposed detection system is given in Figure 14.6.

14.4 EXPERIMENTAL RESULTS AND DISCUSSION

As already pointed out, our focus is devoted to a classification study of subjects as autistic or TD groups. Thus, it is assumed that we have sufficient notations and preliminaries of tensor approximations. Assume that we have a set of K training subjects corresponding to $D=2$ (autistic or control) categories. Note that the numbers of

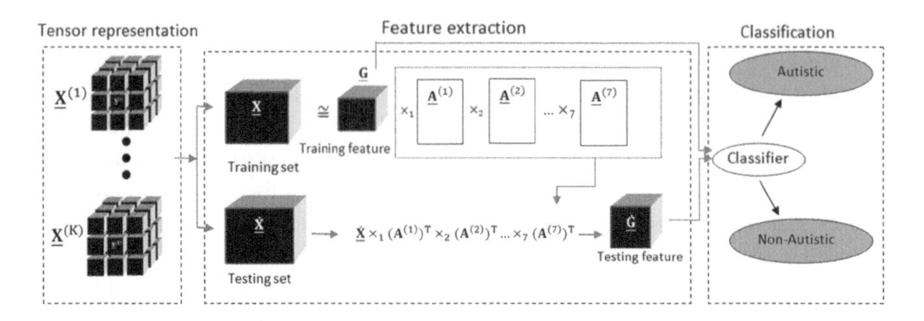

FIGURE 14.6 Conceptual paradigm for the application of the multiway feature extraction of the proposed method on an eye-blink tensor-based classification.

autistic or healthy subjects are k_1 and k_2, respectively. Hence, we assume that the training set is manually classified.

Here we aim to extend the tensor concept for the problem. Thus, to achieve the goal, the K training sample matrices are firstly concatenated along the mode 7 so that the training set is represented by an 7th-order tensor $\underline{X}$defined as

$$\underline{X} = \left(X^{(1)}, X^{(2)}, \ldots, X^{(K)}\right) \in R^{I_1 \times I_2 \times I_3 \times I_4 \times I_5 \times I_6 \times K} \quad (14.13)$$

For the sake of simplicity, we assume $K = I_7$ so $\underline{X} \in R^{I_1 \times I_2 \times I_3 \times I_4 \times I_5 \times I_6 \times I_7}$. Therefore, the training data formed a 7-D spectral tensor with modes *80 Blink number × 2 eyes (right or left) × 7 ROIs areas × 150 Fixation number in each ROI areas × 600 Total fixation number (to save the order) × 2 Conditions (IBEIs or Fixation durations) × 30 Subjects*. That means the training data $\underline{X}$is a 7-dimensional tensor of 30 6-D sub-tensors for our subjects and two classes: 80 × 2 × 7 × 150 × 600 × 2 × 30. The first 15 sub-tensors are for class 1 and the next 30 sub-tensors are for class 2.

$$\underline{X}^{(1)}\underline{G}^{(1)}A^{(n)}$$

The test data $\dot{\underline{X}}$ were also organized in a similar way and consisted of 30 6-D sub-tensors (see Figure 14.7): *30 Blink number× 2 eyes (right or left) × 7 ROI areas × 70 Fixation number in each ROI areas × 300 Total fixation number (to save the order) × 2 Conditions (IBEIs or Fixation durations) × 30 Subjects*. The challenge was to find appropriate labels for the test data with high accuracy. The classification algorithm can be generally performed in the following steps: we first sort the training data into tensors and try to find a set of common factors and corresponding reduced features of the tensors. For example, a Tucker decomposition-based method is then applied to have the approximations

$$\underline{X} = \underline{G} \times_1 A^{(1)} \times_2 A^{(2)} \times_3 A^{(3)} \times_4 A^{(4)} \times_5 A^{(5)} \times_6 A^{(6)} \times_7 A^{(7)} + E^i. \quad (14.14)$$

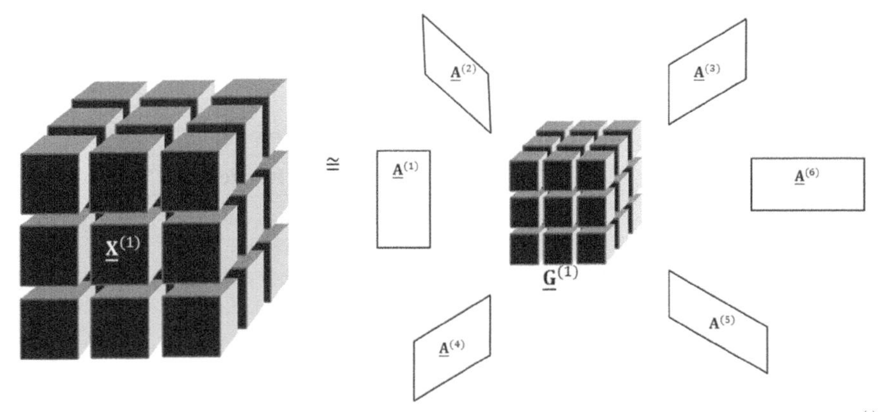

FIGURE 14.7 Illustration for a six-way Tucker decomposition. Decomposition of tensor $\underline{X}^{(1)}$ can be seen as a multiplication in all possible modes of a core tensor $\underline{G}^{(1)}$and a set of basis matrices $A^{(n)}$.

As will become apparent, using the above factors we can extract features for test samples.

To get a better sense of the efficiency of the method presented in the current chapter, let us describe the results in detail. Meanwhile, to measure the performance of the classifiers, accuracy (ACC), specificity (or true negative rate (SPC)) and sensitivity (or true positive rate (TPR)) have been used with the following definitions:

$$ACC = \frac{N_{TP} + N_{TN}}{N_{TP} + N_{TN} + N_{FN} + N_{TN}}, \tag{14.15}$$

$$TRP = \frac{N_{TP}}{N_{TP} + N_{FN}} \tag{14.16}$$

$$SPC = \frac{N_{TN}}{N_{TN} + N_{FP}} \tag{14.17}$$

where N_{TP}, N_{TN}, N_{FP}, and N_{FN} denote the number of true positives, true negatives, false positives, and false negatives, respectively (see Table 14.2).

The numerical implementation and all of the executions are conducted in a MATLAB environment.

14.4.1 Results

Now we turn back to the reminder of the numerical solution to the problem. First of all, in our experiments, the original data were downsampled into a six-order tensor of dimension described in the previous section. Indeed, we tested the performance of the proposed scheme on the binary classification task with different

TABLE 14.2
Confusion Matrix

	Predicted results	
Gold Standard	**Reject H_0 (Autistic)**	**Fail to reject H_0 (Non-Autisitc)**
H_0 is true (Non-Autisitc)	Type I error, False positive (FP)	Correct outcome, True positive (TP)
H_0 is false (Autisitc)	Correct outcome, True negative (TN)	Type II error, False negative (FN)

numbers of training data and test data. The data sets were divided into training and test sets. From the other point of view, the training and test sets were represented by seven-order tensors. Next, the proposed method was performed directly on the tensor-based data sets to seek the reduced set of core tensors consisting of significant features; specifically the method decomposes the training data as $\underline{X} = \underline{G} \times_1 A^{(1)} \times_2 A^{(2)} \times_3 A^{(3)} \times_4 A^{(4)} \times_5 A^{(5)} \times_6 A^{(6)} \times_7 A^{(7)}$. The core tensor $\underline{G}$, which consists of the reduced features of $\underline{X}$ that are of a much lower dimension of $\underline{X}$, was vectorized into the feature vector. Then we selected a prominent set of features by applying the Fisher ranking scores. Obviously, there exists a tradeoff between dimensional reduction and approximation error. Thus, randomly we selected four different values for approximation ratio ε. After ranking the training features based on the Fisher ranking scores, to validate the performance of the proposed method, the leading features in the core tensor were input to the classifiers that are chosen as the kNN, linSVM, two-order polySVM,rbfSVM width of Gaussian kernel from 0.8 to 20, with the step size of 0.2.

As far as the radial basis functions are concerned, several choices are possible, such as the so-called multiquadrics, inverse multiquadrics, Gaussian RBFs, or thin plate splines (see, e.g., [11]). In particular, the multiquadrics, the inverse multiquadrics, and the Gaussian RBFs contain a free shape parameter on which the performance of the RBF approximation strongly depends. Precisely, values of the shape parameters that yield a high spatial resolution (i.e., a high level of accuracy) also lead to severely ill-conditioned linear systems. Therefore, one has to find a value of the shape parameter such that the numerical results are satisfactorily accurate, but at the same time the overall approximation does not blow up (due to ill-conditioning problems). In the technical literature, various approaches for selecting the RBF shape parameter have been proposed (see, e.g., [43, 53, 8, 7, 6, 12, 13, 20, 24]). These algorithms, which are often based on rules of thumbs and on semianalytical relations, can yield satisfactory results in some circumstances. Nevertheless, to the best of our knowledge, a method to choose the RBF shape parameter that is rigorously established and is proven to perform well in the general case is still lacking.

Testifying the accuracy of the classification success rate is of our special interest. The results are given in Figure 14.8. The receiver operating characteristics (ROC) graph is used for better visualization of the performance of our binary classifier system based on the proposed feature extraction algorithms. It should be noted that the ROC

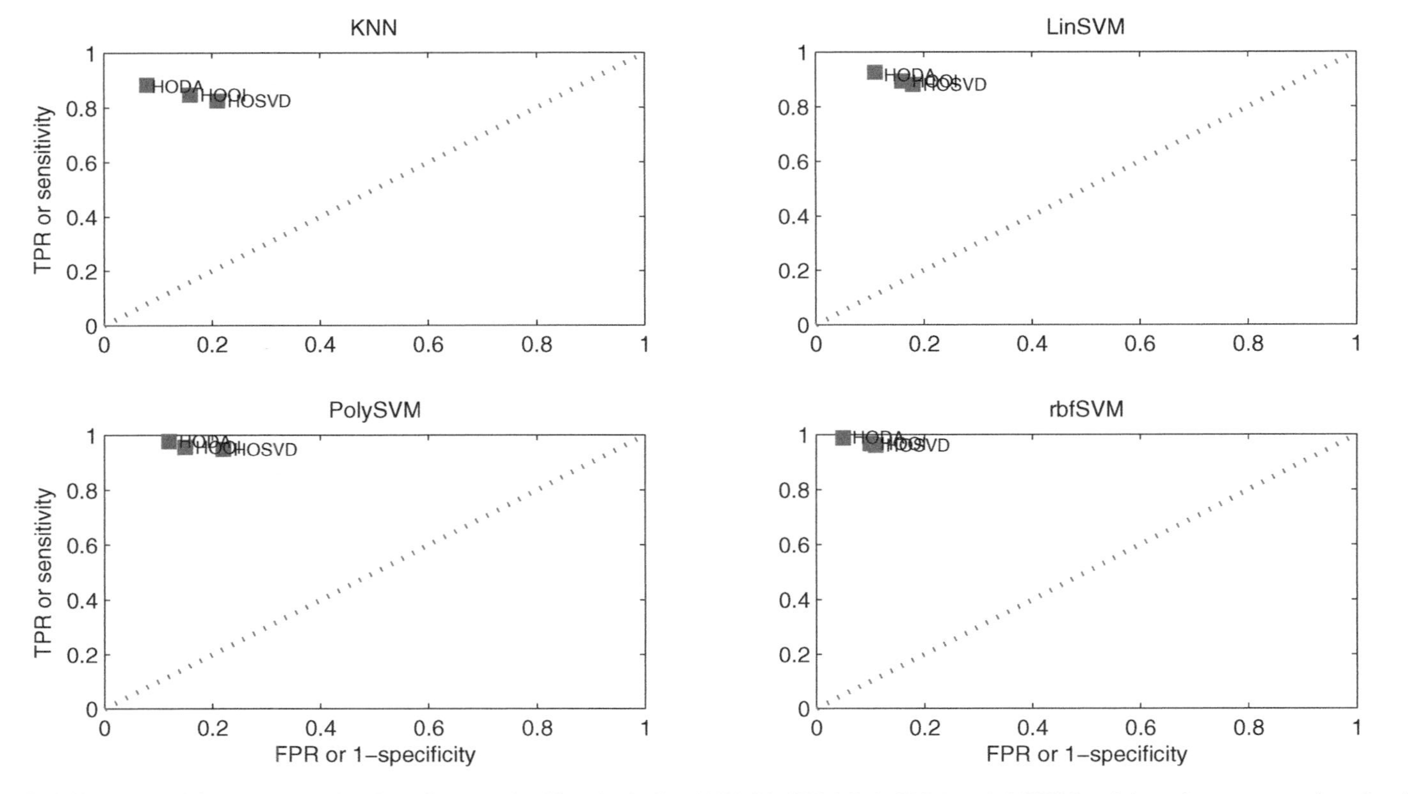

FIGURE 14.8 ROC graphs showing four discrete classifiers including kNN, LinSVM, PolySVM, and rbfSVM and three feature extraction algorithms including HOSVD, HOOI, and HODA.

TABLE 14.3
Best Classification Success Rate Corresponding to Different Feature Extraction Algorithms Obtained by Proposed Scheme

	Performance	kNN	linSVM	PolySVM	RBFSVM
HOSVD	ACC	83.1 ± 1.2	84.2 ± 2.5	93.3 ± 2.1	97.2 ± 1.1
	TRP	82.5 ± 8.4	84.1 ± 2.1	94.8 ± 3.5	96.1 ± 1.4
	SPC	79.0 ± 3.1	82.0 ± 3.2	78.2 ± 1.1	89.1 ± 2.2
HOOI	ACC	86.2 ± 3.5	91.1 ± 3.2	95.4 ± 3.3	97.6 ± 4.4
	TRP	84.7 ± 6.1	89.3 ± 3.5	95.6 ± 7.6	96.7 ± 1.3
	SPC	84.3 ± 7.2	84.0 ± 6.2	85.2 ± 3.6	90.0 ± 8.2
HODA	ACC	88.4 ± 2.4	92.5 ± 3.1	97.4 ± 2.9	99.1 ± 4.6
	TRP	88.3 ± 4.3	92.6 ± 5.4	97.7 ± 2.2	98.8 ± 3.6
	SPC	92.3 ± 7.5	89.4 ± 6.3	88.0 ± 1.2	95.5 ± 7.7

graph is a graph defined by sensitivity and specificity as x and y axes, respectively. In fact, the ROC graph depicts the relative tradeoff between benefits and costs. In each plot, each point in the ROC graphs presented in Figure 14.8 demonstrates the performance of the system using the labeled feature extraction algorithms including HOSVD, HOOI, and HODA for screening of autistic children. From this figure we observe that the system works better applying tensor analysis-based HODA features and rbfSVM classifier. Note that the best classification success rate is used in the tables. The results for the screening of autistic children in terms of mean values of sensitivity and specificity and their standard deviations are shown in Table 14.3. The classification performance was analyzed, and we achieved an average accuracy of 99.1 with a standard deviation of 4.6. The table given in this model obviously confirms these claims.

14.5 CONCLUDING REMARKS

Autism Spectrum Disorder is a cognitive and neurodevelopmental disorder characterized by impairments in social communication, interaction, and imagination. There are many challenges of early autism screening. The relationship between eye-blinking and cognitive state and the effect of eye-blinks on cognition in real-world environments has received limited research attention. In most experimental studies of eye movements, visual scanning and especially cognitive processing the eye-blink data were commonly discarded as artifacts or noise; although the timing of blinks is related to both explicit [38, 19] and implicit [37, 27] attentional pauses in task content. Therefore, in this work we focused on the temporal pattern of eye-blinking during the face processing task in children with ASD when they compared to typically developed children (TD). The senso-Motoric Instrument (SMI) eye-tracking system was used to collect the eye movements at the rate of 120 Hz when the participants were looking at images of human faces. This system reported the average fixation duration. Five ROIs were defined for each face and fixations were assigned to each region. Blinks were recorded as events with a measurable duration, identified by an automated algorithm. We propose a new algorithm to discriminate subjects with ASD from typically developed children (TD) based on their eye-blink patterns using tensor dimensionality reduction and feature extraction algorithms and achieves classification results with error rate lower than 3%. Tensors provide a natural way to multidimensional data objects whose entries are indexed by several continuous or discrete variables and tensor decompositions allow us to extract hidden factors with different dimension in each mode, and investigate interactions among various modalities. The results confirmed that we can use the presented system as an alternative scheme for automatic screening of children with ASD based on their eye-blinking pattern.

REFERENCES

[1] Centers for disease control and prevention. prevalence of autism spectrum disorders-Autismand developmental disabilities monitoring network, 14 sites, United States, 2008. MMWR Surveillance Summaries 61, 119 (2012).

[2] World Health Organization, ICD-10 International Statistical Classification of Diseases and Related Health Problems:10th Revision, second ed., World Health Organization, Geneva, 2007 (Version for 2007).

[3] E. Acar, C. Aykut-Bingol, H. Bingol, R. Bro, B. Yener, Multiway analysis ofepilepsy tensors, Bioinformatics 23 (2007) 10–18.

[4] R. Adolphs, L. Sears, J. Piven, Abnormal processing of social information from faces in autism, J. Cog. Neurosci. 13 (2001) 232–240.

[5] A. Bailey, S. Palferman, L. Heavey, A. L. Couteur, Autism: the phenotype in relatives, J. Autism Dev. Disord. 28 (1998) 369–392.

[6] L. V. Ballestra, G. Pacelli, Computing the survival probability density function in jump-diffusion models: A new approach based on radial basis functions, Eng. Anal. Bound. Elem. 35 (2011) 1075– 1084.

[7] L. V. Ballestra, G. Pacelli, A radial basis function approach to compute the first-passage probability density function in two-dimensional jump-diffusion models for financial and other applications, Eng. Anal. Bound. Elem. 36 (2012) 1546–1554.

[8] L. V. Ballestra, G. Pacelli, Pricing European and American options with two stochastic factors: A highly efficient radial basis function approach, J. Econ. Dyn. Cont. 37 (2013) 1142–1167.

[9] S. Baron-Cohen, S. Wheelwright, T. Jolliffe, Is there a language of the eyes? evidence from normal adults, and adults with autism or asperger syndrome, Vis. Cogn. 4 (1997) 311–331.

[10] A. S. R. Bro, P. Geladi, Multi-Way Analysis: Applications in the Chemical Sciences, Wiley, 2004.

[11] M. D. Buhmann, Radial Basis Functions: Theory and Implementations, Cambridge University Press, New York, 2004.

[12] R. E. Carlson, T. A. Foley, The parameter r^2 in multiquadric interpolation, Comput. Math. Appl. 21 (1991) 29–42.

[13] A. H. D. Cheng, M. A. Golberg, E. J. Kansa, Q. Zammito, Exponential convergence and H-*c* multiquadric collocation method for partial differential equations, Numer. Meth. Part. D. E. 19 (2003) 571–594.

[14] A. Cichocki, R. Zdunek, A. H. Phan, S. Amari, Nonnegative Matrix And Tensor Factorizations, Wiley, 2009.

[15] J. N. Constantino, Autism recurrence in half siblings: strong support for genetic mechanisms of transmission in ASD, Mol. Psychiatry 18 (2013) 137–138.
[16] G. Dawson, L. Carver, A. N. Meltzoff, H. Panagiotides, J. McPartland, S. J. Webb, Neural correlates of face and object recognition in young children with autismspectrum disorder, developmental delay, and typical development, Child Dev. 73 (2002) 700–717.
[17] G. Dawson, A. N. Meltzoff, J. Osterling, J. Rinaldi, E. Brown, Children with autismfail to orient to naturally occurring social stimuli, J. Autism Dev. Disord. 28 (1998) 479–485.
[18] G. Dawson, K. Toth, R. Abbott, J. Osterling, J. Munson, A. E. et al., Early socialattention impairments in autism: social orienting, joint attention, and attentionto distress, Dev. Psychol. 40 (2004) 271–283.
[19] G. C. Drew, Variations in reflex blink-rate during visual-motor tasks, Q. J. Exp. Psychol. 3 (1951) 73–88.
[20] G. Fasshauer, J. Zhang, On choosing "optimal" shape parameters for RBF approximation, Numer. Algorithms 45 (2007) 346–368.
[21] P. A. Filipek, P. J. Accardo, G. T. Baranek, E. H. Cook, G. Dawson, B. Gordon, The screening and diagnosis of autistic spectrum disorders, Journal of Autism and Developmental Disorders 29 (1999) 439–484.
[22] S. E. Folstein, M. L. Rutter, Infantile autism: a genetic study of 21 twin pairs, J. Child Psyco. Psych. 18 (1977) 297–321.
[23] N. A. Fox, R. J. Davidson, Patterns of brain electrical activity during facial signs ofemotion in 10month-old infants, Dev. Psychol. 24 (1988) 230–236.
[24] R. Franke, Scattered data interpolation: test of some methods, Math. Comput. 38 (1982) 181–200.
[25] D. R. Gitelman, ILAB: A program for postexperimental eye movement analysis, Behav. Res. Methods Instrum. Comput. 34 (2002) 605–612.
[26] A. Hall, The origin and purposes of blinking, Bri. J. Opthalmol. 29 (1945) 445–467.
[27] A. Herrmann, The interaction of eye blinks and other prosodic cues in German sign language, Sign. Lang. Linguist. 13 (2010) 3–39.
[28] A. K. Jain, R. P. Duin, J. Mao, Statistical pattern recognition: a review, IEEE Transactions on Pattern Analysis and Machine Intelligence 22 (2000) 4–37.
[29] R. M. Joseph, J. Tanaka, Holistic and part-based face recognition in children with autism, J. Child Psyco. Psych. All. Dis. 44 (2003) 529–542.
[30] C. Karson, Spontaneous eye-blink rates and dopaminergic systems, Brain 106 (1983) 643–653.
[31] P. Kaufman, A. Alm, Adler's Physiology of the Eye: Clinical Applications, Elsevier, 2003.
[32] R. Khosrowabadi, C. Quek, K. K. Ang, A. W. A. Rahmand, A. C. Shen-Hsin, Dynamic screening of autistic children in various mental states using pattern of connectivity between brain regions, Appl. Soft Comput. 32 (2015) 335–346.
[33] A. Klin, W. Jones, R. Schultz, F. Volkmar, D. Cohen, Visual fixation patterns during viewing of naturalistic social situations as predictors of social competence in individuals with autism, Arch. Gen. Psych. 59 (2002) 809–816.
[34] S. R. Leekam, E. Hunnisett, C. Moore, Targets and cues: Gaze following in children with autism, J. Child Psyco. Psych. All. Dis. 39 (1998) 951–962.
[35] P. Mundy, M. Sigman, J. Ungerer, T. Sherman, Defining the social deficits ofautism: the contribution of non-verbal communication measures, J. Child Psycho. Psychiatry 27 (1986) 657–669.

[36] T. Nakano, N. Kato, S. Kitazawa, Lack of eyeblink entrainments in autism spectrum disorders, Neuropsychologia 49 (2011) 2784–2790.
[37] T. Nakano, Y. Yamamoto, K. Kitajo, T. Takahashi, S. Kitazawa, Synchronization of spontaneous eyeblinks while viewing video stories, Proc. Biol. Sci. 276 (2009) 3635–3644.
[38] L. N. Orchard, J. A. Stern, Blinks as an index of cognitive activity during reading, Integr. Physiol. Behav. Sci. 26 (1991) 108–116.
[39] K. A. Pelphrey, N. Sasson, J. Reznick, G. Paul, B. Goldman, J. Piven, Visual scanning of faces in autism, J. Autism Dev. Disord. 32 (2002) 249–261.
[40] A. H. Phan, A. Cichocki, Tensor decompositions for feature extraction and classification of high dimensional datasets, IEICE 1 (2010) 37–68.
[41] E. Ponder, W. P. Kennedy, On the act of blinking, Q. J. Exp. Physiol. 18 (1927) 89–110.
[42] H. Pouretemad, Diagnosis and treatment of joint attention in autistic children (in Persian), Arjmand Book, 2011.
[43] S. Rippa, An algorithm for selecting a good parameter c in radial basis function interpolation, Advan. Comp. Math. 11 (1999) 193–210.
[44] A. Ronald, R. A. Hoekstra, Autism spectrum disorders an autistic traits: a decade of new twin studies, Amer. J. Med. Gen. Part B Neuropsychiatric Genetics 156 (2011) 255–274.
[45] N. Sasson, The development of face processing in autism, J. Autism Dev. Disord. 36 (2006) 381–394.
[46] M. Sigman, P. Mundy, T. Sherman, J. Ungerer, Social interactions of autistic, mentally retarded and normal children and their caregivers, J. Child Psycho. Psychiatry 27 (1986) 647–656.
[47] M. D. Sigman, C. Kasari, J. H. Kwon, N. Yirmiya, Responses to the negative emotions of others by autistic, mentally retarded, and normal children, Child Dev. 63 (1992) 796–807.
[48] D. H. Skuse, Child Psychology and Psychiatry: An Introduction, The Medicine Publishing Company, 2003.
[49] J. Stern, L. Walrath, R. Goldstein, The endogenous eyeblink, Psychophysiology 21 (1984) 22–33.
[50] C. Trevarthen, T. Kokkinaki, G. A. Fiamenghi, What infants' imitation communicate: With mothers, with fathers and with peers, Imitation in infancy (p. 127–185), Cambridge: Cambridge University Press, 1999.
[51] F. VanderWerf, P. Brassinga, D. Reits, M. Aramideh, B. O. de Visser, Eyelid movements: behavioral studies of blinking in humans under diverent stimulus conditions, J. Neurophysiol. 89 (2003) 2784–2796.
[52] P. Vuilleumier, J. L. Armony, J. Driver, R. J. Dolan, Effects of attention and emotionon face processing in the human brain: an event-related fMRI study, Neuron 30 (2001) 829–841.
[53] J. Wang, G. Liu, On the optimal shape parameters of radial basis functions used for $2d$ meshless methods, Comput. Meth. Appl. Mech. Eng. 191 (2002) 2611–2630.
[54] E. Werner, G. Dawson, J. Osterling, N. Dinno, Brief report: recognition of autismspectrum disorder before one year of age: a retrospective study based on homevideotapes, J. Autism Dev. Disord. 30 (2000) 157–162.
[55] B. Wicker, P. Fonlupt, B. Hubert, C. Tardif, B. Gepner, C. Deruelle, Abnormal cerebral effective connectivity during explicit emotional processing inadults with autism spectrum disorder, Soc. Cogn. Affect. Neurosci. 3 (2008) 135–143.
[56] H. Widdel, Theoretical and applied aspects of eye movement research, Gale AG, Johnson F (Elsevier, New York) (1984) 21–29.

[57] L. T. Thanh, N. T. A. Dao, N. V. Dung, N. L. Trung, K. Abed-Meraim, Detection of multichannel EEG spike with new tensor analysis method, J. Neural Eng. 17 (2020) 016023.

[58] F. Duan, H. Jia, Z. Zhang, F. Feng, Y. Tan, Y. Dai, A. Cichocki, Z. Yang, CF Caiafa, Z. Sun, J. Solé-Casals, on EEG robustness methods Completion of Tensor, Science China Technological Sciences 64 (2021) 1828–1842.

[59] S. Ahmadi-Asl, C. F Caiafa, A. Cichocki, A. H. Phan, T. Tanaka, I. Oseledets, J. Wang, Cross Tensor Approximation Methods for Compression and Dimensionality Reduction, IEEE Access 9 (2021) 150809–1508.

[60] Rahman, A. (2020). Statistics for Data Science and Policy Analysis. Springer.

15 Stress-Level Detection Using Smartphone Sensors

*Rekha Pal, Rohith Reddy Vangal, Yashita Watchpillai, Pooja Jain and Tapan Kumar**
Department of Computer Science and Engineering,
Indian Institute of Information Technology, Nagpur, India
*IEEE Senior member, Electronics and Communication Engineering
Indian Institute of Information Technology, Nagpur, India

CONTENTS

15.1 INTRODUCTION

Stress is an important concern for many people. Whether it is based on individual or outside factors, it impacts intellectual and physical aspects of one's self. Usually, when physical or mental changes occur in an individual's life, stress is pretty ordinary [1]. Different emotions like anger, sorrow, frustration, and guilt can cause stress. Not only does stress disrupt a person's daily life, but it also causes health issues. It is also responsible for anger management issues. Stress causes cardiovascular diseases, musculoskeletal diseases, immunological problems, insomnia, obesity, digestive symptoms, skin & hair problems, headaches, sadness, anger, irritability, tension, and depression [2–4]. According to The Indian Express dated 12 Dec 2020, around 74% of Indians are tormented by stress [5].

Stress Level Detection can help people know about their high-stress levels and take appropriate measures to reduce them [6]. The detection of stress has been a

DOI: 10.1201/9781003253051-20

major concern due to its inconsistency. Depending upon age, gender, and occupation, stress may vary [7]. Many types of research have been carried out to detect stress using wearable devices that are capable of obtaining information about the individual such as heart rate, blood pressure, skin conductance, skin temperature, heart rate variability [8], etc., to examine patterns. With the information extracted from wearable sensors, the stress levels of a subject can be identified. Using Voice Analysis, features such as pitch, frequency, harmonics to noise ratio, etc., can be extracted, which are helpful in stress and mood analysis. Surveys like Pittsburgh Sleep Quality Index (PSQI) score, Perceived Stress Scale (PSS), SF-12, and Big Five personality test consist of questions that are to be rated in a specific range. For example, in PSQI, questions are rated on a scale of 0-3 where 0 indicates no difficulty and 3 indicates severe difficulty. Using the sum of the scores obtained in each survey, user's behavior, personality, and sleep patterns can be analyzed.

Smartphone sensors can be used in detecting the stress level of a subject by their activities [9]. Our research data is collected through a mobile application called "Sensors Recorder." It collects the data from various sensors in a smartphone in the background such as 1) Accelerometer – measures linear acceleration along all axes (x,y,z); 2) Gyroscope – measures the rate of rotation around all axes (x,y,z),;3) Ambient light – measures surrounding light. A smartphone also collects data related to call usage, app usage, and the idle state of the user. Learning these patterns will help in reducing stress levels and improving mental health by giving personalized feedback to users with high-stress levels. This informs them about their daily activities that are responsible for high stress and what helps in reducing it [10].

The data and base model were provided to us by Tata Research Development and Design Centre (TRDDC). They experimented with various models such as K-Nearest Neighbors (KNN), Support Vector Machine (SVM), and Long Short Term Memory (LSTM). LSTM gave better results among all the models so it was chosen as our base model. The aim of this project is to build a better machine learning model to identify stress levels with the help of smartphone sensor data and analyze patterns among high-stress and low-stress level individuals.

In Section 15.2 we explain some of the papers and methods from our literature review. Section 15.3 discusses the proposed methodology, which includes 1. Data collection, 2. Data extraction, 3. Exploratory Data Analysis (EDA), and 4. Proposed Model. In Section IV the results we achieved are discussed. Sections 15.4 and 15.5 include the conclusion and future scope, respectively.

15.2 RELATED WORK

In recent times, machine learning and intelligent systems-based techniques have evolved significantly in research with application into different domains [11–14, 15–20]. Chang et al. (2012) designed a feature phone app to identify the user's emotions and find the stress level using voice analysis. The Belfast Dataset, which includes 298 audio clips accrued from television programs, was used for emotion recognition. The SUSAS Dataset is used for stress level detection [21,22]. Feature extraction included rms zero-cross rating, pitch, mel frequency cepstrum coefficients, etc. Linear Support Vector Machine recognizes positive and negative emotion and high & low stress based on feature vectors. Accuracy of 75.5% has been observed in emotion analysis and 93.6% has been observed in identifying stress levels [22].

Ceja et al. (2015) aimed to detect the stress level of individuals while working. Many studies are based on image processing or sound analysis, but this work used an accelerometer sensor. From two different organizations, 30 people participated in this study for 8 weeks excluding weekends. Stress level was reported through a survey users completed three times during working hours. Thirty-four features were taken including mean, standard deviation, variance, and rms. Three models were built. The User Specific model trained with users' data. The General model used leave one person out approach. A similar User model used data from a set of users with similar behaviour. Naive Bayes and Decision Trees models were used on all of them. It was observed that the User Specific model performed the best and the General model gave the least accuracy. Naive Bayes had slightly higher accuracy than the Decision Trees model. It was also observed that maximum stress was reported on Tuesday, then decreased and reached a minimum on Friday [23].

Ferdous et al. (2015) looked into app usage patterns, specific types of apps, and their duration to see if there was a link between these patterns and users' perceived stress levels at work. They analyzed 28 users over the course of six weeks for the amount of time they spent at work using timestamps to collect app usage patterns and felt stress levels via a five-point scale questionnaire. They divided the apps into five groups: (i) entertainment, (ii) social networking, (iii) utility, (iv) browsing, and (v) gaming. They started by creating a Generic Model, which was created by training an SVM classifier with a subset of data for training and the rest for testing. K-fold (k=10) was used for validation. However, the accuracy gained was only 54%. As a result, they developed a User-Centric Model, in which each model was trained using individual app usage patterns and stress levels. For 22 active users, k-fold (k=10) was used to select user-specific models, with an accuracy ranging from 69.2% to 99% [24].

Wang et al. (2014) showed that there are several connections between smartphone sensor data and success in education and mental wellbeing. They developed an app called "StudentLife" that combines Mobile EMA for data collecting. They gathered data from 48 students over the course of ten weeks during the spring semester. The information gathered was sorted into four categories: (i) Automatic Sensing data, which includes activity, conversation, sleep, and location information.;(ii) EMA data, which includes both EMA and PAM responses, and data from surveys (i.e., PHQ-9, PSS, UCLA) to evaluate stress levels; (iii) Survey Instrument data; and (iv) Academic data. To find correlations among sensor data, EMA data, and survey data researchers used the Physical Activity Classifier, Audio and Conversation Classifier, and Sleep Classifier. They observed that the students who slept less were more likely to be depressed. Similarly, there was a negative correlation with conversation frequency, duration, and other factors [25].

Suhara et al. (2017) focused on predicting depressed mood based on the information given by users about their sleep schedule, mood, actions, and behaviour changes. The objective was to use the user's previous 14 days' data to determine if the user experienced any severe depression day during the period of n successive days. SVM was used as the base model, and further LSTM was also used to forecast severe depression. Day-of-the-week variable and the corresponding embedded layer were added to the model. They used two feature sets: i) all and ii) severe only. With AUC-ROC as the evaluation metric and k=14, for different n values (n=1,3,7), (n,k)-day severe depression was achieved. In the case of "all" LSTM-RNN performed better than SVM for every n value [9]. The authors in Ref. [26] worked on the prediction of fault in Hadoop systems using SVM. In Refs. [27, 28] LSTM was used for sentiment analysis and spam detection [29].

Sano et al. (2015) aimed to identify factors that hinder sleep, stress, and educational outcomes. The dataset was created by data from 66 undergraduates who participated in the study for 30 days. Before the start of experiments BFI Personality Test, PSQI, Morningness Eveningness Questionnaire, PSS, and SF-12 Physical and MCS were filled out. A wrist sensor was used to collect the physical data of the user. Also, an application was installed on their phones to get call data, message data, internet usage data, GPS data, and mobile activity time. Participants filled out surveys two times every day. They filled out SF-12, PSS, and STAI surveys, and their email usage was collected after the completion of 30 days. They reported their GPA after the end of the semester. The highest 20% and lowest 20% of participants were grouped for GPA, stress, sleep, and mental health from surveys. These features were compared using two classifiers: i) SVM (linear kernel) and ii) SVM (radial basis function kernel) for determining accuracy. The accuracy of the models ranged from 67 to 92% [10]. The authors in Ref. [30] reported on real-time video summarization using CBVR. The authors in Ref. [7] proposed reinforcement learning for a sustainable environment.

Umematsu et al. (2019) worked to identify how well an automated model can forecast the next day's well-being. The data in this project examined Sleep, Networks, Affect, Performance, Stress, and Health using Objective Techniques, which gathered 30-day multimodal data from college students in one U.S. university, including physiological, mobile phone, and behavioural survey data. Using the preceding seven days of physiological, mobile phone, and behavioural survey data, static models (support vector machine and logistic regression) and time-series models (Long Short Term Memory Neural Network Models (LSTM)) were developed. The results suggest that employing an LSTM instead of static models improves accuracy across all conditions examined. The accuracy achieved was 80.4% [31].

15.3 PROPOSED METHODOLOGY

In this study we performed data collection, preprocessing, exploratory data analysis, developing a new model, and then testing of the proposed model.

15.3.1 Data Collection

The data was collected using the "Sensor Recorder" app as shown in Figure 15.1, which was provided by TRDDC. A total of 51 users, which included interns and employees, participated in data collection for 6 months. There were 39 users from the age range of 18-30 years, 11 from 31-45 range, and 1 user >45 range. Their profession can not be disclosed due to privacy issues. The app collected stress and sensor

TABLE 15.1
Literature Review

Reference Number	Year	Technology	Result
[1]	2015	Two different classifiers: i) SVM linear kernel and ii) SVM radial basis function kernel was used. The best of the two was selected.	Accuracy = 67–92%
[2]	2017	SVM and LSTM-RNN are known to be powerful models for learning from sequential data. (k=14, n= 1,3,7) Two models: severe model, full model	n=7 SVM (all) 0.837 LSTM RNN (severe only) – 0.846 LSTM-RNN (all) – 0.886
[3]	2014	Physical Activity Classifier, Audio, and Conversation Classifier, and Sleep Classifiers.	Negative correlations with sleep, conversation frequency, duration, and other factors.
[4]	2015	SVM Classifier, k-fold validation	Generic Model: 54% User-Centric Model: 69.2 – 99%
[5]	2015	User-specific, General Model, Similar user Models are created using Naive Bayes and Decision Trees	Accuracy of 0.71, 0.52, 0.6 were recorded for User-specific, General, and Similar user models, respectively
[6]	2012	Linear SVM	Emotion Analysis accuracy: 75.% Stress Level accuracy: 93.6
[21]	2019	SVM, Logistic Regression and LSTM (Best was selected)	Accuracy: 80.4%

data every day. When the user installed the app, a unique ID was generated for the user. Later, permissions were given to record the smartphone sensor data. The app sent a notification at 6 pm to enter stress levels and the window was active until 9 pm. The user had to choose a stress level between 0 and 4, where 0 indicates no stress level, 1 indicates mild stress level, 2 indicates moderate stress level, 3 indicates high-stress level, and 4 indicates extreme stress level. On Sunday, we also used the STAI-6 survey to obtain information about users' moods. The State-Trait Anxiety Inventory (STAI-6) Survey has six questions with a score range of one to four. The result is an STAI-6 score that ranges from 6 to 24. These scores can then be used to determine anxiety levels. Calm Value, Tense Value, Upset Value, Relaxed Value, Content Value, and Worried Value are the six questions in our survey, each on a 4-point scale: Not at all, Somewhat, Moderately, and Very much. Secondly, sensor data, which includes

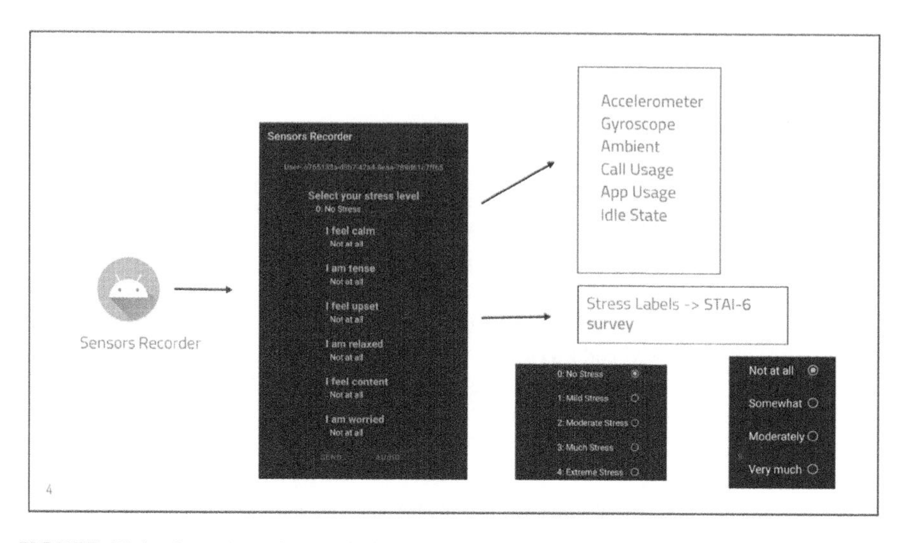

FIGURE 15.1 Snapshot of Data Collection App.

Timestamp	Stress Value (0-4)	UserName	Calm Value	Tense Value	Upset Value
03/08/2021 18:05:35	1	User- 43e6d935-3119-44f7-95d8-2bce44dfbd62	Somewhat	Somewhat	Somewhat
03/08/2021 18:08:26	2	User- b4ca14a7-ad0e-4608-aa24-56e45b02122a	Not at all	Moderately	Somewhat
03/08/2021 18:10:50	2	User- 5ca4b3a8-7c8e-49ce-9f3b-4ec1da135db8	Moderately	Somewhat	Somewhat
03/08/2021 18:30:45	2	User- d3799dc7-a41b-4d72-8af1-1f7e10f89ba6	Somewhat	Not at all	Moderately
04/08/2021 18:01:47	2	User- d601e640-ae69-42e4-9b11-a4a1d176bac9	Not at all	Somewhat	Somewhat
04/08/2021 18:10:29	0	User- ea0844eb-a4d4-4b53-be56-2ea3d2c9ace7	Very much	Somewhat	Not at all

FIGURE 15.2 Stress-Level Sheet.

Accelerometer, Gyroscope, Ambient Light, App data, and Call usage data, were also collected from the app, which runs in the background 24*7.

Accelerometer sensor measures acceleration in three axes. By tracking rotation or twists, the gyroscope provides another dimension to the information provided by the accelerometer. The Ambient Light Sensor adjusts the brightness of the computer's LCD panel automatically based on ambient light levels. The Idle state is when an Android device is left without being used, first it will dim the screen, then turn off the screen, and finally, turn off the CPU. The novelty of this data collection is that, unlike other research methods, the user does not have to fill in a large amount of information in surveys or wear wearable sensors throughout the day.

15.3.2 Data Extraction

Initially, the stress sheet shown in Figure 15.2 and sensor datasheets shown in Figure 15.3 is loaded, which contain the raw data. The stress datasheet has the Timestamp, Username, Stress Value, Calm Value, Tense Value, Upset Value, Relaxed Value, Content Value, and Worried Value information. The Username is in the form

Hardware Sensor Data	Call Usage	App Usage Data	Idle State Data
			Off- 16:43:50
15:47:24; ; ; 4.7925; ;	06-Oct-2019 17:51 Outgoing 0	Start	On- 16:51:21
15:47:24; ; ; ; ; 5.0	06-Oct-2019 17:51 Outgoing 0	------------------------------	Off- 16:54:07
15:47:24; -0.55575407, 6.1472626, 7.5239286; ; ; ;	07-Oct-2019 09:08 Missed 0	Start	On- 16:56:35
15:47:28; -0.5000889, 6.037728, 7.704092; ; ; ;	07-Oct-2019 12:31 Missed 0	youtube: 17 Sep 2021 09:42:27:887 +0530 2021723	Off- 17:03:58
15:47:28; ; -0.13498816, -0.110226184, -0.047441646; ; ;	07-Oct-2019 14:20 Missed 0	truecaller: 16 Sep 2021 21:16:13:873 +0530 701011	On- 17:14:27
15:47:28; -0.32052374, 5.866243, 7.9177747; ; ; ;	09-Oct-2019 11:53 Incoming 6	instashot: 27 Aug 2021 14:25:22:022 +0530 367171	Off- 17:14:28
15:47:28; -0.25049335, 5.7602997, 7.701548; ; ; ;	10-Oct-2019 09:12 Missed 0	googlequicksearchbox: 16 Sep 2021 09:12:38:704 +	On- 17:23:21
15:47:28; -0.44083238, 5.9081416, 7.893533; ; ; ;	14-Oct-2019 09:11 Incoming 4	messenger: 13 Sep 2021 12:53:14:032 +0530 1873	Off- 17:23:27
15:47:28; 0.054917, 5.746084, 7.8531313; ; ; ;	15-Oct-2019 13:03 Incoming 5	snapseed: 09 Aug 2021 20:07:09:764 +0530 87924	On- 17:23:32
15:47:28; -0.054468088, 5.7595515, 7.881712; ; ; ;	16-Oct-2019 11:51 Incoming 1	myairtelapp: 14 Sep 2021 12:40:47:035 +0530 1023	Off- 17:23:33
15:47:28; ; 0.07423012, -0.015631063, 0.07144135; ; ;	17-Oct-2019 09:12 Missed 0	whatsapp: 17 Sep 2021 15:46:40:586 +0530 145081	On- 17:34:56
15:47:28; 0.047135845, 5.850681, 7.679402; ; ; ;	18-Oct-2019 09:09 Missed 0	messaging: 16 Sep 2021 16:11:19:549 +0530 34767	Off- 17:34:59
15:47:28; 0.058358666, 5.9287915, 8.033444; ; ; ;	19-Oct-2019 09:11 Missed 0	vending: 17 Sep 2021 15:46:49:249 +0530 2965673	On- 17:35:36
15:47:28; -0.0068833297, 5.86759, 7.7428484; ; ; ;	20-Oct-2019 12:07 Missed 0	tcsappstore: 14 Sep 2021 12:02:46:329 +0530 3518	Off- 17:35:36
15:47:28; -0.013617022, 5.915773, 7.844153; ; ; ;	21-Oct-2019 09:10 Missed 0	duomobile: 26 Aug 2021 11:59:52:990 +0530 8557 c	On- 17:39:38
15:47:29; -0.11971009, 5.8527756, 7.697209; ; ; ;	21-Oct-2019 13:16 Missed 0	android: 17 Sep 2021 08:59:54:841 +0530 119775 a	Off- 17:39:38
15:47:29; -0.18629882, 5.8461914, 7.884256; ; ; ;	21-Oct-2019 15:13 Missed 0	launcher3: 17 Sep 2021 15:46:35:001 +0530 434097	On- 17:47:35
15:47:29; -0.17372926, 5.961562, 7.616853; ; ; ;	22-Oct-2019 12:07 Missed 0	android: 17 Sep 2021 13:13:02:816 +0530 8393077	Off- 17:47:35
15:47:29; -0.28236616, 5.8916817, 7.9566803; ; ; ;	23-Oct-2019 12:12 Missed 0	camera2: 16 Sep 2021 20:38:33:113 +0530 721563	On- 17:52:44
15:47:29; -0.25004444, 5.9681463, 7.6226892; ; ; ;	24-Oct-2019 12:05 Missed 0	deskclock: 13 Sep 2021 18:10:10:345 +0530 43564	Off- 17:52:45
15:47:29; -0.25617957, 5.873725, 7.807492; ; ; ;	25-Oct-2019 09:10 Missed 0	hotstar: 25 Jul 2021 01:31:31:959 +0530 27373 in st	On- 17:52:45

FIGURE 15.3 Raw Data.

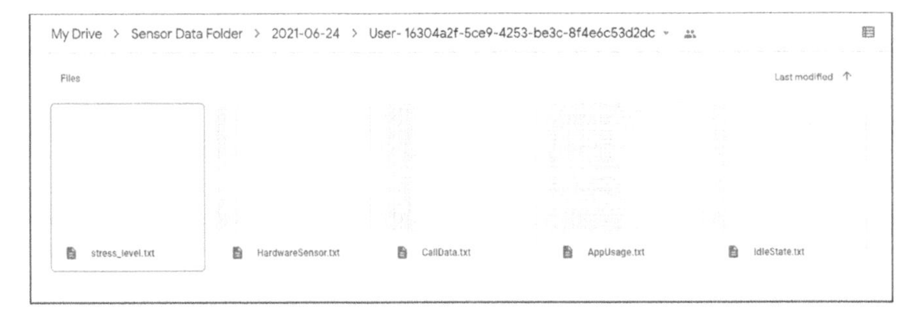

FIGURE 15.4 Data Extraction.

of "User-" appended to a unique id, which is generated by the Sensors Recorder app. The sensor datasheet consists of hardware sensor data, Call Usage Data, App Usage Data, and Idle State Data. The hardware sensor data consists of timestamp, accelerometer, gyroscope, ambient light, ambient temperature, and proximity.

The Call Usage Data consists of a timestamp, category (Incoming, Outgoing, and Missed), and duration. The App Usage Data consists of the app name, current timestamp, the timestamp when it was previously used, duration, and package name. The Idle state data consists of the state (On or Off) and timestamp. Next, we create a sensor data folder and create the date directories. In each of these date directories, we create folders for each user. These user folders have five text files as shown in Figure 15.4 into which the sensor and stress data are loaded. Now our aim is to create minute-level data for each user and store it in a data frame. The same stress level, which is submitted once a day, is used for all the minute timestamps of the day. If the stress level is not submitted, it is considered to be -1.

To calculate idle count and idle duration per minute, the following steps are followed. If the idle state is Off we increment the idle count for that minute. The latest Off timestamp and idle state is saved. If the idle state is On and the previous state is Off then we calculate the idle duration and increment the idle count until we reach the current On state. The maximum value of idle duration for a minute can be 60.

For app data, we store the runtime and category of the app used. There is a total of 56 app categories (business, communication, education, entertainment, etc.). We can find the category of the app used by the genre data from the Play store. The Hardware sensor data requires some refining. The Accelerometer and Gyroscope data values are quite random but we can normalize these by assigning a value of 1 if data is present; else 0. Finally, our data frame is saved in an Excel file. For each of the dates, the users were divided into low-stress and high-stress folders where 0,1 were considered low-stress levels while 2,3,4 were considered high-stress levels.

15.3.3 Exploratory Data Analysis

The correlation between sleeping hours on stress is important. Therefore, we used the Idle state sensor to calculate the sleeping hours. Time differences between consecutive on and off timestamps were calculated and the highest value was considered as the sleeping time. There are times during our sleep when we receive notifications and our Idle state becomes On; this lasts up to a few seconds. Hence, the time-difference < 1 minute was removed and the adjacent time differences merged to ensure that the fake rise time situation was avoided. Later, we divided the sleeping time into two categories: <6 and >=6 hours. For each of these categories, we found the percentages of stressed and non-stressed subjects.

Later the app categories, app durations, and stress were also compared as can be seen in Figures 15.9 and 15.10. The weekday factor was also taken into consideration where 0-6 values were assigned to Monday to Sunday, respectively. The App data was taken for each of the apps and its app category found. The 56 app categories were brought down to 10 app categories as follows: Education, Entertainment, FaceApps, Finance, Sports, Communication, Necessities, Games, Family, and Miscellaneous. Later we found the percentage of significant app-based categories used by stressed and non-stressed subjects on an hourly basis. U.S researchers suggest that the amount of time we should spend daily using the phone should not be more than 2 hours outside work [6]. Therefore, to analyze this we calculated the active time spent on the phone using the Idle state sensor, which was initially used to calculate sleeping hours.

15.3.4 Proposed Model

A Recurrent Neural Network (RNN) is a type of neural network that excels at learning from sequential input in deep learning. LSTM is a type of RNN that has found practical uses since it is robust to the challenges of long-term dependency [32]. LSTMs differ from more standard feedforward neural networks because they have feedback connections. This trait allows LSTMs to process complete data sequences without having to handle each point in the sequence separately, instead preserving important knowledge about prior data in the sequence to aid in the processing of incoming data points [33].

Figure 15.5 shows the workflow of our proposed model. Initially, meaningful information is extracted from raw data. EDA is performed to understand the dependency of sensor data features and stress levels. Later, different permutations of features were found. The LSTM model is iteratively trained on each of these feature sets. If the addition of a feature into the feature set increases the accuracy of the model then

TABLE 15.2
Feature Set

Features (17)
Accelerometer_x_mean, Accelerometer_y_mean, Accelerometer_z_mean
Gyroscope_x_mean, Gyroscope_y_mean, Gyroscope_z_mean
Accelerometer_x_std, Accelerometer_y_std, Accelerometer_z_std
Gyroscope_x_std, Gyroscope_y_std, Gyroscope_z_std
Ambient_Light_mean, Ambient_Light_std
Weekday Factor
App usage Duration
Call usage Duration

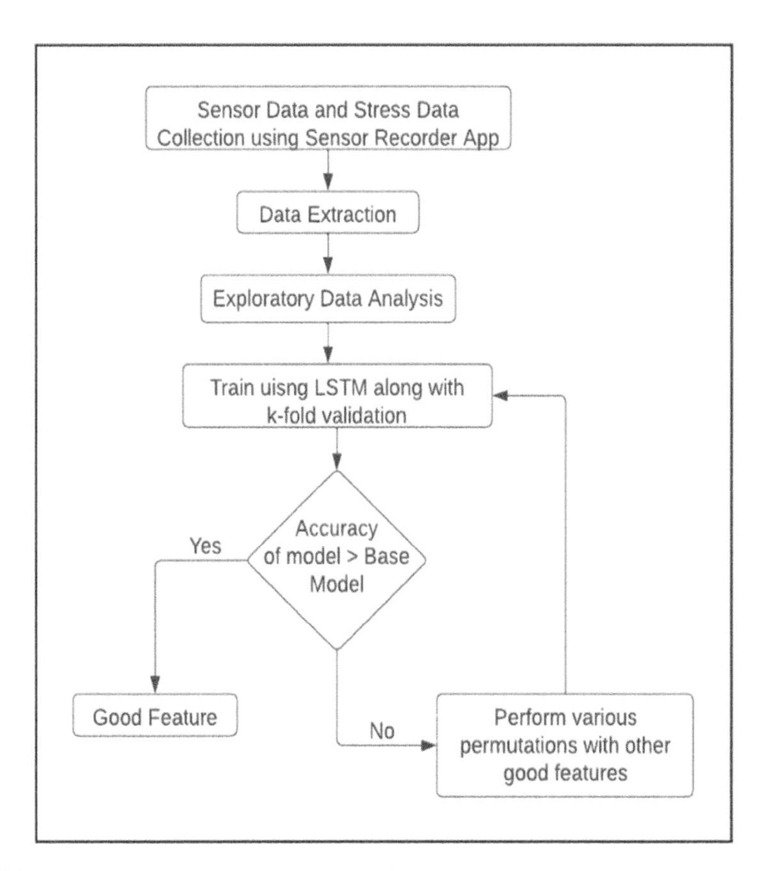

FIGURE 15.5 Workflow of the proposed model.

it is considered to be a Good feature; otherwise, instead of directly discarding the feature we find new permutations of that feature along with other Good features and retrain the model.

The base LSTM model was trained on the minute-level data of the Accelerometer, Gyroscope, and Ambient Light sensors. The mean and standard deviation of the Accelerometer (3 axes), Gyroscope (3 axes), and Ambient Light were used as features in the base model (14 features). In our four proposed models, we additionally used different permutations of App data, Call data, and Weekday factors. In model 1, we included the Weekday factor in the base model. For building model 2 we calculated the sum of app durations for every minute and added this feature to model 1. Similarly, for model 3 we calculated the sum of call durations for every minute and added this feature to model 1. For model 4 all the features were combined; the feature set can be seen in Table 15.2. (The models are compared in Section 15.4.) The LSTM model was used to train the data as it is effective for time series data. Here we collected the data for each day minute-wise. LSTM has memory cells that are useful in remembering a user's history.

A Keras sequential model was built by adding sixteen units of LSTM layers. Eight fully connected layers with an activation function of ReLU were added to the model, and in the last layer, the number of dense layers was equal to the shape of the output vector, which uses the Softmax activation function. It was compiled using categorical cross-entropy as a loss parameter, Adam optimizer, and accuracy as a metric. The model was fit on the data and ran for 200 epochs using a batch size of 4 and a validation split of 0.1.

15.4 RESULTS AND DISCUSSION

We made an hourly pie chart, as seen in Figure 15.11, which consists of the percentage use of each app category. As per minute data was available, we found the per hour usage of each of these app categories by adding up the duration (in ms) and found the percentage used. For 20 hours out of 24 hours, the maximum used app category was communication in the low-stress category while 14 hours out of 24 hours were spent in the case of stressed subjects. This indicates that communication is an important factor affecting stress. The time period of 3:00 am – 5:00 am was mostly spent on education and communication in the case of non-stressed subjects while stressed subjects spent it on entertainment. The percentages of low-stressed and high-stressed subjects with respect to their sleeping hours were measured. If the sleeping hours were less than 6 hours, 42% of participants had low stress levels while 57% had high-stress levels. If the sleeping hours were greater than 6 hours, 87% of participants had low stress levels while only 12% had high stress levels. These results match with the previous studies that say low sleeping hours cause more stress.

Coming to the active hours on the phone, about 87% and 12%, respectively, are the percentages of low stress and high stress level subjects for the active hours of less than equal to 3 hours; while 63% and 36% are the percentages of low stress and high stress levels in subjects whose active hours is greater than 3. As the active hours increase the percentage of highly stressed subjects also increased. Figure 15.9 and Figure 15.10 show the correlation between app data and stress. The app category entertainment was used a lot by users and the app category of finance correlates with the maximum stress.

TABLE 15.3
Result Comparison

Model	Validation Accuracy
Base Model	79%
Base Model Weekday Factor	81%
Base Model Weekday Factor App Duration	76%
Base Model Weekday Factor Call Duration	81%
Base Model Weekday Factor App Duration Call Duration	82%

In Table 15.3 we can see the accuracy comparison of the models designed by us to the base model. The model that incorporates the base model, weekday factor, app usage, and call usage data gives us the highest accuracy among all the models. Though app data could not contribute with an individual feature, the overall accuracy of the model increased when all the features were included. In Figures 15.6 and 15.7 we can see the graphs for accuracy and loss. Figure 15.8 is a plot for stress vs. weekday. We can see that it correlates with the studies. The stress increases from Monday to Tuesday and drops down on Wednesday. The stress levels again increase from Wednesday to Thursday and start decreasing until Sunday. Saturday and Sunday have the lowest stress levels.

Similar to our model, in 2019 Terumi et al. achieved higher accuracies in LSTM models over static methods. Physiology data from wearable sensors was collected. They achieved an accuracy of 80.4% in stress forecasting. In 2019, Yasin et al. used the StudentLife dataset (which consists of mobile sensor data) and trained a CNN-LSTM model to obtain an accuracy of 62.83%. Our model is better than the recent research because a higher accuracy of 82.81% was achieved, and the model is trained on a vast dataset of 6 months and our method of data collection uses only smartphone sensor data, which is user friendly.

15.5 CHALLENGES AND FUTURE DIRECTIONS

One of the difficulties we encountered was a lack of continuous data. Because the ID issued by the data-gathering app is randomized owing to privacy concerns, re-installing the application leads to the development of a new ID, causing data continuity to be lost. Because of its enormous size, maintaining the raw data proved

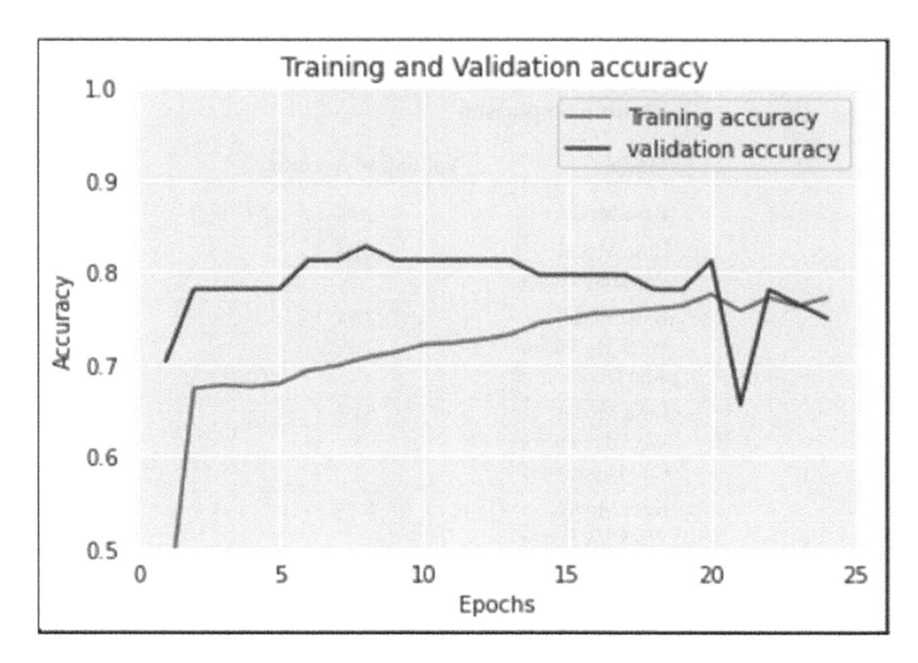

FIGURE 15.6 Training and Validation Accuracy.

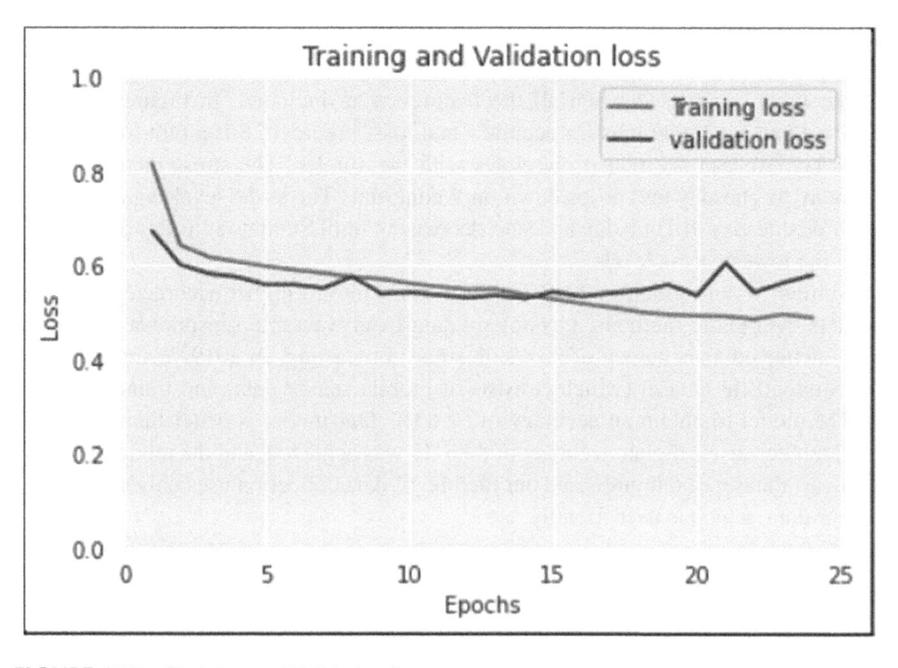

FIGURE 15.7 Training and Validation Loss.

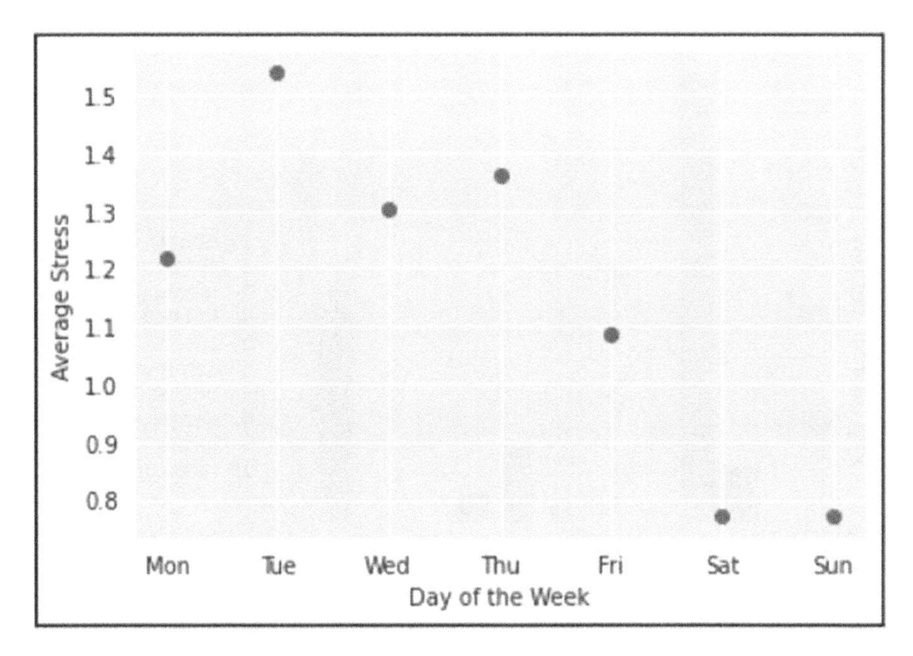

FIGURE 15.8 Stress vs. Weekday.

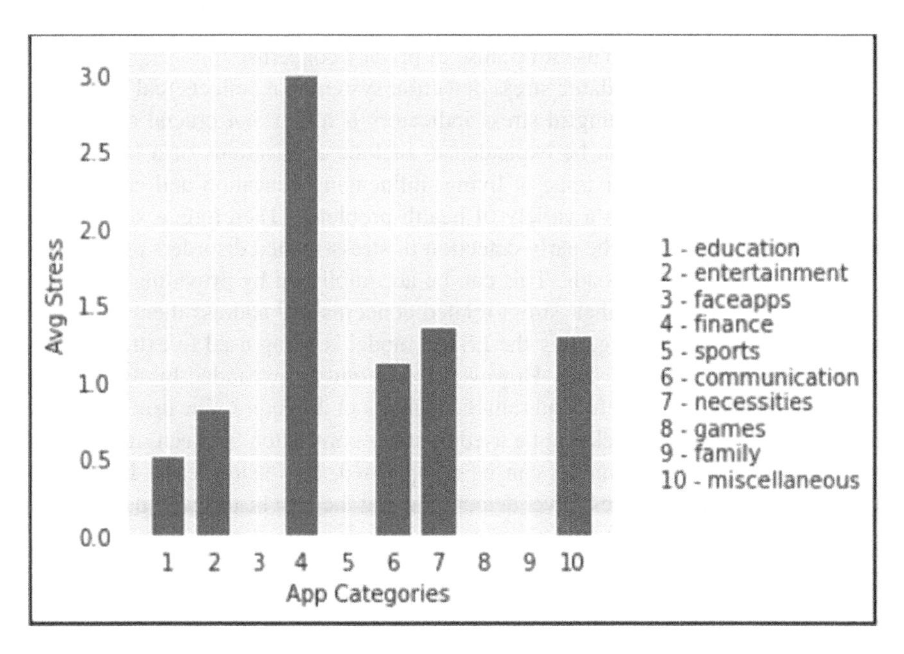

FIGURE 15.9 App Categories vs. Stress.

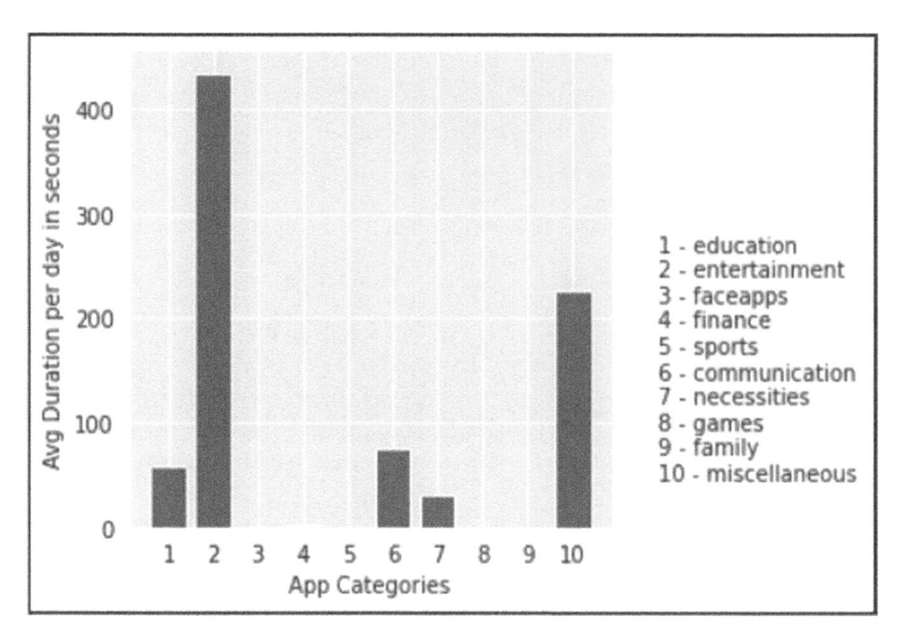

FIGURE 15.10 App Categories vs. App Duration.

problematic. Despite the fact that the software did not collect personal information, some users were hesitant to use it because of privacy concerns.

The creation of a dependable stress detection system that collects real-world data and translates it into meaningful stress indicators is a first, but crucial step toward long-term wellness and can be expanded to include applications of a large range. Stress has become a major issue at living, influencing education and employment productivity and leading to a variety of health problems. Therefore, a stress detection system can help with the early detection of stress-related disorders as well as in supporting balanced workloads. This can be accomplished by providing users with feedback to keep track of their stress-related concerns and address them as soon as possible. For the time being, only the LSTM model is being used to extract various insights from the data collected. However, this might be expanded by using multi-models to acquire more useful and reliable insights in the future. The user mood data that is collected once a week can be used to study correlation between mood, sleep, and stress. Our proposed model can be compressed, thus using lesser RAM. This would be helpful in live stress level detection using the past histories of users.

15.6 CONCLUSION

In this research, we use an app called Sensor Recorder, which captures data from smartphone sensors such as the accelerometer, gyroscope, ambient light, call, and app usage, and uses that data to determine the user's stress level. This app is made for smartphones, and no personal information is collected to preserve users' privacy. We gathered the data for 6 months. Later, we used Exploratory Data Analysis, which

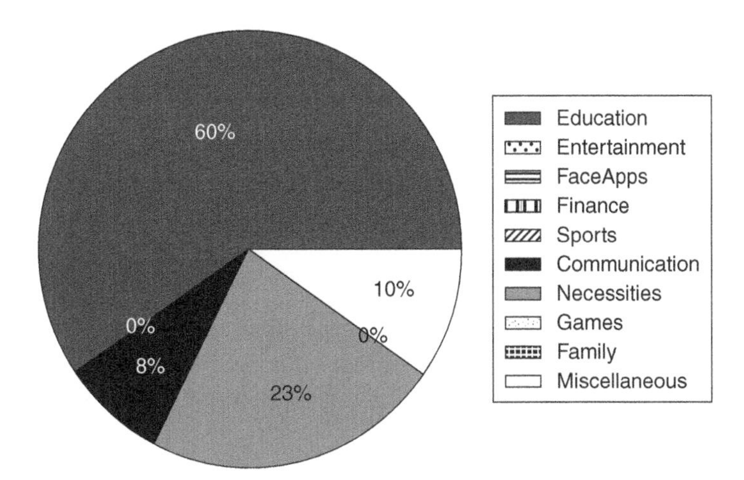

FIGURE 15.11 An example of percentage division of app category usage at 3 AM for non-stressed subjects.

has not been done in previous research, to discover links between stress and sleep, app usage, call data, active hours, and other factors. We chose an LSTM model to extract features from the data because it works best with time series data. According to the literature review, data of 14 successive days is required to predict future stress levels with good accuracy, but due to the lack of continuous data, we trained our model on the entire 6 months of data. Our research method is better than other studies since the data collection process is simple and the user does not feel obligated to provide information. Other valuable characteristics are derived directly from sensor data. Additionally, once a week, we collected the user's mood values, but we have not utilized that data. Our model outperforms the baseline model by 6%. Detecting stress levels has always been a challenging issue. By combining user and sensor data, we can get a little closer to accurately detecting stress levels.

ACKNOWLEDGMENTS

We appreciate Tata Research and Development Design Center's (TRDDC) assistance in running the project and gathering data. We also like to thank everyone who took part in the survey.

REFERENCES

[1] "Cleveland Clinic. Stress. https://my.clevelandclinic.org/health/articles/11874-stress (Jan. 2021)"

[2] "National Institute of Mental Health. Transforming the understanding and treatment of mental illnesses. www.nimh.nih.gov/health/publications/stress "

[3] "Eleesha Lockett. Emotional Signs of Too Much Stress. www.healthline.com/health/emotional-symptoms-of-stress (Mar. 2022)"

[4] "Hedy Marks. Stress Symptoms. www.webmd.com/balance/stress-management/stress-symptoms-effects_of-stress-on-the-body (Aug. 2021)"
[5] "The Indian Express. 74% Indians suffering from stress, 88% from anxiety: Study. https://indianexpress.com/article/lifestyle/health/indians-stress-anxiety-mental-health-study-7101237/ (Dec. 2020)"
[6] "Reid Health. How Much Screen Time is Too Much For Adults?. www.reidhealth.org/blog/screen-time-for-adults"
[7] Selukar, Mayur, Pooja Jain, and Tapan Kumar. "Inventory control of multiple perishable goods using deep reinforcement learning for sustainable environment." Sustainable Energy Technologies and Assessments 52 (2022): 102038.
[8] R. K. Dishman, Y. Nakamura, M. E. Garcia, R. W. Thompson, A. L. Dunn, and S. N. Blair, "Heart rate variability, trait anxiety, and perceived stress among physically fit men and women," International Journal of Psychophysiology, vol. 37, no. 2, pp. 121–133, Aug. 2000.
[9] Yoshihiko Suhara, Yinzhan Xu, and Alex Pentland, "Forecasting Depressed Mood Based on Self-Reported Histories via Recurrent Neural Networks", 2017.
[10] Akane Sano, Andrew J. Phillips, Amy Z. Yu, Andrew W. McHill, Sara Taylor, Natasha Jaques, Charles A. Czeisler, Elizabeth B. Klerman, and Rosalind W. Picard, "Recognizing Academic Performance, Sleep Quality, Stress Level, and Mental Health using Personality Traits, Wearable Sensors and Mobile Phones", 2015.
[11] Agarwal, B., Agarwal, A., Harjule, P., & Rahman, A. (2022). Understanding the intent behind sharing misinformation on social media. Journal of Experimental and Theoretical Artificial Intelligence, 1–15. https://doi.org/10.1080/0952813X.2021.1960637
[12] Agarwal, B., Rahman, A., Patnaik, S., & Poonia, R. C. (Eds.) (2022). Proceedings of International Conference on Intelligent Cyber-Physical Systems. (Algorithms for Intelligent Systems). Springer. www.springer.com/gp/book/9789811671357.
[13] Chowdhury, M.M.H., Rahman, A., & Islam, M. R. (2018). Protecting data from malware threats using machine learning technique. In Proceedings of the 2017 12th IEEE Conference on Industrial Electronics and Applications (ICIEA) (pp. 1691–1694). IEEE, Institute of Electrical and Electronics Engineers. https://doi.org/10.1109/ICIEA.2017.8283111
[14] Rahman, A. (2020). Statistics for Data Science and Policy Analysis. Springer.
[15] Rahman, A., Nimmy, S. F., & Sarowar, G. (2019). Developing an automated machine learning approach to test discontinuity in DNA for detecting tuberculosis. In J. P. Davim (Ed.), Proceedings of the Twelfth International Conference on Management Science and Engineering Management (pp. 277–286). Springer. https://doi.org/10.1007/978-3-319-93351-1_23
[16] Uddin, M. G., Nash, S., Rahman, A., & Olbert, A. I. (2023). A novel approach for estimating and predicting uncertainty in water quality index model using machine learning approaches. Water Research, 1–24. https://doi.org/10.1016/j.watres.2022.119422
[17] Rahman, A., & Harding, A. (2017). Small area estimation and microsimulation modeling. CRC Press. https://doi.org/10.1201/9781315372143
[18] Chowdhury, M., Rahman, A., & Islam, R. (2018). Malware analysis and detection using data mining and machine learning classification. In J. Abawajy, K-K. R. Choo, & R. Islam (Eds.), International Conference on Applications and Techniques in Cyber Security and Intelligence: Applications and Techniques in Cyber Security and Intelligence (Vol. 580, pp. 266–274). (Advances in Intelligent Systems and Computing; Vol. 580). Springer. https://doi.org/10.1007/978-3-319-67071-3_33
[19] Rahman, A., & Upadhyay, S. K. (2015). A Bayesian reweighting technique for small area estimation. In U. Singh, A. Loganathan, S. K. Upadhyay, & D. K. Dey (Eds.),

Current trends in Bayesian methodology with applications (1st ed., pp. 503–519). CRC Press.

[20] Rahman, A. (2019). Statistics-based data preprocessing methods and machine learning algorithms for big data analysis. *International Journal of Artificial Intelligence*, 17(2): 44–65.

[21] Hansen, J.H.L. (1999). SUSAS LDC99S78. Web Download. Philadelphia: Linguistic Data Consortium. https://doi.org/10.35111/chqa-vd56

[22] Keng-hao Chang, Drew Fisher, and John Canny, "AMMON: A Speech Analysis Library for Analyzing Affect, Stress, and Mental Health on Mobile Phones", 2012.

[23] Enrique Garcia-Ceja, Venet Osmani and Oscar Mayora, "Automatic Stress Detection in Working Environments from Smartphones' Accelerometer Data: A First Step", 2015.

[24] Raihana Ferdous, Venet Osmani and Oscar Mayora, "Smartphone app usage as a predictor of perceived stress levels at workplace", 2015.

[25] Rui Wang, Fanglin Chen, Zhenyu Chen, Tianxing Li, Gabriella Harari, Stefanie Tignor, Xia Zhou, Dror Ben-Zeev, and Andrew T. Campbell, "StudentLife: Assessing Mental Health, Academic Performance and Behavioral Trends of College Students using Smartphones", 2014.

[26] Pinto, Joey, Pooja Jain, and Tapan Kumar. "Fault prediction for distributed computing Hadoop clusters using real-time higher order differential inputs to SVM: Zedacross." International Journal of Information and Computer Security 12, no. 2-3 (2020): 181–198.

[27] Sreesurya, Ilayaraja, Himani Rathi, Pooja Jain, and Tapan Kumar Jain. "Hypex: A Tool for Extracting Business Intelligence from Sentiment Analysis using Enhanced LSTM." Multimedia Tools and Applications 79, no. 47 (2020): 35641–35663.

[28] Agarwal, B. Financial sentiment analysis model utilizing knowledge-base and domain-specific representation. Multimed Tools Appl (2022). https://doi.org/10.1007/s11042-022-12181-y.

[29] Gauri Jain, Manisha Sharma, Basant Agarwal, "Optimizing Semantic LSTM for Spam Detection", In International Journal of Information Technology, Springer, DOI:10.1007/s41870-018-0157-5, 2018.

[30] Jain, Rahul, Pooja Jain, Tapan Kumar, and Gaurav Dhiman. "Real time video summarizing using image semantic segmentation for CBVR." Journal of Real-Time Image Processing 18, no. 5 (2021): 1827–1836.

[31] Umematsu, Terumi, Sano, Akane and Picard, Rosalind W. 2019. "Daytime Data and LSTM can Forecast Tomorrow's Stress, Health, and Happiness." Proceedings of the Annual International Conference of the IEEE Engineering in Medicine and Biology Society, EMBS, 2019

[32] "Rowel Atienza. LSTM by Example using Tensorflow. https://towardsdatascience.com/lstm-by-example-using-tensorflow-feb0c1968537 (Mar. 2017)"

[33] "Rian Dolphin. LSTM Networks | A Detailed Explanation. https://towardsdatascience.com/lstm-networks-a-detailed-explanation-8fae6aefc7f9 (Oct. 2020)"

16 Antecedents and Inhibitors for Use of Primary Healthcare

A Case Study of Mohalla Clinics in Delhi

Kaushal Kumar[1] *and Rahul Kumar*[2]
[1]Department of Operational Research, Faculty of Mathematical Sciences, University of Delhi, Delhi, India
E-mail: kkumar@or.du.ac.in
[2] Faculty of Management Studies, University of Delhi, Delhi, India
E-mail: rahulkumar886049@gmail.com

CONTENTS

DOI: 10.1201/9781003253051-21

16.1 INTRODUCTION

The health status of India has improved substantially in recent decades, despite formidable challenges (Agrawal, Bhattacharya, & Lahariya, 2020). Though primary care has played a vital role in this success, the publicly funded healthcare systems require substantial improvement to be of internationally acceptable standards. The facilities are unevenly utilized, patients visit private institutions and incur out-of-pocket expenses, the tertiary care hospitals are burdened with large queues at OPDs (Out-patient Departments), and there are many more such deterrents that discourage patients from seeking primary care from public-funded facilities (Balarajan, Selvaraj, & Subramanian, 2011).

As is said, prevention is better than cure. For India to become a developed economy any time soon, this adage has to be the guiding principle in the development of health policy. The health and wellness of the population is a necessary condition for the progress and growth of a country. The improvements in health status achieved in recent years need to be maintained and accelerated in the future. India has a huge population and it will be a nightmarish situation if the burden of diseases deteriorates. This brings primary care into focus, the area of inferential and normative investigation in the proposed study.

The services under primary care are manifold; they range from promotion and prevention to treatment, rehabilitation, and palliative care (WHO). If primary care gets the right importance that it deserves, the load on secondary and tertiary care would reduce and the performance of the healthcare system would improve. The government of India has recently launched the ambitious 'health care for all' program called Ayushman Bharat – Pradhan Mantri Jan Arogya Yojana (AB-PMJAY) (Lahariya C., 2018). The objective of the program echoes the policies of the National Health Mission to deliver equitable, affordable, and quality health care to all.

The program gives equal importance to both promotive and curative health care through proportionate support to all three hierarchies of primary, secondary, and tertiary care. The curative tertiary care for the underprivileged would be financed through government-sponsored insurance. Whereas for comprehensive primary care, 150,000 health and wellness centres are to be created (Lahariya C., 2018). These would supplement existing community health centers. For the success of this program, it is imperative to identify the determinants for use of primary care. The purpose of our study is to help policymakers make better decisions. We have utilized primary research (question survey) to determine antecedents for higher utilization of primary health centers and derive policy implications.

To strengthen community-level primary healthcare services, the Government of Delhi launched *Mohalla clinics* or community/neighborhood clinics in 2015 (Lahariya C., 2017). The concept of Mobile Medical Units (MMUs) stimulated the idea for this project (Parmar, Singh, & Jadhav, 2021). MMUs have a team of doctors and paramedics with medical supplies and testing kits to provide healthcare services to underserved areas. More than 500 *Mohalla clinics* are functional in the city.[1] These clinics operate through portable containers called 'porta cabin' or rented accommodation called 'rented premises' with approximately INR 2,000,000 as set-up cost per clinic (Komal & Rai, 2017). These health centers provide free medical consultation, diagnostic services, and medicines. These community clinics assist in rendering door-to-door primary health services and also help in reducing the primary care load on tertiary care facilities in Delhi (Lahariya C., 2016). This study is based on the responses received from patients visiting *Mohalla clinics* in Delhi, India.

16.2 BACKGROUND

Past studies have found inequity related to the distribution of ill health across geographies, age, gender, wealth, and socioeconomic strata (Powell-Jackson et al., 2013). Governments have the liability for ensuring health services in countries having a large economically weaker population. The public-funded facilities have limited capacity and it is generally observed that patients visit hospitals even for primary care. The government hospitals as a result experience overcrowding and thereby have high workloads (Kumar & Bardhan, 2020). A patient is required to wait for a long time to get services there and often must skip work as well. The poor and unprivileged section of society is therefore forced to visit private facilities for outpatient care and spend extra money in the form of out-of-pocket expenses (Balarajan et al., 2011). A primary care system complementing secondary and tertiary care systems is required to achieve the goals of health promotion and education, health care for all, and prevention of diseases (Marten et al., 2014).

Das et al. (2016) have found that despite spending four times more on *per patient interaction*, the public healthcare system in India lags in performance in comparison to the private sector. People from lower socio-economic strata make poor choices while selecting healthcare options. They have also found public facilities on paper have better-qualified care providers, but due to a variety of reasons patients may not get better care in them, and there is a lack of empirical evidence on the quality of clinical interactions.

India has seen significant improvements in health outcomes over the years with increments in life expectancy, reduction in infant and maternal mortality rates, and control of diseases like polio and HIV/AIDS, yet we witness instances that show the presence of inadequate and low-quality healthcare services in India (Patel et al., 2015). Despite many efforts by the government, millions in India can not even meet basic needs, malnutrition is uncontrollable, and vaccination coverage is not adequate (Jacob, 2007). India accounts for 21% of the global disease burden and ranks third highest in the world for HIV-infected persons (Yeravdekar et al., 2013). Implementation of universal health coverage requires good quality and accessible primary care services (Beran et al., 2016).

Sabde et al. (2018) conducted a study in Madhya Pradesh, India to identify individual and facility determinants of bypassing public-funded health facilities for childbirth and discussed its implication for the *Janani Suraksha Yojana* (JSY). Perceptions of quality of care and the attributes of a facility (i.e., number of doctors, availability of drugs and equipment, number of services offered, etc.) are deciding factors associated with bypassing behavior. They found that the likelihood of bypassing for childbirth can be reduced by strengthening the availability of basic emergency obstetric care. Their results support and add to the knowledge from similar studies conducted in multiple geographies.

Health is a state subject in India. There have been many initiatives by state governments to boost primary care in the geographies under their jurisdictions. There are many successes and learning from these used cases can lead to the formulation of a better program at a larger scale. Mohanan et al. (2016) have suggested adopting a cautious approach to scaling up. After the implementation of recommendations of the fourteenth finance commission,[2] states have more financial resources, yet these are limited considering the large population they need to support. Therefore, quantum of effectiveness needs to be ascertained before scaling up a program. A large number of health and wellness centers are going to be established under AB-PMJAY. For optimal utilization of limited resources, it is pertinent to generate robust empirical evidence on program effectiveness.

16.3 RELATED WORK

Primary care is the first point of contact for a patient seeking health services and is linked to health facilities providing higher-level services. Primary care is an achievable alternative for addressing the healthcare needs of society in countries with limited resources (Rahman & Kuddus, 2020, 2021; Abdulla et al., 2021; Rahman et al., 2021). Primary care not only helps in reducing illness and death but also helps in ensuring equitable distribution of healthcare services (Starfield et al., 2005). Primary health care is aimed at providing better health services for everyone by reducing social disparities and ensuring equitable and quality services at a lesser cost (Stange et al., 2014). A functional healthcare system requires a high-quality primary care system but low- and middle-income nations have poor primary care systems with high rates of low-quality treatments and longer waiting times (Gage et al., 2017). The accomplishments related to health care are lower than the expectations and the primary care system needs to be reinvented by integrating it into the remaining health system to strengthen the healthcare system as a whole (Frenk, 2009).

Sharma & Jyoti (2019) assessed the quality of services and dimensions of healthcare services offered by *Mohalla clinics*. Bhandari et al. (2017) conducted an evaluative study of newly established *Mohalla clinics* (primary care facilities operated by the government of Delhi). They found that the users of these centers have benefitted from a substantial reduction in transportation costs, costs due to loss of work, and out-of-pocket expenses. Nearly 65% of the patients at the surveyed *Mohalla clinics* were female, indicating the easy access has encouraged female patients to seek care that they could have otherwise skipped. One problem area they identified was the lack of proper communication between caregivers and patients. Other studies have also pointed out the short duration that doctors devote to each patient. It is necessary to understand the reasons for this so that solutions can be found.

Alkuwaiti et al (2020) carried out a cross-sectional study in Saudi Arabia to identify the determinants of satisfaction associated with outpatient services using exploratory factor analysis. They found professional care, service availability, waiting time, and laboratory services to be the major drivers of patient satisfaction. The study conducted by Liu & Fang (2019) in China also utilized exploratory factor analysis and found service quality, expenditure, and convenience as major drivers of patient satisfaction. Hush et al. (2013) performed a cross-cultural comparison of determinants of satisfaction of physiotherapy patients in Australia and Korea and reported effective communication, care, and respect from the therapist as major factors of patient satisfaction.

Natesan et al (2019) conducted a study in Lebanon and revealed various satisfaction factors associated with patient visits to emergency department: admission and discharge process, length of stay, confidentiality, and overall disturbance of the emergency department. Khudair & Raza (2013) administered a study in Qatar for determining factors influencing patient satisfaction associated with pharmaceutical services. The factors identified include the waiting space, location of the pharmacy, counseling of medication, service agility, and behavior of the pharmacist.

From the above discussion, it can be concluded that patient satisfaction is influenced by various factors including availability of services, the behavior of staff, convenience, communication, waiting time, diagnosis time, etc. In this study, we have considered these attributes to understand the patient satisfaction level associated with *Mohalla clinics* in Delhi, India. A list of all the variables considered can be found in Section 16.5.

16.4 RESEARCH GAP

Diseases have their determinants in the environments where people live, therefore interventions need to be planned around communities. The primary care system under the promising program AB-PMJAY shuns the siloed approach of developing plans for groups of diseases. For the program to be successful, it is imperative to identify the determinants for use of primary care. A limited number of studies have been conducted on such problems in India. Our proposed research intends to address the gap through primary research (questionnaire survey). In this regard, our research seeks to find patient-choice determinants for use of primary care. There have been many studies on patient behavior in choosing treatment for specific types of conditions. Perceptions about and consequent use of primary care as a holistic care service have not been adequately measured. This is another gap our research intends to address. Researchers in India have not given adequate attention to issues related to the quality of care and its impact on program performance in the health system. For evidence-based policymaking, it is necessary to collect more data on the quality attributes of healthcare facilities and their relative importance to patients.

16.5 OBJECTIVES

The objective of the study is to identify factors that influence the intention to use or to avoid primary care by people. This will assist policymakers in making judicious decisions for the success of health and wellness centers under the recently launched Pradhan Mantri Jan Arogya Yojana (PM-JAY).

Specific objectives: The framework and scope of the study are as follows:

(a) Highlight the importance of primary care in health outcomes.
(b) Identify determinants of patient choice for a primary care facility.

(c) Identify reasons for patients skipping primary care facilities.
(d) Understand how holistic approaches under the PM-JAY are being implemented.
(e) Generate management and policy implications.

16.6 METHODOLOGY

The study is based on collecting primary data through a survey over a period of two months. The flow diagram of this cross-sectional study is presented in Figure 16.1.

16.6.1 Design of the Study

A survey-based cross-sectional study was conducted for the identification of factors influencing choice of primary care services. A structured questionnaire was utilized to collect the responses from the visitors of *Mohalla clinics* of selected districts of Delhi. The responses were collected during Jan-Feb 2021.

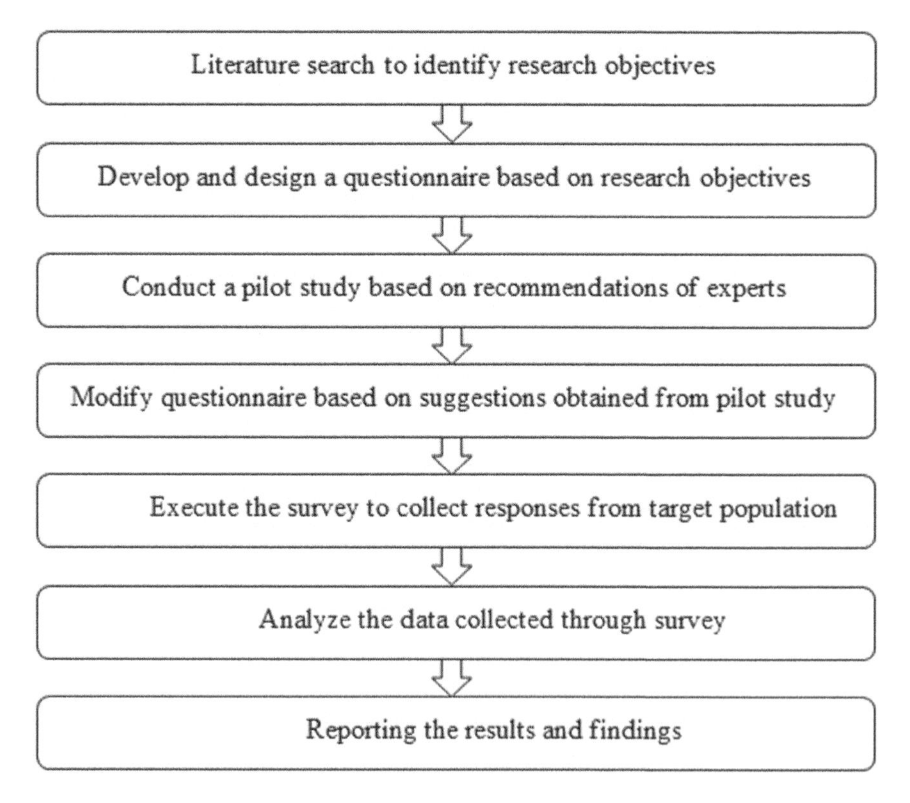

FIGURE 16.1 Overview of the proposed methodology.

16.6.2 Sampling

The study aimed to ascertain factors and criteria that influence the use (selection and choice) of primary care. Based on the discussions with experts working in government, private sector, NGOs, agencies like WHO, UNICEF, PHFI, etc., a questionnaire was created for conducting a survey-based study to identify the healthcare needs of the respondents and their choice behavior for primary care. A convenience sampling method was used to collect data from 309 respondents. Equation (16.1) was used to compute the sample size:

$$n = \frac{z^2 * p(p-1)}{e^2} \quad (16.1)$$

The notations used in the formula are:

n = sample size
z = critical value of the standard normal distribution
p = expected population proportion (0.75 for this study)
e = margin of error (5%)

To begin with, a pilot study was carried out that provided sample proportion as 75% with a 0.05 margin of error and a 95% confidence level. Using these values in equation (16.1), 'n' was computed as 288.12 or 288 approximately. This is the minimum sample size for carrying out the study and any sample size greater than or equal to 288 is acceptable. We were able to collect 309 responses, which becomes the final sample size for our study.

16.6.3 Instrument Deployed

A questionnaire was created for describing the choice attributes for primary care. The attributes/variables considered were:

- A1: Availability of healthcare workforce
- A2: Availability of information, education, and communication materials
- A3: The behavior of reception desk
- A4: The behavior of medical assistants
- A5: The behavior of doctor (s)
- A6: Doctor's attentiveness
- A7: Effectiveness of remedy
- A8: Explanation of treatment given by the doctor
- A9: Functioning of the healthcare workforce
- A10: Sanitation
- A11: Sufficient diagnosis time

All these attributes were rated using a 5-point Likert scale. The rating indicates the degree of satisfaction of patients visiting *Mohalla Clinics* associated with respective attributes in increasing order: 1 = Extremely dissatisfied and 5 = Extremely satisfied.

16.6.4 Statistical Analysis

IBM SPSS Statistics 21 was used to carry out statistical analysis. Exploratory factor analysis (EFA) (Alkuwaiti et al, 2020; Liu & Fang, 2019) was conducted for recognizing the factors associated with the satisfaction of patients of *Mohalla clinics*. The extraction was based on the *principal components method* and the rotation technique deployed was *varimax* rotation. Indirectly, this study identifies the factors influencing the choice behavior of patients for primary care. Further, the Chi-square test of association was used to identify the associations among various categorical variables. This study also revealed various useful guidelines for policymakers and managers working towards strengthening primary care networks.

16.7 RESULTS AND DISCUSSION

A total of 309 responses were received from the visitors of *Mohalla clinics* located in five districts of Delhi: North, North East, North West, Shahdara, and West Delhi. The proportion of respondents from North, North East, North West, Shahdara, and West Delhi was 7.12%, 25.89%, 14.24%, 23.95%, and 28.80%, respectively. The major population of these districts has a low socio-economic profile (Dutta, et al., 2020). Out of 309 respondents, 44.66% correspond to porta cabins while 55.34% corresponded to rented premises. Respondents were asked to rate 11 attributes/variables (A1, A2... A11) to assess their satisfaction in visiting *Mohalla clinics* using a 5-point Likert scale. In other words, this study attempted to find the factors influencing patient choice for primary care.

Attributes A1, A6, and A9 were excluded from analysis because the majority of respondents did not respond to these attributes. For the remaining variables, *Cronbach's alpha* value was computed as 0.739. This value was computed from 291 valid responses as the remaining 18 were excluded due to missing values. A *Cronbach's alpha* value greater than 0.7 was considered acceptable (Ahmed, Vveinhardt, Štreimikienė, Ashraf, & Channar, 2017). Thus, the variables considered were reliable and exploratory factor analysis could be carried out.

16.7.1 Exploratory Factor Analysis

Next, exploratory factor analysis (EFA) (Alkuwaiti et al, 2020; Liu & Fang, 2019) was carried out to reduce the attributes listed above into fewer factors. *IBM SPSS Statistics 21* was used to perform EFA. The extraction was based on *the principal components method* and the rotation technique deployed was *varimax* rotation. The factors were extracted by setting the eigenvalue threshold to be greater than 1. The cut-off point for factor loadings was taken as 0.4 (i.e., absolute value of factor loading less than 0.4 should be ignored). As a result of EFA, two factors were obtained, and these explain 60.3% of the total variance. As seen in Table 16.2, factor 1 comprises the attributes concerned with the services received at the healthcare facility. Thus, factor 1 is named as *Service quality*. Similarly, factor 2 is named as *the Physical appearance of the facility* based on its variables.

TABLE 16.1
KMO and Bartletts's Tests

KMO measure – sample adequacy		0.819
Bartlett's Test	Chi-Square value	961.707
	deg of freedom	28
	Sig.value	.000

TABLE 16.2
Description of factors

Label	Variable/Attribute name	Factor loading
Factor 1: Service quality % of variance explained: 45.845		
A3	The behavior of the reception desk	0.792
A4	The behavior of medical assistants	0.814
A5	The behavior of doctor (s)	0.734
A7	Effectiveness of remedy	0.544
A8	Explanation of treatment given by a doctor	0.858
A11	Sufficient diagnosis time	0.818
Factor 2: Physical appearance of the facility % of variance explained: 14.439		
A2	Availability of information, education, and communication materials	0.898
A10	Sanitation	0.814

As seen in Table 16.1, both the KMO test (KMO value is 0.819) and Bartlett's Test of Sphericity are significant. The results of the EFA are given in Tables 16.1 and 16.2.

16.7.2 Chi-Square Testing for Association

The Chi-square test is widely used to test the association between categorical variables (Verma, 2013). We utilized this test to evaluate the relationship between different useful variables considered in our study.

Hypothesis 1:
H_0: There is no association between district and medium of awareness about *Mohalla clinic.*
H_1: There is a significant relationship between district and medium of awareness.

In research hypothesis 1, there are two variables involved each having five categories. The variables are *district* (North, Northeast, Northwest, Shahdara, West) and *medium of awareness* (Advertisement/Hoardings, Patient/staff at a health facility, Relative/neighbor, Witnessed the construction, Word of mouth). The results of the analysis are given in Table 16.3. From the results, it is inferred that the relationship between the *district* and *medium of awareness* is statistically significant because of the p-value being < 0.001. But the association was low (Cramer's V = 0.215) (Sayassatov & Cho, 2020). That means that even though the medium of awareness about *Mohalla clinic* was dependent on its district the degree of association was modest.

Hypothesis 2:

H_0: There is no association between the type of *Mohalla clinic* and patient travel time.

H_1: There is a significant association between the type of *Mohalla clinic* and patient travel time.

In research hypothesis 2, there are two variables involved: *type of Mohalla clinic* having two categories (Porta cabin and Rented premises) and *patient travel time* having three categories (< 2 km, 2–5 km, and > 5 km). The results of the analysis are given in Table 16.3. The results infer that there is no connection between the type of *Mohalla clinic* and patient travel time. The strength of association is also very low (Cramer's V = 0.15) (Sayassatov & Cho, 2020). That means that time taken by patients to reach the *Mohalla clinic* is independent of its type.

Hypothesis 3:

H_0: There is no association between type of doctor and doctor-patient time.

H_1: There is a significant association between type of doctor and doctor-patient time.

In research hypothesis 3, there are two variables involved: *type of doctor* having two categories (Empanelled and 'On deputation from NHM/DGHS') and *doctor-patient time* having three categories (< 2 min, 2–5 min, and > 5 min). The results of the analysis are given in Table 16.3. We can infer from the results that *type of doctor* and *doctor-patient time* are statistically connected because of the p-value being < 0.001. The strength of association is medium (Cramer's V = 0.327) (Sayassatov & Cho, 2020). That means that the average time spent by the doctor to treat patients is dependent on the type of doctor. It was observed that the majority of the doctors who spent more than 5 minutes for patient consultation were 'empanelled' while only a few doctors on deputation spent more than 5 minutes on consultation.

Hypothesis 4:

H_0: There is no association between district and patient waiting time.

H_1: There is a significant association between district and patient waiting time.

In research hypothesis 4, there are two variables involved: *district* having five categories (North, North East, North West, Shahdara, West) and *patient waiting time* having five categories (< 5 min, 5–10 min, >10 min & ≤ 30 min, > 30 min & ≤ 1-hour, and > 1 hour). The results of the analysis are given in Table 16.3. It can be derived from the results that there is a statistically significant association between *district* and *patient waiting time* because of the p-value being < 0.001. The strength of association is low (Cramer's V = 0.253) (Sayassatov & Cho, 2020). That means patient waiting time is dependent on the district. It was observed that majorly patients from the North East and Shahdara districts of Delhi had to wait for more than 30 minutes before getting healthcare services.

Hypothesis 5:

H_0: There is no association between the type of *Mohalla clinic* and patient waiting time.

H_1: There is a significant association between the type of *Mohalla clinic* and patient waiting time.

In research hypothesis 5, there are two variables involved: *type of Mohalla clinic* having two categories (Porta cabin and Rented premises) and *patient waiting time* having five categories (< 5 min, 5–10 min, >10 min & ≤30 min, >30 min & ≤ 1-hour min, and > 1 hour). The results of the analysis are given in Table 16.3. From the analysis, we may say that *type of Mohalla clinic* and *patient waiting time* are associated because of the p-value being < 0.001. The strength of association was also high (Cramer's V = 0.541) (Sayassatov & Cho, 2020). That means that patient waiting time is dependent on the type of *Mohalla clinic*. It was observed that the majority of patients visiting *Mohalla clinics* operating through 'Rented premises' had to wait for more than 30 minutes before getting healthcare services while waiting time for only a few patients visiting 'Porta cabin' was more than 30 minutes. Moreover, many patients visited 'Porta cabin' and had to wait for a meager less than 5 min while only a few visiting 'Rented premises' had a waiting time of fewer than 5 minutes.

Hypothesis 6:

H_0: There is no association between the type of *Mohalla clinic* and patient travel cost.

H_1: There is a significant association between the type of *Mohalla clinic* and patient travel cost.

In research hypothesis 1, there are two variables involved: *type of Mohalla clinic* having two categories (Porta cabin and Rented premises) and *patient travel cost* having four categories (Nil, 1–20 rupees, >20 & ≤50 rupees, >50 rupees). The results of the analysis are given in Table 16.3. We can say that the variables *type of Mohalla clinic* and *patient travel cost* are not related because p-value > 0.001. The strength of association was also very low (Cramer's V = 0.171) (Sayassatov & Cho, 2020). That means that patient travel cost is independent of the type of *Mohalla clinic*.

Hypothesis 7:

H_0: There is no association between *the type of facility visited for primary care in the absence of Mohalla clinics* and *Savings on consultation and medicines*.

H_1: There is a significant association between the *type of facility visited for primary care in the absence of Mohalla clinics* and *Savings on consultation and medicines.*

In research hypothesis 7, there are two variables involved: *type of facility visited for primary care in the absence of Mohalla clinics* having four categories (Government, private, both govt. & private and others) and *Savings on consultation and medicines* having five categories (Nil, 1–100 rupees, >100 & ≤200 rupees, >200 & ≤500 rupees, >500 rupees). The results of the analysis are given in Table 16.3. It is concluded that variables *type of facility visited for primary care in the absence of Mohalla clinics* and *Savings on consultation and medicines* are related statistically because p-value < 0.001. The strength of association was mediocre (Cramer's V = 0.291) (Sayassatov & Cho, 2020). It was observed that savings of more than 200 on consultation and medicines were primarily attributed to earlier visits to private healthcare facilities.

Hypothesis 8:

H_0: There is no association between *the type of facility visited for primary care in the absence of Mohalla clinics* and *Savings on travel costs.*

H_1: There is a significant association between *the type of facility visited for primary care in the absence of Mohalla clinics* and *Savings on travel costs.*

In research hypothesis 8, there are two variables involved: *type of facility visited for primary care in the absence of Mohalla clinics* having four categories (Government, private, both govt. & private and others) and *Savings on travel cost* having five categories (Nil, 1–20 rupees, >20 & ≤50 rupees, >50 & ≤100 rupees, >100 rupees). The results of the analysis are given in Table 16.3. We can conclude that the two variables considered, *type of facility visited for primary care in the absence of Mohalla clinics* and *Savings on travel cost,* are not associated because of the p-value being > 0.001. The strength of association was very low (Cramer's V = 0.075) (Sayassatov & Cho, 2020).

Hypothesis 9:

H_0: There is no association between *the type of facility visited for primary care in the absence of Mohalla clinics* and *Savings on pathology tests.*

H_1: There is a significant association between the *type of facility visited for primary care in the absence of Mohalla clinics* and *Savings on pathology tests.*

In research hypothesis 9, there are two variables involved: *type of facility visited for primary care in the absence of Mohalla clinics* having four categories (Government, private, both govt. & private and others) and *Savings on pathology tests* having five categories (Nil, 1–200 rupees, >200 & ≤500 rupees, >500 & ≤1000 rupees, >1000 rupees). The results of the analysis are given in Table 16.3. From the results, we can say that *type of facility visited for primary care in the absence of Mohalla clinics* and *Savings on pathology tests* are not associated because p-value > 0.001. The strength of association was low (Cramer's V = 0.179) (Sayassatov & Cho, 2020).

TABLE 16.3
Chi-square test results

	Pearson chi-square			Likelihood ratio		
Hypothesis	**Value**	**deg of freedom**	**Asymptotic Sig. (2-sided)**	**Value**	**deg of freedom**	**Asymptotic Sig. (2-sided)**
Hypothesis 1	57.103	16	.000	58.454	16	.000
Hypothesis 2	6.907	2	.032	7.525	2	.023
Hypothesis 3	33.095	2	.000	38.612	2	.000
Hypothesis 4	79.091	16	.000	92.912	16	.000
Hypothesis 5	90.458	4	.000	105.521	4	.000
Hypothesis 6	9.032	3	.029	10.768	3	.013
Hypothesis 7	78.726	12	.000	90.997	12	.000
Hypothesis 8	5.206	12	.951	6.237	12	.904
Hypothesis 9	29.776	12	.003	22.732	12	.030
Number of valid responses: 309						

From the above discussion, it can be concluded that choice for primary care is attributed to two main factors: service quality and physical appearance of the facility. Alkuwaiti et al. (2020) and Liu & Fang (2019) also found similar results. The study by Alkuwaiti et al. (2020) reported that professional care, service availability, waiting time, and laboratory services are major drivers of patient satisfaction while Liu & Fang (2019) found patient satisfaction to be impacted by three factors: service quality, healthcare expense, and convenience.

Additional interesting findings are out-of-pocket expenditure incurred by patients is majorly attributed to the expense incurred on consultation and medicines and patient waiting time is dependent on the type of healthcare facility.

16.8 CONCLUSION

The use of data and insights gained from their analysis, in policymaking, has increased in recent times. There is a growing appreciation for evidence-based decision-making. The government, therefore, is putting efforts into the collection and dissemination of data, especially in the health sector. Data will play an important role in the success of the *Ayushman Bharat* program. It is presently used in fraud detection and some resource allocation decisions.[3] Though the trend is encouraging, this is a modest start. For exploiting the potential of data to the fullest all three practices of analytics, descriptive, predictive, and normative need, to be employed.

In the proposed research, we intend to work on a framework that provides a practical template for evidence-based decision-making. Presently descriptive and predictive analysis feeds decision-making, which is subjective and based on human judgments. We have collected primary data to identify qualitative inputs on social and behavioral factors of patients seeking primary care. The approach would assist in improving decision-making.

Some antecedents for use of primary care such as distance, cost, etc., are well known. We have identified some new antecedents including service quality (Alolayyan, et al., 2018) and physical appearance of the facility and found patient choice as primarily driven by these two factors. Additional interesting findings are out-of-pocket expenditure incurred by patients is majorly attributed to the expense incurred on consultation and medicines and patient waiting time is dependent on the type of healthcare facility.

Ensuring participation of primary beneficiaries, the members of society, can not only reduce the costs, it would also enhance the effectiveness of the primary care program. This is an objective set in the operational doctrine of AB-PMJAY. We have identified factors that encourage people to participate in a primary care system.

As a future research direction, a new optimization model would be developed for designing a primary care network. This would lead to better decision-making regarding the location of facilities. The study would suggest ways to augment existing capacities, increase efficiency of existing system and wherever required model of engagement of private sector. We will use location analysis, determine antecedents for higher utilization of primary health centers, develop an optimal location model, and derive policy implications.

ACKNOWLEDGMENTS

This research was supported by the FRP grant 2020–21 received from the Institute of Eminence, University of Delhi, India. The authors are thankful to Prof. Amit Kumar Bardhan for extending his valuable support in writing and improving this study. The authors would also like to thank the editor(s) and reviewer(s) for their valuable inputs.

NOTES

1 www.financialexpress.com/lifestyle/health/mpd-2041-suggestions-to-include-mohalla-clinics-more-maternity-hospitals-in-old-delhi/2383647/
2 www.indiabudget.gov.in/budget2015-2016/es2014-15/echapvol1-10.pdf
3 www.thehindu.com/news/cities/mumbai/5-data-analytics-firms-shortlisted-for-fraud-detection-in-pmjay/article25728554.e

REFERENCES

Abdulla, F., Nain, Z., Karimuzzaman, M., Hossain, M. M., & Rahman, A. (2021). A non-linear biostatistical graphical modeling of preventive actions and healthcare factors in controlling COVID-19 pandemic. *International Journal of Environmental Research and Public Health, 18*(9), 4491. https://doi.org/10.3390/ijerph18094491

Agrawal, T., Bhattacharya, S., & Lahariya, C. (2020). Pattern of use and determinants of return visits at community or Mohalla clinics of Delhi, India. *Indian journal of community medicine: official publication of Indian Association of Preventive & Social Medicine, 45*(1), 77–82.

Ahmed, R. R., Vveinhardt, J., Štreimikienė, D., Ashraf, M., & Channar, Z. A. (2017). Modified SERVQUAL model and effects of customer attitude and technology on customer satisfaction in banking industry: mediation, moderation and conditional process analysis. *Journal of Business Economics and Management, 18*(5), 974–1004.

Alkuwaiti, A., Maruthamuthu, T., & Akgun, S. (2020). Factors associated with the quality of outpatient service: The application of factor analysis – A case study. *International Journal of Healthcare Management, 13*(sup1), 88–93.

Alolayyan, M. N., Al-Hawary, S. I., Mohammad, A. A., & Al-Nady, B. A.-H. (2018). Banking service quality provided by commercial banks and customer satisfaction. A structural equation modelling approaches. *International Journal of Productivity and Quality Management, 24*(4), 543–565.

Balarajan, Y., Selvaraj, S., & Subramanian, S. (2011). Health care and equity in India. *The Lancet, 377*(9764), 505–515.

Beran, D., Chappuis, F., Cattacin, S., Damasceno, A., Jha, N., Somerville, C., ... Miranda, J. (2016). The need to focus on primary health care for chronic diseases. *The lancet Diabetes & endocrinology, 4*(9), 731–732.

Bhandari, A., Bhardwaj, I., & Sawhney, S. (2017). *Quality and Accessibility of Government Hospitals and Mohalla Clinics, Delhi Citizen's Handbook - 2017.* New Delhi: Centre for Civil Society.

Das, J., Holla, A., Mohpal, A., & Muralidharan, K. (2016). Quality and Accountability in Health Care delivery: audit-study evidence from primary care in India. *American Economic Review, 106*(12), 3765–99.

Dutta, C. K., Chibber, K. R., Srivastava, A. K., Srivastava, P. K., Mathur, A. K., Saini, N., & Gupta, S. (2020). *Socio-Economic profile of residents of Delhi, Part-I: Households Characteristics.* Delhi: Directorate of Economics & Statistics, Government of National Capital Territory of Delhi.

Frenk, J. (2009). Reinventing primary health care: the need for systems integration. *The Lancet, 374*(9684), 170–173.

Gage, A. D., Leslie, H. H., Bitton, A., Jerome, J. G., Thermidor, R., Joseph, J. P., & Kruk, M. E. (2017). Assessing the quality of primary care in Haiti. *Bulletin of the World Health Organization, 95*(3), 182.

Hush, J. M., Lee, H., Yung, V., Adams, R., Mackey, M., Wand, B. M., ... Beattie, P. (2013). Intercultural comparison of patient satisfaction with physiotherapy care in Australia and Korea: an exploratory factor analysis. *Journal of Manual & Manipulative Therapy, 21*(2), 103–112.

Jacob, K. S. (2007). Public health in India and the developing world: beyond medicine and primary healthcare. *Journal of Epidemiology and Community Health, 617*, 562–563.

Khudair, I. F., & Raza, S. A. (2013). Measuring patients' satisfaction with pharmaceutical services at a public hospital in Qatar. *International Journal of Health Care Quality Assurance, 26*(5), 398–419.

Komal, & Rai, P. (2017). HEALTH INFRASTRUCTURE: A STUDY OF MOHALLA CLINICS. *International Journal of Research in Economics and Social Sciences, 7*(5), 133–135.

Kumar, K., & Bardhan, A. K. (2020). A choice-based model to reduce primary care load on tertiary hospitals. *International Journal of Management Science and Engineering Management, 15*(3), 155–164.

Lahariya, C. (2016). Delhi's Mohalla Clinics: Maximising Potential. *Economic and Political Weekly, 51*(4), 15–17.

Lahariya, C. (2017). Mohalla Clinics of Delhi, India: Could these become platform to strengthen primary healthcare? *Journal of Family Medicine and Primary Care, 6*(1), 1–10.

Lahariya, C. (2018). Ayushman Bharat'program and universal health coverage in India. *Indian pediatrics, 55*(6), 495–506.

Liu, L., & Fang, J. (2019). Study On Potential Factors Of Patient Satisfaction: Based On Exploratory Factor Analysis. *Patient Preference and Adherence, 13*, 1983–1994.

Marten, R., McIntyre, D., Travassos, C., Shishkin, S., Longde, W., Reddy, S., & Vega, J. (2014). An assessment of progress towards universal health coverage in Brazil, Russia, India, China, and South Africa (BRICS). *The Lancet, 384*(9960), 2164–2171.

Mohanan, M., Hay, K., & Mor, N. (2016). Quality of health care in India: challenges, priorities, and the road ahead. *Health Affairs,, 35*(10), 1753–1758.

Natesan, P., Hadid, D., Harb, Y. A., & Hitti, E. (2019). Comparing patients and families perceptions of satisfaction and predictors of overall satisfaction in the emergency department. *PLOS ONE, 14*(8), e0221087.

Parmar, A., Singh, A., & Jadhav, S. (2021). Satisfaction Affecting Factors of Mohalla Clinic Services in the Times of COVID -19 Pandemic: A Cross-Sectional Study. *Journal of Pharmaceutical Research International, 33*(36B), 187–194.

Patel, V., Parikh, R., Nandraj, S., Balasubramaniam, P., Narayan, K., Paul, V. K., ... Reddy, K. S. (2015). Assuring health coverage for all in India. *The Lancet, 386*(10011), 2422–2435.

Powell-Jackson, T., Acharya, A., & Mills, A. (2013). An assessment of the quality of primary health care in India. *Economic and Political Weekly, 48*(19), 53–61.

Rahman, A., & Kuddus, M. A. (2020). Cost-effective modeling of the transmission dynamics of malaria: A case study in Bangladesh. *Communications in Statistics: Case Studies, Data Analysis and Applications*, *6*(2), 270–286. https://doi.org/10.1080/23737484.2020.1731724

Rahman, A., & Kuddus, M. A. (2021). Modelling the transmission dynamics of COVID-19 in six high-burden countries. *BioMed Research International*, 2021, 1–17. [5089184]. https://doi.org/10.1155/2021/5089184

Rahman, A., Kuddus, M. A., Ip, H. L., & Bewong, M. (2021). A review of COVID-19 modelling strategies in three countries to develop a research framework for regional areas. *Viruses*, *13*(11), 2185. http://10.3390/v13112185

Sabde, Y., Chaturvedi, S., Randive, B., Sidney, K., Salazar, M., De Costa, A., & al, e. (2018). Bypassing health facilities for childbirth in the context of the JSY cash transfer program to promote institutional birth: A cross-sectional study from Madhya Pradesh, India. *PLoS ONE, 13*(1), e0189364. Retrieved from https://doi.org/10.1371/journal.pone.0189364

Sayassatov, D., & Cho, N. (2020). The Analysis of Association between Learning Styles and a Model of IoT-based Education: Chi-Square Test for Association. *Journal of Information Technology Applications and Management, 27*(3), 19–36.

Sharma, S., & Jyoti. (2019). *Moholla Clinics: Providing Quality Healthcare to the deprived?* Delhi: Institute of Economic Growth, University Enclave, University of Delhi.

Stange, K., Etz, R., Gullett, H., Sweeney, S., Miller, W., Jaén, C. R., ... Glasgow, R. (2014). Metrics for assessing improvements in primary health care. *Annual review of public health, 35*, 423–442.

Starfield, B., Shi, L., & Macinko, J. (2005). Contribution of primary care to health systems and health. *The milbank quarterly, 83*(3), 457–502.

Verma, J. (2013). Chi-Square Test and Its Application. In J. Verma, *Data Analysis in Management with SPSS Software* (pp. 69–101). India: Springer.

Yeravdekar, R., Yeravdekar, V., Tutakne, M. A., Bhatia, N. P., & Tambe, M. (2013). Strengthening of Primary Health Care: Key to Deliver Inclusive Health Care . *Indian Journal of Public Health, 57*(2), 59–64.

Index

Note: Page locators followed by 'f' and 't' represents figures and tables.

C

D

K

L

M

N

O

P

R